Annual Reports

on Fermentation Processes

Volume 7

Annual Reports on Fermentation Processes

VOLUME 7

EDITED BY

GEORGE T. TSAO

Laboratory of Renewable Resources Engineering
A. A. Potter Engineering Center
Purdue University
West Lafayette, Indiana

ASSOCIATE EDITORS

M. C. FLICKINGER

National Cancer Institute
Frederick Cancer Research Facility
Frederick, Maryland

ROBERT K. FINN

School of Chemical Engineering
Cornell University
Ithaca, New York

1984

ACADEMIC PRESS, INC.
(Harcourt Brace Jovanovich, Publishers)
Orlando   San Diego   San Francisco   New York   London
Toronto   Montreal   Sydney   Tokyo   São Paulo

ACADEMIC PRESS, INC.
Orlando, Florida 32887

United Kingdom Edition published by
ACADEMIC PRESS, INC. (LONDON) LTD.
24/28 Oval Road, London NW1 7DX

LIBRARY OF CONGRESS CATALOG CARD NUMBER: 0-275-3790

ISBN 0-12-040307-2

PRINTED IN THE UNITED STATES OF AMERICA

84 85 86 87   9 8 7 6 5 4 3 2 1

Contents

Contributors

Numbers in parentheses indicate the pages on which the authors' contributions begin.

Harvey W. Blanch (81), *Lawrence Berkeley Laboratory and Department of Chemical Engineering, University of California, Berkeley, California 94720*

William E. Brown (135), *The Squibb Institute for Medical Research, Princeton, New Jersey 08540*

Bruce E. Dale (107), *Department of Agricultural and Chemical Engineering, Colorado State University, Ft. Collins, Colorado 80523*

Nikos K. Harakas (159), *Monsanto Company, Central Research Laboratories, Chemistry and Applied Sciences, Biochemistry Department, St. Louis, Missouri 63167*

Ching T. Hou (21), *Exxon Research and Engineering Company, Corporate Research Laboratories, Annandale, New Jersey 08801*

James C. Linden (107), *Department of Agricultural and Chemical Engineering, Colorado State University, Ft. Collins, Colorado 80523*

John H. Luong[1] (45), *John Labatt Ltd., London, Ontario N6A 4M3, Canada*

Mary Mandels (1), *U.S. Army Natick Research and Development Laboratories, Science and Advanced Technology Laboratory, Natick, Massachusetts 01760*

Elwyn T. Reese (1), *U.S. Army Natick Research and Development Laboratories, Science and Advanced Technology Laboratory, Natick, Massachusetts 01760*

Carlos Rolz (213), *Central American Research Institute for Industry (ICAITI), Guatemala City, Guatemala*

Murray C. Tseng (45), *International Minerals and Chemical Corporation, Terre Haute, Indiana 47808*

[1]Present address: National Research Council of Canada, Biotechnology Center, Ottawa, Ontario K1A OR6, Canada.

Preface

As in the earlier volumes of "Annual Reports on Fermentation Processes," we have tried to present a balanced coverage of various subjects of interest to our readers. We feel fortunate that this publication is becoming well known and is stimulating feedback; readers' comments continue to guide us in the selection of topics for future volumes.

Again we are privileged to include award-winning contributions to the field of fermentation research. Dr. Mary Mandels, as the winner of the Marvin Johnson Award of 1982, presented her lecture at the Annual Meeting of the Microbial and Biochemical Technology Division of the American Chemical Society in Kansas City, Missouri, and it is reprinted here in Chapter 1. Dr. William E. Brown, the winner of the David Perlman Award of 1983, presented his lecture at the Annual Meeting of the Microbial and Biochemical Technology Division of the American Chemical Society in Washington, D.C., and it is reprinted here in Chapter 6.

ROLLING WITH THE TIMES: PRODUCTION AND APPLICATIONS
OF *Trichoderma reesei* CELLULASE[1]

Elwyn T. Reese
Mary Mandels

U.S. Army Natick Research and Development Laboratories
Natick, Massachusetts

Elwyn and I are pleased, grateful, and frankly over-whelmed at the honor you are bestowing on us today. We have had some good times together over the past 30 years with *Trichoderma reesei*. In our experience, it is the best available source of extracellular cellulase; and, after a little hands-on experience, most other workers have come to agree with us. The parent strain, QM6a, was isolated from a shelter half from Bougaineville Island after World War II and was originally identified as a *Trichoderma viride*. In 1977 it was recognized as a new species and named *Trichoderma reesei* by Emory Simmons.

We are a pair of old timers and have seen the field of biotechnology develop during our working years. When we were in school we did not know what DNA was for, nor did anyone else. I took Biochemistry from J.B. Sumner who had crystallized the first enzyme, Urease, in 1926. He was still rather annoyed because chemists had been so slow to accept his results and agree that enzymes really were proteins. Elwyn worked on penicillin production at J.T. Baker Company in the early 1940's and turned to Marvin Johnson at the University of Wisconsin and to the group at Peoria for guidance on submerged fermentation. World War II led to a golden age for scientists and we were generously supported by a public still grateful for the technological miracles supplied during the war. Science was the "endless frontier" that would wipe out want and misery forever. We were indeed fortunate to be involved with microbial degradation. The fundamental problems are extremely interesting and the work can always be justified as having a potential for practical

---

[1]*Third Annual Marvin Johnson Memorial Lecture Presented by Mary Mandels in Kansas City, Missouri 15 September 1982.*

Figure 1.   Distribution of Cellulases in Nature

Animals *(rare)*	*Molluscs* *Snails* *Marine Borers* *Insects (Silverfish, Cockroaches)* *Nematodes*
Bacteria *(uncommon)*	*Cellvibrio* *Cellulomonas* *Ruminococcus* *Clostridium thermocellum*
Actinomycetes *(common)*	*Streptomyces* *Thermomonospora*
Phycomycetes *(rare)*	*Karlingia rosea* *Rhizopus arrhizus*
Ascomycetes *(50%)*	*Trichoderma* * *reesei, viride,* *koningii, lignorum* *Chaetomium* *Fusarium* * *solani* *Penicillium* * *iriensis, funiculosum*
Basidiomycetes *(very common)*	*White rotters: Trametes, Sporotrichum* *Irpex (Polyporus)* *Brown rotters: Poria, Lenzites* *Sclerotium* * *rolfsii*

* *Complete cellulase*

Figure 2.   Cellulase Components Produced by Fungi Which are
Actively Degrading Cellulose in Shake Flask

Organism	*Cellulase Component,* μ*/ml* C_1	$\beta1{\to}4$ *glucanase*	*Hydrolysis* *Cotton,%*
Trichoderma viride QM6a	*50.0*	*50.0*	*58*
Pestalotiopsis *westerdijkii QM381*	*0.7*	*60.0*	*4*
Chaetomium globosum QM459	*0.2*	*0.5*	*0*

(Mandels and Reese, 1964)

applications. We began by fighting tropical deterioration.
Later we looked for applications of the enzymes such as
production of sugar, and single cell protein, and the utilization
of cellulose as a resource for chemicals and fuel, and most
recently for production of edible cell free protein.

Now I would like to tell you a little about how our
work developed over the years and to indicate some of the
colleagues who worked with us and deserve most of the credit.
During the war (World War II) in the South Pacific, the Army
lost large stocks of cellulosic materials (tents, clothing,
sand bags, etc.) to microbial deterioration. These items
were difficult to replace because of an acute shortage of
shipping. The Quartermaster Corps asked Professor William
Weston, a Harvard mycologist, to set up a long range basic
research laboratory to investigate the nature of the rotting,
the causal organisms, their mechanism of action, and the
development of methods of control not requiring fungicides.
Lawrence White went to the battle area and collected deteriorated
items from which he isolated the fungi that became the
nucleus of the QM collection at the Philadelphia Quartermaster
Depot (Reese, 1976). The collection later moved to Natick
in 1954 and the USDA Laboratory in Peoria in 1979. It is
the largest culture collection focussed on decomposition of
materiel. Thousands of transfers from it have been supplied
to investigators all over the world. A group of eager young
scientists including Elwyn, Ralph Siu, Gabriel Mandels, and
Hillel Levinson used these fungi to investigate the microbiology
and biochemistry of cellulose decomposition. Cultures were
grown on cotton duck and then the strips were tested for
breaking strength. Based on a 10% loss in tensile strength
after two weeks growth, 50% of the cultures were cellulolytic.
The more active cultures were then evaluated in shake flask
growth studies (Figure 1).

Some interesting facts quickly became apparent. Most
cellulolytic fungi grew well on cellulose and rapidly consumed
it. When filtrates from such cultures were tested, they
usually contained enzymes which produced soluble sugars from
carboxymethyl cellulose. Yet these same culture filtrates
rarely had the ability to produce any significant quantity
of soluble sugar from insoluble cellulose (Figure 2). Elwyn
and his colleagues proposed that the reason for this was
that several enzymes are required to hydrolyze insoluble
cellulose and the filtrates of cellulose cultures may not
contain all of these. This is the well known C_1 C_x hypothesis
(Reese, Siu and Levinson, 1950). This theory greatly stimulated
research and interest in cellulolytic enzymes.

$$\text{Crystalline Cellulose} \xrightarrow{C_1} \text{Reactive Cellulose} \xrightarrow{C_x} \text{Cellobiose + Glucose} \xrightarrow{\text{Cellobiase}} \text{Glucose}$$

In the 1960's we learned much more about the biochemistry of cellulose degradation and made our first approaches to developing a saccharification process. In 1964 we demonstrated the near total hydrolysis of fibrous cotton to glucose by *Trichoderma* cellulase. The proteins in the culture filtrate were separated on DEAE Sephadex and shown to be different enzymes, C_1, C_x, and cellobiase, that individually had little or no action on cellulose, but acted synergistically to hydrolyze it when recombined (Mandels and Reese, 1964). This is beautifully shown in data collected by Wood at the Rowett Research Institute (Figure 3). More recently Wood has demonstrated cross synergism between cellulase components from different fungi (Wood and McCrae, 1979).

Figure 3. Relative Cellulase Activities of the Components of Trichoderma koningii *Cellulase Alone and in Combination (as a Synergistic Effect)*

Enzyme	Relative Cellulase Activity %
C_1	< 1
C_x - 1	< 1
C_x - 2	< 1
β-Glucosidase - 1	NIL
β-Glucosidase - 2	NIL
C_1 + β-Glucosidase (1 + 2)	5
C_x (1 + 2) + β-Glucosidase (1 + 2)	4
C_1 + C_x (1 + 2) + β-Glucosidase (1 + 2)	103
Original Culture Filtrate	100

Cellulase activity = hydrolysis of cotton

Components added at level equal to original filtrate

(Wood, 1975)

Today the cellulases are recognized as endo- and exo-β-glucanases. The endo-enzymes attack cellulose randomly, analogous to α amylase, producing free chain ends. The exo-enzymes are chiefly cellobiohydrolases that remove cellobiose units sequentially from the non-reducing chain ends, analogous to β-amylase. The cellulases are strongly inhibited by cellobiose. Therefore cellobiase is also a critical enzyme for cellulose hydrolysis. Unfortunately for process development, cellobiase is in turn inhibited by its product, glucose. Nature provided *Trichoderma reesei* with cellobiase adequate for its growth on cellulose, but inadequate for use in a hydrolysis reactor where glucose concentrations quickly reach 10% or more. We can solve this problem by adding supplemental cellobiase from another source such as *Aspergillus phoenicis* (Sternberg, Vijayakumar and Reese, 1977; Bissett and Sternberg, 1978; Allen and Sternberg, 1980; Allen and Andreotti, 1981).

In 1968 Elwyn and Moshe Katz produced 30% glucose from a 50% slurry of ball-milled cellulose in a test tube (Figure 4). This inspired Tarun Ghose to come from India for two years as a visiting scientist. He is still working on cellulase. Tarun fed a 10% slurry of milled cellulose in culture filtrate into a one-liter continuous stirred reactor, and with a retention time of 40 hours, maintained 5% sugar in the effluent (Figure 5). The situation was improved with the membrane reactor. With 30% milled cellulose and higher concentrations of enzyme he produced 14% sugar in a semi-continuous effluent. Enzymes and undigested cellulose were retained in the reactor, volume was replenished with buffer or dilute enzyme, and fresh milled cellulose was added to maintain substrate concentration (Ghose and Kostick, 1970). The next innovation was the adsorption reactor. Since cellulase is strongly adsorbed by cellulose (Peitersen et al, 1977) we could dispense with the ultrafiltration membrane and use a coarser filter that retained the cellulose. As long as a high cellulose concentration was maintained, the enzyme was adsorbed and retained in the vessel (Mandels, Kostick and Parizek, 1971).

The Army drafted Dixon Brandt from Hercules in 1971 and we began to look at the engineering aspects of a practical process. Newspaper was used as a model substrate. Since 30% of the newspaper consists of non-hydrolyzable components, the continuous membrane and adsorption reactors were not useable. However, the newspaper was readily hydrolyzed in batch reactors. Five to 10% sugar syrups were produced from 10-15% ball-milled newspaper. The ink and clay did not interfere (Figure 6). The products from newspaper, as from pure cellulose, were

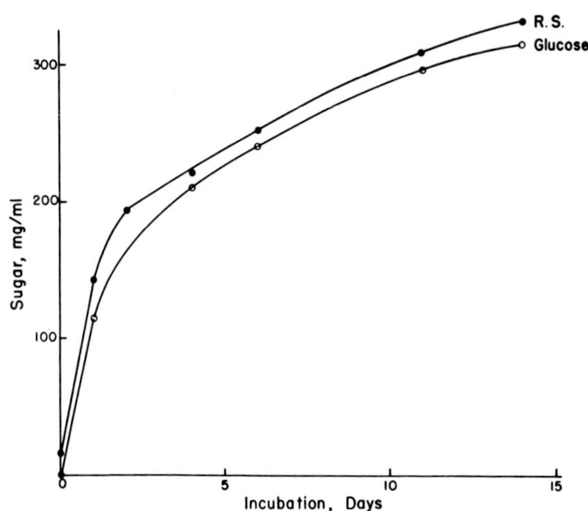

Figure 4.   *The Production of 30% Glucose Syrup from Heated,
Milled Cellulose by* Trichoderma reesei *Cellulase*

*Test tube experiment (1 ml) 50% milled cellulose plus
cellulase* (Trichoderma) *and β-glucosidase* (Aspergillus)
●——● Reducing Sugar (Dinitrosalicylic Acid)
0——0 Glucose (glucose oxidase)   (Katz and Reese, 1968)

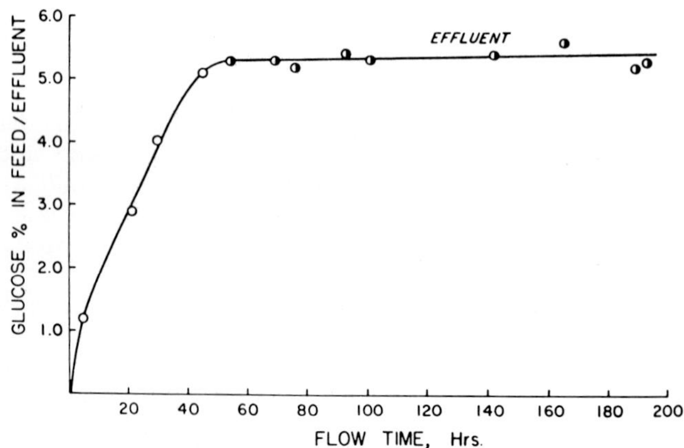

Figure 5. *Continuous Saccharification of Heated Milled
Cellulose in a Stirred Tank Reactor*

Feed slurry 10% cellulose in Trichoderma reesei *cellulose
culture filtrate 4L STR - 50°C, pH 4.8.*
0——0 *Batch;*   ●——● *Continuous D = 0.25 hr*[1]

Enzyme Protein mg/ml	Newspaper %	Temp C	Glucose				Sacchar-ification %
			1 hr %	4 hr %	24 hr %	48 hr %	
0.7	5	50	1.0	2.0	2.8	–	50
0.7	5	50	1.0	2.0	2.3	–	42
1.0	10	50	2.1	3.1	5.5	7.3	66
1.6	10	45	2.0	3.6	5.4	6.5	59
1.6	10	50	2.3	4.2	6.4	6.3	57
0.8	15	50	0.8	2.8	6.1	6.3	38
1.8	15	50	3.2	6.0	8.6	10.0	60

Reactor Volume   1 Liter   Stirred 60 RPM   pH 4.8

(Brandt, Hontz and Mandels, 1973)

Figure 6. *Hydrolysis of Milled Newspaper in Stirred Reactors*

chiefly glucose, cellobiose and xylose. More recently Curtis
Blodgett has produced 10% sugar syrups in eight hours from 30%
ball-milled newspaper on a pilot plant scale. This required
active cellulase from *Trichoderma* supplemented with β-glucosidase from *Aspergillus* (Figure 7).

The engineering studies revealed that the most important
cost factor was the enzyme. Therefore we began a mutation program. Dick Parizek, another draftee, isolated the first cellulase enhanced mutants, QM9123 and QM9414. These mutants have
been widely distributed and used. Later Bland Montenecourt and
Douglas Eveleigh at Rutgers and Benedict Gallo at Natick produced even better mutants. Now there are world wide mutation
programs to enhance the cellulase productivity of *Trichoderma
reesei* (Figure 8). Only QM6a and its descendants are properly
named *Trichoderma reesei*. *Trichoderma viride* is alive, well,
and in use, particularly in Japan.

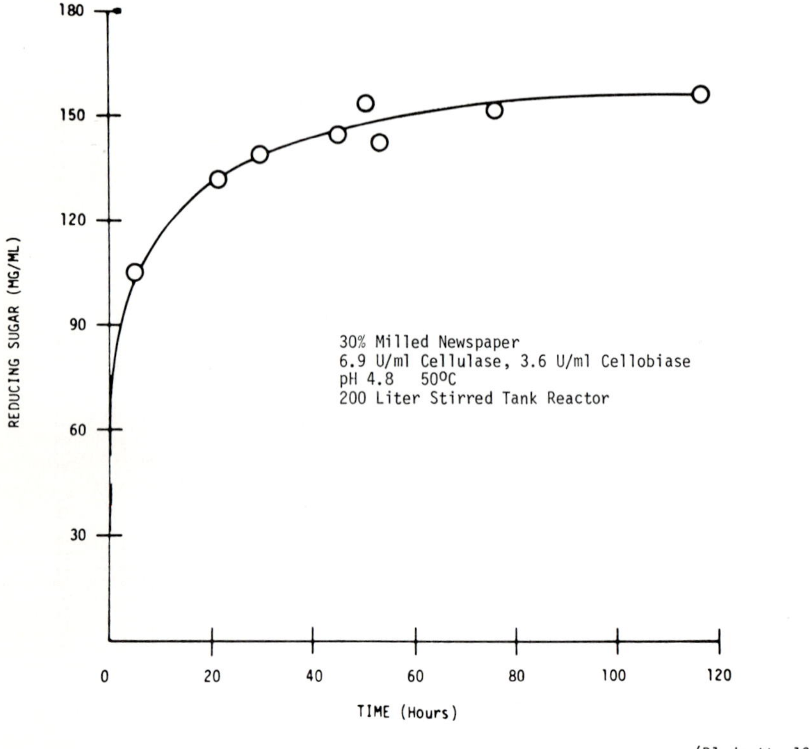

(Blodgett, 1981)

Figure 7. Hydrolysis of Newspaper by Trichoderma reesei
  Cellulase

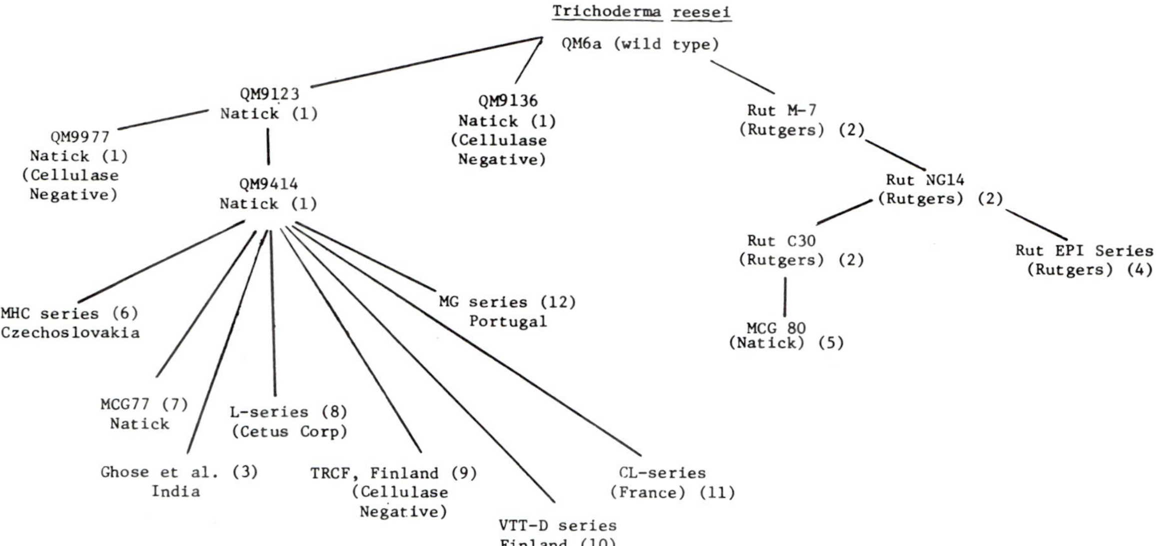

Figure 8. Worldwide Mutant Selection Programs for Cellulolytic Trichoderma reesei Strains

1) Mandels et al., 1971-1975
2) Montenecourt and Eveleigh, 1977-1979
3) Ghose et al., 1982
4) Cuskey et al., 1980-1981
5) Gallo, 1982
6) Farkas et al., 1981
7) Gallo et al., 1979
8) Shoemaker et al., 1981
9) Nevalainen and Palva, 1978
10) Nevalainen et al., 1980
11) Durand and Tiraby, 1980
12) Beja de Costa and Van Uden, 1980

(Cuskey, Montenecourt and Eveleigh, 1982)

In the past ten years we have had support from the Department of Energy in addition to our Army support. This gave us a bigger budget and attracted a larger number of bright young colleagues. Mr. Leo Spano became our program manager and our interest turned to process development and economics.

The availability of improved strains encouraged us to optimize the fermentation. The objectives are clear. The more the better, the faster the better, and the purer the better. Achieving them is more difficult. The enzyme is induced, so practically it must be produced on cellulose, and the specific activity is low, so a high cellulose concentration is required to achieve a good yield. This creates difficulty for the bioengineer because, above 6% cellulose, problems with stirring, foam and oxygen transfer become serious (Figure 9). The enzyme is a secondary metabolite. When 80% of the cellulose or nitrogen has been consumed, only 20% of the enzyme is released (Figure 10). Conditions which favor growth reduce enzyme productivity. However, cellulase is extracellular and is secreted from living cells (Figure 11).

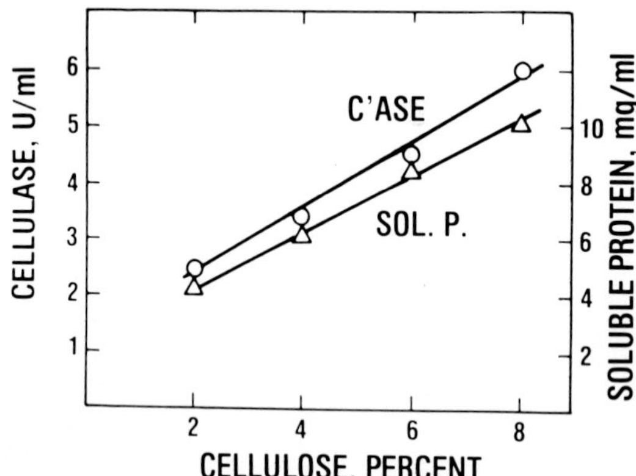

Figure 9. Effect of Cellulose Concentration in the Medium on Production of Cellulase (0) and Soluble Protein (Δ) by Trichoderma reesei QM9414. (Sternberg, 1978)

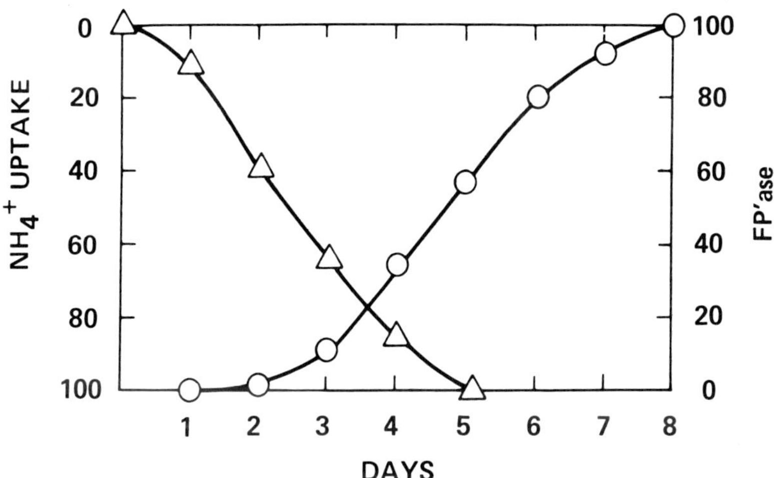

Figure 10. *Relative Values for Ammonia Uptake (Δ) and Cellulase Appearance in the Filtrate (O) for* Trichoderma reesei *QM9414 Grown on 8% Cellulose. (Sternberg and Dorval 1979).*

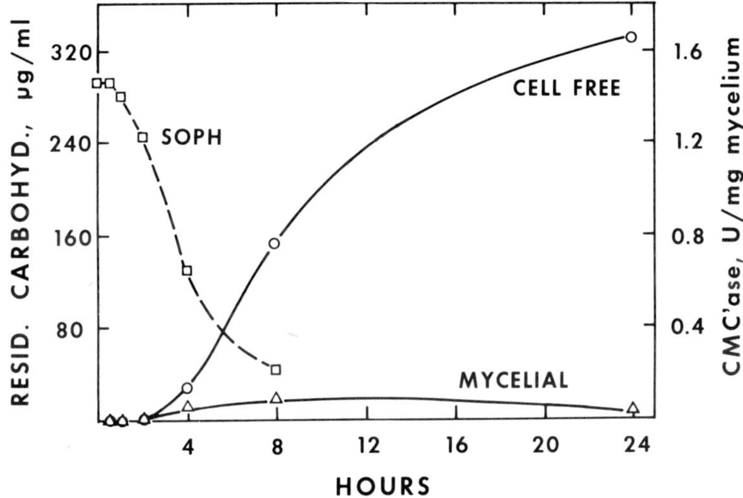

Figure 11. *Location of Cellulase (CMC'ase) and Uptake of Sophorose During Induction at 28°C, pH 2.8 With 300μg Sophorose/ml. (Sternberg and G. Mandels 1979).*

Substrate	%	Fermentation	Protein g/liter	Cellulase FPU/ml	Volumetric P* U/L Hr^{-1}	Specific P* U/g hr^{-1}
Cellulose	9	Batch	22.7	15.0	93.	–
	0.8	Continuous D=0.044	1.2	0.85	37.	17.8
Lactose	2	Batch	–	0.7	–	–
	5	Continuous D=0.028	10.2	6.0	168	17
	6	Fed Batch (Repeated)	19.8	10.1	105	6-11
Hydrolysis	0.8	Continuous D=0.046	1.5	1.1	51	19
Syrup	5	Fed Batch (Repeated)	16.4	9.7	101	7.2
Glucose	0.8	Continuous D=0.044	–	0.2	9	3

*P = Productivity

Figure 12. Cellulase Production by Trichoderma reesei MCG80

(Allen, 1982)
(Andreotti et al, 1981)

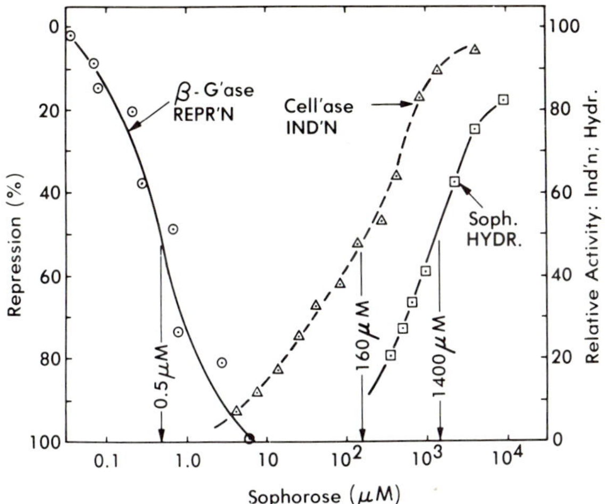

Figure 13.   Effect of Sophorose Concentration on its Role in
        Enzyme Production by Trichoderma reesei QM6a.
                        (Sternberg and G. Mandels, 1980)
        ⊙ - ⊙ Repression of β-Glucosidase Induction (Methyl-β-
            Glucoside Inducer)
        △ - △ Induction of Cellulase
        ⊡ - ⊡ Sophorose as a Substrate for β-Glucosidase,
            K_m = 1.4 mmolar

Figure 14.   Mutation versus Gene Expression
        (Effect of inducer, strain of Trichoderma reesei, and
        fermentation conditions on cellulase yield.)

Fermentation	Growth Substrate	Wild Type (QM6a) Filter Paper Cellulase	Best Mutant Units per ml
Shake Flask	1% lactose	0.07	1.0 (C30)
	1% BW200	0.7	2.2 (MCG80)
Batch			
Fermentor	6% Lactose	0.3	1.8 (MCG77)
10 Liter	6% BW200	0.6	11.9 (C30)
	6% FB Cotton	5.0	14.8 (NG14)
Fed Batch	Lactose	0.2	10.1 (MCG80)
Fermentor	Hydrolysis Sugar	11.6	9.7 (MCG80)

BW200 = Ball Milled Cellulose Pulp
FB Cotton = Two Roll (Compression Milled Cotton)
Hydrolysis Sugar = Cellulase Digest of BW200
                        (Allen, 1982)
                        (Andreotti et al, 1981)

We have investigated the cellulase fermentation by various approaches attempting to maximize yields and productivities. Fortunately, we had some excellent equipment that allowed us to monitor and control such parameters as pH, gas analysis and feed rates of various nutrients, and we also had the bright young engineers who knew how to use both the equipment and the data (Nystrom et al, 1975, 1976, 1977, 1978; Peitersen, 1977; Ryu et al, 1979, 1980). One of them is Alfred Allen who collected most of this data for MCG80, one of our best mutants (Figure 12). In batch cultures on cellulose we can obtain more than 2% extracellular protein. This is 4-5 times the level of mycelial protein. It has occurred to us that an alternate use for this cellulase might be as an edible protein, a soy substitute. It has a good amino acid profile and is free of unwanted cell components such as walls and nucleotides. In continuous cultures on cellulose, yields are low. Lactose, 4-0-β-D-galactosyl-D-glucose, is a good soluble inducer particularly for MCG80 in continuous or fed batch cultures. Recently Allen has discovered that hydrolysis sugars (produced from cellulose by cellulase) also induce, particularly in the fed batch system where soluble sugar levels are kept low to minimize repression. Artificial hydrolysis sugars containing similar levels of glucose, cellobiose and xylose do not induce. Induction must be due to sophorose which is present at low level in the hydrolysis syrup (Reilly, 1982), apparently a reversion product due to transglucosylation by cellulase or cellobiase.

In 1962 we found that sophorose, 2-0-β-D-glucopyranosyl-D-glucose, an impurity present in glucose produced by acid hydrolysis of corn starch is a cellulase inducer for *Trichoderma reesei* even at 10^{-5} M. Until recently, we regarded this as an interesting curiosity, but now we believe that sophorose may plat a role in regulation of the enzymes of the cellulase complex. As noted above it is present in hydrolysis syrups. The role of sophorose in induction and repression has been re-investigated by David Sternberg and Gabriel Mandels (1979,1980,1982). Sophorose is a substrate for cellobiase with a Km of about 10^{-3}M. At very low concentrations, 10^{-6}-10^{-7}M, it represses the induction of cellobiase by methyl-β-glucoside (Figure 13).

Our best mutants normally yield 3-20 times as much cellulase as the wild parent strain, QM6a, and cellulase concentrations for the wild strain grown on lactose or cellulose are usually less than one filter paper cellulase unit per ml. However, on two roll milled cotton, a resistant growth substrate that is very slowly consumed, it reaches a much higher enzyme level. Recently Allen has achieved almost 12 units per ml with QM6a in a fed batch fermentation on hydrolysis sugar (Figure 14).

This is almost equal to the best levels obtained under any
conditions by the mutants. It suggests that the mutants are
altered not in the cellulase genes, but in some factor related
to gene expression. Data like these should be seriously considered by anyone planning to transfer *Trichoderma* cellulase
genes to another organism.

In fact, we have known for some time that the mutants
produced the same cellulase complex as the wild strain. Frank
Bissett (1979, 1981) has developed an HPLC procedure for
separating the cellulase proteins on DEAE glass beads. The
fractions can be recovered and identified. When equal levels
of protein are applied, the wild strain and the mutants give
very similar HPLC profiles. About 70% of the protein is
included in the exo-glucanase areas. About 30% of the protein
is endo-glucanase. The cellobiase represents 1% or less of the
protein and is included in the first endo-β-glucanase peak
(Figure 15).

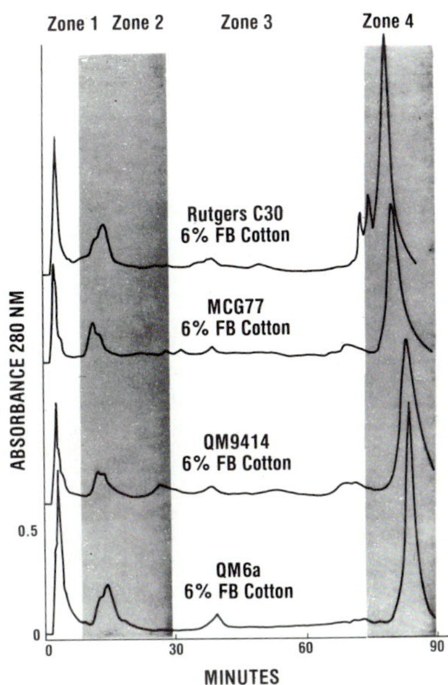

Figure 15. Separation of Endo- and Exo-β-Glucanases of
     Trichoderma reesei *by HPLC.*

    Cultures Grown on 6% Compression Milled (FB) Cotton
    Zone 1 β-Glucosidase + Endo-β-Glucanase A
    Zone 2 (Shaded) Exo-β-Glucanase A
    Zone 3 (Endo-β-Glucanase B
    Zone 4 (Shaded) Exo-β-Blucanase B

*Based on 25 x 10^6 gallons/year from 5 x 10^8 Kg Urban Waste**

(Cyclone → Light Fraction = 3.7 x 10^8 Kg Hydrolyzable Substrate)
24 Hour Hydrolysis → 1.67 x 10^8 Kg Glucose

Cost Factor	Assumption	$/Gallon 95% Ethanol
Substrate*	30% Charge to Reactor	0.13
Pretreatment	Two Roll Mill	0.22
Enzyme	10 FPU/g Substrate	0.45
Hydrolysis	45% Yield (10% syrup)	0.13
Ethanol Production	40% Yield from Sugar	0.28
Total Factory Cost	No Credit for By-Products	1.21

Enzyme productivity = 125 filter paper cellulase units/liter/hour
Enzyme utilization efficiency (24 hours) = 13%

**Separated fiber fraction at $6/ton*

Figure 16. Cost Analysis - Ethanol From Cellulose

(Spano et al, 1980)

Finally, I will say a few words about process development and economics. We have discussed the work on enzyme production. Despite all the advances, this is still the largest cost because of the large quantities required. It is not unduly expensive per kilogram. Substrate is also costly. We have investigated over 200 wastes and other cellulosic materials as possible substrates for a commercial process (Andren et al, 1975, 1976). The most promising are municipal wastes, agricultural residues, and cellulosic materials such as poplar grown on an energy plantation. Many such materials are readily hydrolyzed but fortunately most require pretreatment (Mandels et al, 1974). One pretreatment that we have investigated in some detail at Natick is the two roll or compression mill (Tassinari et al, 1977, 1980, 1982; Ryu et al, 1982). It is very effective and fairly energy efficient, about 0.2 - 0.3 Kw hours per pound of substrate. Capital costs may be high.

These are the economics developed by the engineers at Natick (Figure 16). The cost, $1.21 per gallon of ethanol, is not commercial without a subsidy. In my opinion the process will not be commercial until the enzyme requirement is substantially reduced (Mandels et al, 1981). Since the specific activity of cellulase is only 0.6 units per mg of protein, the requirement of 10 units per gram equals 15 mg of protein or 15 kg per 1000 kg of substrate. One thousand kilograms of substrate would yield 50 gallons of ethanol - in other words, 300 g of enzyme protein would be required per gallon of ethanol produced. This is roughly 100 times the amount of enzyme required to produce a gallon of ethanol from starch (corn).

At the moment there is plenty of oil and interest in ethanol or chemicals from cellulose has declined. No doubt, the pendulum will swing back soon. Challenging practical and fundamental research opportunities still exist. The critical areas are:

    a. <u>Enzyme production</u> including the search for new strains, improvement of existing strains, greater understanding of the controls over enzyme production and secretion, the role of proteases in enzyme secretion, stability and multiplicity, and further optimization of fermentation to produce enzymes.

    b. <u>Hydrolysis</u> including a search for better pretreatments that are efficient and economical, enzyme kinetics and the nature of synergism, and enzyme stabilization and desorption from hydrolysis residues to permit greater enzyme recovery and reuse (Reese et al, 1980, 1982).

Such research will be interesting to carry out and may have valuable commercial and social implications.

REFERENCES

Allen, A. L. and D. Sternberg, *Biotechnol. Bioeng. Symp. 10*, 189-197 (1980).

Allen, A. L. and R. E. Andreotti, Proceedings International Symp. on Wood and Pulping Chemistry, Stockholm, Sweden. The Ekman Days, Volume 4. Chemistry and Biochemistry of Alternative Usage of Wood Components (Biomass), 14-17 (June 1981).

Allen, A. L., Unpublished data (1982).

Andreotti, R. E., J. E. Medeiros, C. R. Roche, and M. Mandels, Proceedings Second International Symposium on Bioconversion. New Delhi, India, February 17-March 6, 1980. Ed., T. K. Ghose, Vol. 1, 353-372 (1980).

Andren, R. K., M. Mandels, and J. E. Medeiros, *Appl. Polymer Symp. 28,* 205-219 (1975).

Andren, R. K., R. J. Erickson, and J. Medeiros, *Biotechnol. Bioeng. Symp. 6,* 177-203 (1976).

Andren, R. K., M. Mandels, and J. Medeiros, *Process Biochem. 11,* 2-11 (1976).

Bissett, F. and D. Sternberg, *Appl. Env. Microbiol. 35,* 750-755 (1978).

Bissett, F. H., *J. Chromatog. 178,* 515-523 (1979).

Bissett, F. H., R. E. Andreotti, and M. Mandels, Proceedings Second International Symp. on Bioconversion, New Delhi, India. Ed. T. K. Ghose, Vol. 1, 373-388 (March 6, 1980).

Blodgett, C., Unpublished data (1981).

Brandt, D., L. Hontz, and M. Mandels, AIChE Symp. Series *69,* 127-133 (1973).

Cusky, S. M., B. S. Montenecourt, and D. E. Eveleigh, *Organic Fuels,* Ed. D. L. Wise. CRC (in press).

Gallo, B. J., R. Andreotti, C. Roche, D. Ryu, and M. Mandels, Proceedings Symp. on Biotechnology in Energy Production and Conservation, Gatlinburg, TN. *Biotechnol. Bioeng. Symp. 8,* 89-101 (May 10-12, 1978).

Ghose, T. K. and J. Kostick, in Cellulases and their Applications. *Adv. in Chem. Series, 95.* Washington, DC. Eds. G. Hajny, and E. T. Reese, 415-446 (1969).

Ghose, T. K. and J. A. Kostick, *Biotechnol. Bioeng. 12,* 921-946 (1970).

Katz, M. and E. T. Reese, *Appl. Microbiol. 16,* 419-420 (1968).

Mandels, M., F. W. Parrish, and E. T. Reese, *J. Bact. 83,* 400-408 (1962).

Mandels, M. and E. T. Reese, *Dev. Ind. Microbiol. 5,* 5-20, (1964).

Mandels, M., J. Kostick, and R. Parizek, *J. Polymer Science, Part C (36)* 445-459 (1971).

Mandels, M., J. Weber, and R. Parizek, *Appl. Microbiol. 21,* 152-154 (1971).

Mandels, M., L. Hontz, and J. Nystrom, *Biotechnol. Bioeng. 16,* 1471-1493 (1974).

Mandels, M., J. E. Medeiros, R. E. Andreotti, and F. H. Bissett, *Biotechnol. Bioeng. 23,* 2009-2026 (1981).

Montenecourt, B. S. and D. E. Eveleigh, *Appl. Env. Microbiol. 34,* 777-782 (1977).

Montenecourt, B. S. and D. E. Eveleigh, Proceedings Second Annual Symp. on Fuels From Biomass. Troy, NY, Vol. II, 613-625 (June 1978).

Montenecourt, B. S. and D. E. Eveleigh, *TAPPI*, Proceedings Annual Meeting, New York, 101-108 (March 1979).

Nystrom, J. M. and K. A. Kornuta, Proceedings Symp. on Enzymatic Hydrolysis of Cellulose. Aulanko, Finland, 181-192. (March 1975).

Nystrom, J. M. and A. L. Allen, *Biotechnol. Bioeng. Symp. 6,* 55-74 (1976).

Nystrom, J. M. and P. H. DiLuca, *Proc. Bioconversion Symp. IIT,* Delhi, India, 293-304 (1977).

Nystrom, J. M., R. K. Andren, and A. L. Allen, AIChE Symp. Series *74,* 82-88 (1978).

Peiterson, N., *Biotechnol. Bioeng. 19,* 337-348 (1977).

Peiterson, N., J. Medeiros, and M. Mandels, *Biotechnol. Bioeng. 19,* 1091-1094 (1977).

Reese, E. T., R.G.H. Siu, and H. S. Levinson, *J. Bacteriol. 59,* 485-497 (1950).

Reese, E. T., *Biotechnol. Bioeng. Symp. 6,* 9-20 (1976).

Reese, E. T., *J. Appl. Biochem. 2,* 36-39 (1980).

Reese, E. T. and D. Y. Ryu, *Enzyme Microb. Technol. 2,* 239-240 (1980).

Reese, E. T. and M. Mandels, *Biotechnol. Bioeng. 22,* 323-335 (1980).

Reese, E. T. in Solution Behavior of Surfactants, Vol. 2, Eds. Mittel and Fendler. Plenum Press, 1487-1504 (1982).

Reese, E. T., *Proc. Biochem. 17(3):2* (1982).

Reilly, P., Personal communication (1982).

Ryu, D., R. Andreotti, M. Mandels, B. Gallo, and E. T. Reese, *Biotechnol. Bioeng. 21,* 1887-1903 (1979).

Ryu, D. Y., R. Andreotti, J. Medeiros, and M. Mandels, *Enzyme Engineering 5,* 33-40 (1980).

Ryu, D. Y., S. B. Lee, T. Tassinari, and C. Macy, *Biotechnol. Bioeng. 24,* 1047-1067 (1982).

Simmons, E. G., Abstract Second Intnl. Mycological Congress, Tampa, FL 618 (1977).

Spano, L., T. Tassinari, D. Ryu, A. Allen, and M. Mandels, Proceedings Biogas and Alcohol Fuels Production Seminar, Chicago, IL, October 1979, Ed., J. Goldstein. The J.G. Press, Emmaus, PA 62-81 (1980).

Sternberg, D., *Biotechnol. Bioeng. 18,* 1751-1760 (1976).

Sternberg, D., P. Vijayakumar, and E. T. Reese, *Can. J. Microbiol. 23,* 139-147 (1977).

Sternberg, D. and S. Dorval, *Biotechnol. Bioeng. 21,* 181-191 (1979).
Sternberg, D. and G. R. Mandels, *J. Bact.* 761-769 (1979).
Sternberg, D. and G. R. Mandels, *J. Bact.* 1197-1199 (1980).
Sternberg, D. and G. R. Mandels, *Experimental Mycology 6,* 115-124 (1982).
Tassinari, T. and C. Macy, *Biotechnol. Bioeng. 19,* 1321-1330 (1977).
Tassinari, T. C. Macy, L. Spano, and D. Y. Ryu, *Biotechnol. Bioeng. 22,* 1689-1705 (1980).
Tassinari, T. H., C. F. Macy, and L. A. Spano, *Biotechnol. Bioeng. 24,* 1495-1505 (1982).
Wood, T. M., *Biotechnol. Bioeng. Symp. 5,* 111-137 (1975).
Wood, T. M. and S. McCrae, in Hydrolysis of Cellulose:Mechanism of Enzymatic and Acid Catalysis. *Adv. in Chem.Series 181,* Eds. R. D. Brown and L. Jurasek, 181-209 (1979).

Acknowledgment

We are grateful to the Departments of Defense and Energy for their generous support of this research.

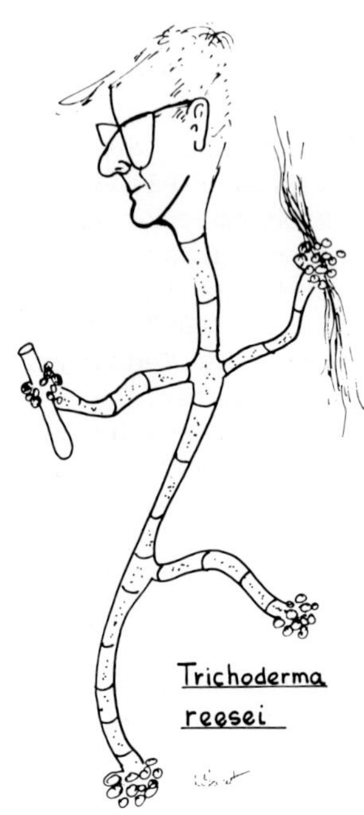

CHAPTER 2

MICROBIAL UTILIZATION OF GASEOUS ALKANES

Ching T. Hou

Exxon Research and Engineering Company
Linden, New Jersey

I. INTRODUCTION

Gaseous (C_1-C_4) hydrocarbons are abundant in nature and in various fractions produced in oil refineries. The advantages in employing gaseous alkanes as substrates for the growth of microorganisms in fermentation processes are their relatively low cost, abundance, and in the high degree of purity that is possible to obtain. There are also a number of disadvantages in using gaseous hydrocarbons. Generally, microorganisms that utilize gaseous hydrocarbons have long doubling times and low total growth yields. Furthermore, microbial growth on hydrocarbons requires a large amount of oxygen, and mixtures of gaseous hydrocarbons and oxygen can represent an explosion hazard. These disadvantages need to be overcome in order to establish an industrial process.

The progress of research in the fermentation of gaseous hydrocarbons has been relatively slow. Prior to 1970, most of the research activity on microbial oxidation of gaseous hydrocarbons was concentrated in Foster's laboratory in Texas and in Quayle's laboratory in Sheffield, England. Leadbetter and Foster (1,2) studied the oxidation products formed from gaseous alkanes by the bacterium *Pseudomonas methanica*. During the last decade, considerable industrial interest has developed in gaseous alkane-utilizing microorganisms, initially for single-cell protein production and more recently as biocatalysts or for useful biopolymers. Gaseous hydrocarbons may also be utilized to produce metabolic intermediates, e.g. organic acids, amino acids, purines, pyrimidines, and vitamins.

Methane and propane constitute the largest volumes available of gaseous hydrocarbons. Methane is the main ingredient of natural gas and is a by-product of many oil refining processes. Propane is also of widespread occurrence in petroleum deposits, and represens 1-2% of natural gas. Microorganisms capable of utilizing methane or propane are common in soils and lake water.

Recently, there have been a number of reports and reviews written regarding gaseous hydrocarbons as substrates for the growth of microorganisms (3-7). Most of these deal with microbial transformation of the hydrocarbons. The purpose of this contribution is to review information on techniques used in isolation and growth of microorganisms that utilize gaseous hydrocarbons as sole energy and carbon sources. Their industrial potential is also discussed.

II. ISOLATION OF MICROORGANISMS

Methane-utilizing microorganisms (methanotrophs) are distinct from C_2-C_4 gaseous alkane-utilizers in their ability to grow on compounds that contain no carbon-carbon bonds and to assimilate carbon as formaldehyde or a mixture of formaldehyde and carbon dioxide. However, the technique employed for the isolation of gaseous alkane-utilizers is more or less the same. The principles, media, gas mixtures and other conditions employed for this purpose have been reviewed by Dworkin and Foster (8), Quayle (9), and Whittenbury et al. (10).

Small amounts of soil or water samples are inoculated into mineral salts liquid media and then incubated under a gaseous alkane-air mixture. This usually results in the appearance of turbidity in the culture media, indicating microbial growth, in a few days to two weeks.

A. Methane Utilizers

The incubation of an inoculated, unshaken liquid-salt-medium in a methane-air atmosphere favored the growth of pink bacterium *Pseudomonas methanica* (8). When loopfuls of the liquid culture were streaked onto agar and incubated under methane and air at 30°C, pink colonies appeared in five to seven days. This technique produced pure cultures of *P. methanica*. *Pseudomonas methanitrificans* was isolated by Davis et al. (11) from soils exposed to natural gas by successive transfer in liquid media without combined nitrogen. *Methylococcus capsulatus* was isolated by Foster and Davis (12) by incubating liquid culture at 50-55°C and then streaking onto mineral-salts agar. Visible colonies formed in 10-14 days at 37°C under methane-air atmosphere.

Whittenbury et at.. (10) inoculated 25 ml of a mineral-salts medium (with NO_3 or NH_3 as nitrogen sources) in 250 ml screw-cap bottles with approximately 1 gm of soil or water samples. The bottles were sealed with Subaseal caps and 20 ml of methane were injected. The enrichments were incubated statically at

30, 45, and 55°C. Turbidity, often accompanied by a pellicle
that appeared after three to four days of incubation was taken
as evidence of the growth of methane utilizers.

These cultures were serially diluted in sterile tap water
and spread onto mineral-salt agar plates. The plates were
incubated in a methane-air mixture in vacuum desiccators or
polyethylene containers. Colonies of non-methane-utilizing
bacteria reached their maximum size in about three days.
Colonies of methanotrophs began to appear after about five to
seven days and were about 10 to 100-fold fewer in number than
nonmethane-utilizers. These colonies were picked with the aid
of a plate microscope and were transferred with a straight wire
to mineral-salt agar slopes, which were incubated for two weeks.
Isolation was most successful at the small colony stage (0.2 mm
diam.). With this technique, Whittenbury et al. (10) successfully obtained pure cultures of more than 100 methanotrophs.
The isolation of cultures from microcolonies as soon as they
were visible under the plate microscope and frequent observa-
tion of the plates seem to be the keys to the success.

Galchenko (13) applied silica gel plates for the isolation
of methanotrophs. Soil and water samples were inoculated into
a mineral-salts medium and incubated, without shaking, with a
methane-air mixture at 30°C. These enrichment cultures were
serially diluted and were used to inoculate media solidified
with silica gel. Microcolonies were picked from silica gel
plates with the aid of a plate microscope and were transferred
to silica gel slopes. The use of silica gel rather than agar
has been reported to be more effective because of its reduced
toxicity for methanotrophs and because fewer colonies of
nonmethane-utilizers developed on the plates.

Another effective technique involved the use of nitro-
cellulose filters which were wetted with a mineral-salt
medium (14). Bacteria were streaked on the surfaces of filters
with inoculating loops to obtain clones. The filters were
incubated under a methane-air atmosphere. This technique also
reduced the amount of contamination by nonmethane-utilizing
bacteria and permitted growth of strains that were sensitive
to substances in agar.

Recently, Hou et al (15) isolated from soil and water
samples more than 20 new strains of methanotrophs employing an
enrichment technique similar in principle to that of
Whittenbury et al. (10). Approximately 1 gm of soil or water
sample was inoculated into a 50 ml mineral-salts medium (12)
in a 300 ml flask (Figure 1). Flasks were fitted with a
rubber stopper with a glass tube and clamps for gassing. The

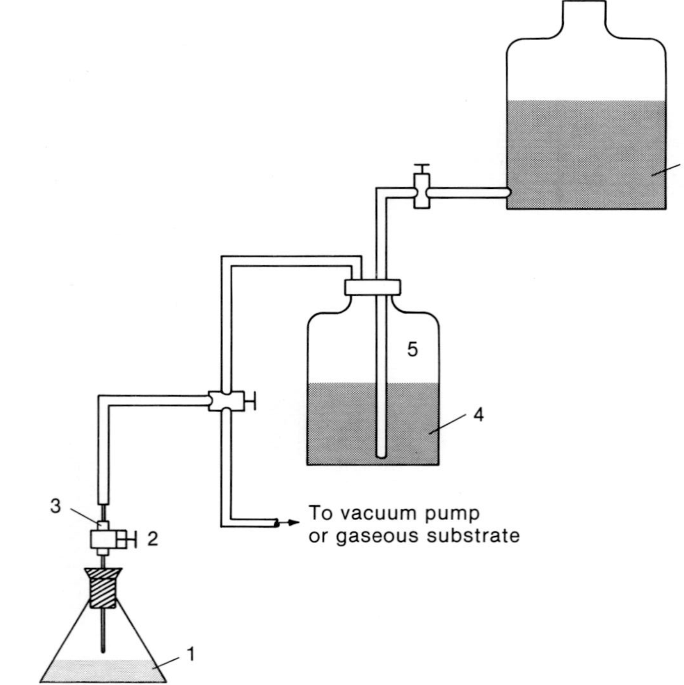

Figure 1. Culture flask and gassing device used in studies on gaseous hydrocarbon-utilizing microorganisms. 1) mineral salts medium; 2) clamp; 3) rubber tube filled with glass wool or cotton; 4) water; 5) mixture gases.

gaseous phase of the flask was evacuated and replaced with a
mixture of methane-air (1:1 v/v). A simple gas replacement
device utilizing the house vacuum system was constructed in the
laboratory (Fig. 1). In each operation, the vacuum cycle took
about three minutes, and the refill of methane-air mixture gas
took about two minutes. Two cycles of gas exchange were performed per flask routinely. The inoculated flasks were incubated in a shaker at 30°C. After three to four days of
incubation, turbidity appeared, indicating growth of microorganisms. Transfers were made to new flasks three times to
encourage the growth of methanotrophs. The cultures in the
final flask were streaked onto plates of a mineral-salts agar
medium and incubated under methane-air atmosphere to obtain
single colonies. Colonies of methanotrophs were observed in
five to ten days. These colonies were picked and were transferred to another plate. Serial transfers of a single colony
resulted in successful isolation of pure cultures. These pure
cultures were re-inoculated into liquid flasks to make sure of
their ability in growth on methane. The organisms were maintained on mineral salt plates in a desiccator jar under an
atmosphere of methane and air (1:1, v/v) at 30°C. It is
important to keep the desiccator jar dry by placing silica gel
at the bottom of the jar to help avoid fungal contamination.
By using this isolation technique, over 20 pure cultures of
new methanotrophic strains were isolated by Exxon's group (15)
(Table 1). These methanotrophs were identified by the classification given by Whittenbury et al. (10) and were deposited
at the Northern Regional Research Laboratory in Peoria, Illinois.

B. C_2-C_4 Alkane-Utilizers

Gaseous alkane-utilizing bacteria can be isolated readily
from soil or water samples by enrichment techniques similar to
those described for the isolation of methanotrophs. Davis et
al. (16) employed an atmosphere of ethane-oxygen-nitrogen
(40-20-40 v/v) for isolating *Mycobacterium paraffinicum*. An
atmosphere of propane-air (1:3 v/v) was used by Bokova (17)
for the isolation of several propane-oxidizing bacteria, and
she stated that 0.01% propane in air was sufficient to obtain
measurable growth. Kuznetsov and Telegina (18) also used an
atmosphere of propane-air (1:3 v/v) for isolation of propane-
utilizers from soil taken around gas wells. The optimum
growth atmosphere for *Mycobacterium vaccae* JOB5 (19) was
reported to be propane-oxygen-nitrogen (50:40:10 v/v). In
general, a range of 30 to 50% of gaseous alkane in the atmosphere of a growth container is sufficient to enrich for
gaseous alkane-utilizing microorganisms.

Table 1. *New methanotrophs isolated by Exxon's*
          research group (15)

Obligate type I

Methylomonas *sp.* CRL 4	NRRL B-11,205
Methylomonas *sp.* CRL 8	NRRL B-11,207
Methylomonas *sp.* CRL 10	NRRL B-11,212
Methylomonas *sp.* CRL 17	NRRL B-11,208
Methylomonas *sp.* CRL 20	NRRL B-11,211
Methylomonas *sp.* CRL 21	
Methylomonas *sp.* CRL 22	NRRL B-11,209
Methylomonas *sp.* CRL M6P	
Methylobacter *sp.* CRL 5	NRRL B-11,218
Methylobacter *sp.* CRL M6	NRRL B-11,214
Methylobacter *sp.* M1Y	NRRL B-11,216
Methylobacter *sp.* CRL M19	NRRL B-11,217
Methylobacter *sp.* CRL 23	NRRL B-11,215
Methylococcus *capsulatus* CRL M1	NRRL B-11,219
Methylococcus *sp.* CRL 24	NRRL B-11,220
Methylococcus *sp.* CRL 25	NRRL B-11,221

Obligate type II

Methylosinus *sp.* CRL 15	NRRL B-11,202
Methylosinus *sp.* CRL 16	
Methylocystis *sp.* CRL 18	NRRL B-11,204

Facultative

Methylobacterium *sp.* CRL 26	NRRL B-11,222

A basal-salts medium with a broad array of trace elements
described by Leadbetter and Foster (20) was modified by
Perry (4). He lowered the level of NaNO$_3$ from 0.2% to 0.1%
added NH$_4$Cl at 0.1% to the basal-salts medium of Leadbetter
and Foster and found this to be a more effective basal-salts
medium.

McLee et al. (21) isolated 15 bacterial cultures that
utilized ethane, propane, or n-butane as a source of carbon
and energy. The bacteria belonged to two genera: *Brevibacter*
and *Arthrobacter*. Smirnova and Taptykova (22) examined 22
nocardial stains belonging to 15 different species and found
that two strains grew on ethane (poorly) and nine on propane.
Kuznetsov and Telegina (18) found that mycobacteria and
psuedomonads were major propane oxidizers among several
isolates. Cerniglia and Perry (23) isolated propane-utilizing
fungi: *Cunninghamella elegans* and *Penicillium zonatum*.

Davis et al. (24) isolated 12 fungi from raw sewage by enrichment with ethane as substrate. All were *Hyphomycetes* and grew with ethane, propane, or n-butane as the sole carbon and energy sources. A number of propane-utilizing mycobacteria were isolated by Perry and his co-workers: *Mycobacterium rhodochrous* OFS (25); *Mycobacterium convolutum* R-22 *and* M. rhodochrous A-78 (26); *M. rhodochrous* OC2A (19); *M. convolutum* NPA-1 (27); and *M. vaccae* JOB5 and *M. rhodochrus* 7EIC (28).

Kormendy and Waymen (29) observed the formation of intracellular structures in n-butane-grown and not in glucose-grown *Arthrobacter* species and *Candida utilis*. These substrate-specific structures consisted of electron-dense bodies nearly surrounded by a round electron-transparent area, and connected to the main cytoplasm by a bridge. These structures may represent the sites of hydrocarbon oxidation and are named "oxisomes."

Recently, Hou et al. (30, 31) used an enrichment technique similar to that employed for methanotrophs to isolate C_2- C_4 gaseous alkane-utilizing bacteria from both soil and water samples. The atmosphere used in these cases was a mixture of propane-air (1:2 v/v) provided through the device shown in Fig. 1. They isolated 20 pure cultures of new C_2-C_4 gaseous alkane utilizers covering a broad spectrum of genera (Table 2). Eighteen cultures known for their ability to utilize liquid alkane were selected from the culture collections of either the Northern Regional Research Laboratory (Peoria, Illinois) or the American Type Culture Collection (Rockville, Maryland) to examine their ability to utilize gaseous alkanes. All of them were found able to grow on propane as the sole carbon and energy source (Table 3). These findings suggested that C_2-C_4 gaseous alkane utilizers are of widespread occurrence in microbial populations.

III. GROWTH OF MICROORGANISMS

Small-scale cultures of gaseous alkane-utilizing organisms were grown at 30°C in 300-ml flasks containing 50 ml of mineral salts medium with gaseous alkane and air (1:2 v/v) as the sole carbon and energy source. Cultures were incubated at 30°C on a rotary shaker at 200 rpm. Maximum cell yield was reached after two to three days of incubation. For larger volumes, a 50-ml two-day culture was used to inoculate a 2.8-liter flask containing 800 ml of mineral salts medium with an atmosphere of gaseous alkane and air (1:2 v/v). Still larger-scale cultures were grown in a 30-liter explosion-proof fermentor containing 22 liters of mineral salts medium with a continuous supply of the gas mixture which consisted of gaseous alkane

*Table 2.   C₂-C₄ gaseous alkane utilizers newly isolated from
           soil and water samples by Hou et al. (30, 31)*

Ethane utilizers

Pseudomonas *sp.*	*NRRL B-11,329*
Arthrobacter *sp.*	*NRRL B-11,316*
Brevibacterium *sp.*	*NRRL B-11, 320*
Mycobacterium *sp.*	*NRRL B-11,324*

Propane utilizers

Acinetobacter *sp.*	*NRRL B-11,313*
Acinetobacter *sp.*	*NRRL B-11,314*
Arthrobacter *sp.*	*NRRL B-11,315*
Brevibacterium *sp.*	*NRRL B-11,318*
Brevibacterium *sp.*	*NRRL B-11,319*
Corynebacterium *sp.*	*NRRL B 11,321*
Mycobacterium *sp.*	*NRRL B-11,322*
Mycobacterium *sp.*	*NRRL B-11,323*
Nocardia *sp.*	*NRRL B-11,325*
Nocardia *sp.*	*NRRL B-11,326*
Nocardia *sp.*	*NRRL B-11,327*
Pseudomonas *sp.*	*NRRL B-11,330*
Pseudomonas *sp.*	*NRRL B-11,331*
Pseudomonas *sp.*	*NRRL B-11,332*

Butane utilizers

Arthrobacter *sp.*	*NRRL B-11,317*
Pseudomonas *sp.*	*NRRL B-11,333*
Brevibacterium *sp.*	*NRRL B-11,320*

Table 3. Cultures originally known to grow on liquid alkanes
also showed ability to grow on C_2-C_4 gaseous alkanes

Ethane utilizers	
Brevibacterium fuscum	ATCC 15993
Mycobacterium rhodochrous	ATCC 29670
Rhodococcus rhodochrous	ATCC 21198
Pseudomonas aeruginosa	ATCC 15525
Propane utilizers	
Arthrobacter petroleophagus	ATCC 21494
Arthrobacter simplex	ATCC 21032
Brevibacterium *sp.*	ATCC 14649
Brevibacterium fuscum	ATCC 15993
Hydrogenomonas eutropha	ATCC 17697
Mycobacterium album	ATCC 29676
Pseudomonas aeruginosa	ATCC 15525
Pseudomonas aeruginosa	ATCC 15528
Pseudomonas crucurae	NRRL B-1021
Pseudomonas fluorescens	NRRL B-1244
Pseudomonas cepacia	ATCC 17616
Pseudomonas putida	ATCC 17453
Rhodococcus rhodochrous	ATCC 19140
	ATCC 29670
	ATCC 29672
	ATCC 21198
Rhodococcus *sp.*	ATCC 21499
Butane utilizers	
Brevibacterium fuscum	ATCC 15993
Rhodococcus rhodochrous	ATCC 21198
Rhodococcus rhodochrous	ATCC 29670

Table 4. Cell yields from various alkane substrates

Organisms	Substrate and amount	Cell Yield (g/liter) or* (g/g substrate)	Reference
Nocardia sp.	propane gas flow, 15% in air	12.8	11
	n-butane gas flow, 15% in air	12.3	11
HR mixed culture	methane 8% in air	0.62*	33
Nocardia spp.	propane 16% in air	0.3–0.9	34
M 45 mixed culture	methane, fed continuously	0.62*	35
Arthrobacter simplex *B-129*	propane 16% in air	0.3	36
Norcardia paraffinica *Ky 4334*	propane 3 ℓ/min	30.0	37
	n-butane 3 ℓ/min	22.0	37
Candida rigida *Mo 113*	n-butane 0.1%, continuous	29.0	38
Brevibacterium ketoglutamican	n-butane	13.0	38
Mycobacterium cuneatum	methane 33% in air	0.10	39
Mycobacterium petroleophilum	ethane 33% in air	0.19	39
Mycobacterium cuneatum	6% propane + 25% butane in air	0.30	39
Mycobacterium petroleophilum	propane 33% in air	0.25	39
Methylomonas *FERM P-2400*	methane 15% in air, 200 mℓ/min	16.3	40
Methylmonas sp.	Methane 19.5% air 77.5% CO_2 3%	0.55	41
Methylococcus capsulatus	methane-air (1:3), continuous 35 Nm^3/hr	10.5	42
Mixed culture	methane-air (6:24 v/v hr)	3.4	43
Pseudomonas methanica 102/71	methane-air (1:0.2)	30.0	44
Pseudomonas butanovora *sp.*	butane-Co_2-air (40:30:400)	0.83*	45
Graphium *sp.*	natural gas-air (60:40)	0.08	46
Pseudomonas fluorescens *NRRL B-1244*	propane-air-(15:85)	11	
Methylococcus *sp. CRL 31*	methane-air-CO_2 (15:80:5)	5.4	

and air (15:85 v/v). In many cases, the presence of a small
amount of carbon dioxide in the gas mixture (for example,
gaseous alkane:air:CO_2, 15:80:5 v/v), particularly for those
methanotrophs which utilize the serine pathway for their carbon
assimilation, proved to be useful for better cell yield.
Typical cell yields for methane- and propane-utilizers grown in
the 30-liter fermentor with continuous flow of substrate gases
at one liter per min. were 5.4 and 11 g cells/liter, respectively.
In order to achieve maximum rates of gas-liquid mass transfer,
the gases are dispersed as small bubbles throughout the fermentation broth. It is important to achieve high gaseous substrate conversion efficiencies in the fermentor. During the
growth of cells from gaseous hydrocarbon, either single or
double gaseous substrate limitation can occur. The yield
coefficient, productivity, and mass transfer for SCP production
from methane were discussed by Hamer et al. (32). The yield
coefficient is affected by oxygen consumption and carbon
dioxide production. For example, to achieve a yield coefficient
of 0.9 g cells/g methane, it requires 0.065 moles oxygen and
the production of 0.03 moles carbon dioxide per g cells. Yield
coefficient is also affected by productivity, minimum oxygen
transfer rate and the heat production.

Cell yields from various gaseous alkane substrates are
listed in Table 4. They varied from laboratory to laboratory
due to different growth conditions. Data also showed that
considerably greater growth can be obtained with a continuous
flow of gaseous hydrocarbon into the system.

IV. BIOCHEMISTRY

1. Methanotrophs
There is evidence that obligate methanotrophs oxidize
methane by a series of two-electron oxidation steps, through
methanol, formaldehyde, formate, and carbon dioxide. The
initial enzymatic attack on methane involves a monooxygenase
that converts methane to methanol using molecular oxygen.
Methane monooxygenase has been purified from different types
of methanotrophs. The enzyme obtained from a type 1 methanotroph, *Methylococcus capsulatus* (Bath) comprises three components
(47,48). Protein A is an iron protein with a molecular weight
of about 200,000 daltons and contains sub-units of 68,000 and
47,000 daltons. It probably binds substrate after being
reduced by protein C which has a molecular weight of 44,000
daltons and contains one FAD, 2 g atoms of nonheme iron and
2 moles of acid-labile sulfide per mole of protein. Protein B
is colorless, with a molecular weight of 15,000 daltons. Its
role in the reaction is uncertain. The methane monooxygenase
purified from a type II methanotroph, *Methylosinus trichosporium*
OB3b, also comprises three components (49): a high potential
CO-binding cytochrome, a copper protein, and a third protein;

the molecular weights are 13,000, 47,000, and 9,400 daltons,
respectively. This enzyme cannot be reduced directly by NADH
but utilizes ascorbate as an artificial electron donor. The
methane monooxygenase purified from a facultative methanotroph,
Methylobacterium sp. CRL 26 resembles that obtained from the
Bath strain (50).

In methanotrophs, methanol is oxidized to formaldehyde by
methanol dehydrogenase. This enzyme has been purified from at
least 16 strains. Colby et al. (51) classified these enzymes
into four groups based on their molecular weight and substrate
specificity. The prosthetic group of methanol dehydrogenase
has recently been identified as an unusual quinone, 4,5-dihydro-
4,5-dioxo-1H-pyrolo 2,3-quinoline-2,7,9-tricarboxylic acid.

Formaldehyde occupies a key position in the oxidative
pathway as it is either assimilated into cell mass or dissimilated
to carbon dioxide to generate energy and growth. In the diss-
imilation pathway, formaldehyde is oxidized to formate by
either methanol dehydrogenase or one of several different
aldehyde dehydrogenases. NAD-linked formaldehyde dehydrogenases
are flavoproteins of two general types, glutathione-dependent
or independent (51). Another type of formaldehyde dehydrogenase
requires dichlorophenol indophenol *in vitro* rather than NAD.

In methanotrophs, NAD-linked formate dehydrogenase is the
last enzyme in the dissimilative pathway of methane or methanol.
Formate dehydrogenase from methanol-grown *Achromobacter
parvulus* (52) was purified recently. Its molecular weight is
relatively small, about 80,000, and it consists of two identical
subunits. This enzyme is specific for formate and NAD as the
electron acceptor. Formate dehydrogenase from methane-utilizing
bacteria has not been purified due to its extreme instability.

In the carbon assimilation pathway of methanotrophs,
formaldehyde is assimilated into cell constituents through
either the ribulose monophosphate pathway (51) or the serine
pathway (53). The net synthesis of ribulose monophosphate
pathway is one molecule of triose phosphate from three
molecules of formaldehyde:

$$3HCHO + ATP \rightarrow triose\ phosphate + ADP$$

The net synthesis in the serine pathway is one molecule of
phosphoglycerate from two molecules of formaldehyde and one of
carbon dioxide:

$$HCHO + CO_2 \rightarrow CH_2OP.CHOHCOOH$$

2. C_2-C_4 Gaseous Alkane-Utilizers

Alkanes can be metabolized through either terminal or subterminal oxidation. In terminal oxidation, alkanes are oxidized into primary alcohols, aldehydes, and acids. The acids are metabolized through acetyl CoA or acyl CoA. Blevins and Perry (26) reported that propionic acid was metabolized via the methyl malonate pathway (Fig. 2) by *Mycobacterium vaccae* JOB5.

Subterminal oxidation was also reported for propane and butane by *M. smegmatis* 422 (54) and *M. vaccae* (55). The presence of isocitrate lyase and malate synthase in propane-, isopropanol-, and acetone-grown cells and the absence of detectable levels of isocitrate lyase in cells following growth on propionate or glucose suggests a two-carbon intermediate in propane metabolism (Fig. 3).

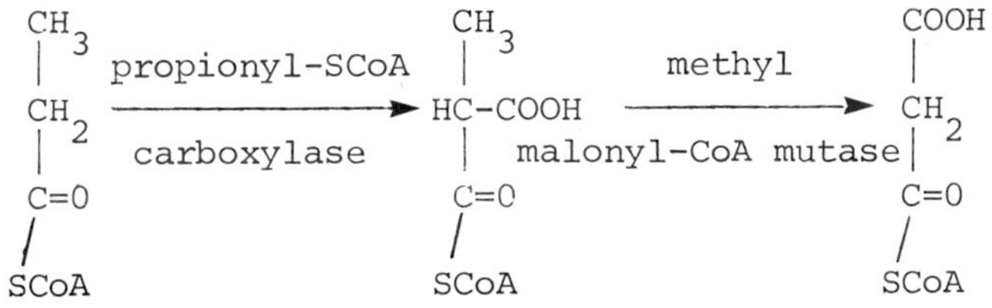

$$
\begin{array}{ccccc}
CH_3 & & CH_3 & & COOH \\
| & propionyl\text{-}SCoA & | & methyl & | \\
CH_2 & \xrightarrow{\hspace{2cm}} & HC\text{-}COOH & \xrightarrow{\hspace{2cm}} & CH_2 \\
| & carboxylase & | & malonyl\text{-}CoA\ mutase & | \\
C{=}0 & & C{=}0 & & C{=}0 \\
/ & & / & & / \\
SCoA & & SCoA & & SCoA
\end{array}
$$

Figure 2. Pathway for propionic acid utilization in
            Mycobacterium vaccae *strain JOB5 (4)*

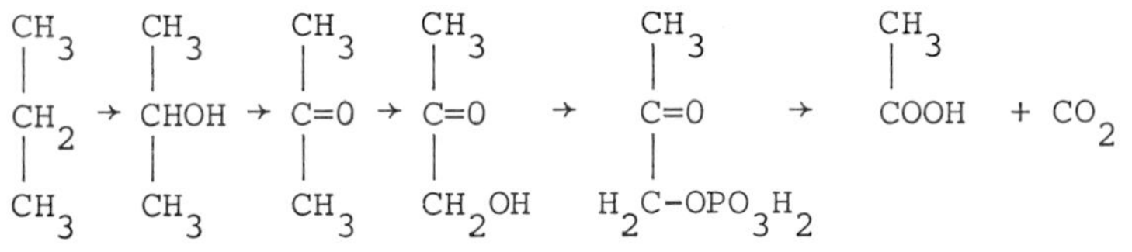

$$
\begin{array}{ccccccc}
CH_3 & CH_3 & CH_3 & CH_3 & & CH_3 & CH_3 \\
| & | & | & | & & | & | \\
CH_2 & \to CHOH & \to C{=}0 & \to C{=}0 & \to & C{=}0 & \to COOH + CO_2 \\
| & | & | & | & & | & \\
CH_3 & CH_3 & CH_3 & CH_2OH & & H_2C\text{-}OPO_3H_2 &
\end{array}
$$

Figure 3. Subterminal oxidation of propane in Mycobacterium
            vaccae *JOB5 (4)*

Table 5. Amino acid analyses of SCP produced from various gaseous alkanes (percent by weight)

	Methylococcus capsulatus(32)[a]	*Pseudomonas butanovora(56)[c]*	*Mycobacterium petroleophilum(39)[b]*	*Arthrobacter Simplex B129(36)[b]*
Leucine	*8.10*	*9.0*	*6.62*	*7.4*
Isoleucine	*4.28*	*4.8*	*3.74*	*3.5*
Valine	*6.46*	*7.0*	*5.17*	*5.5*
Threonine	*4.56*	*5.0*	*4.21*	*3.4*
Methionine	*2.67*	*2.9*	*1.38*	*0.5*
Lysine	*5.71*	*5.7*	*4.15*	*4.4*
Phenylalanine	*4.58*	*4.7*	*3.75*	*3.8*
Arginine	*6.22*	*7.9*	*5.58*	*NG*
Histidine	*2.22*	*2.2*	*1.98*	*NG*
Cystine	*0.60*	*0.3*	*0.32*	*NG*
Serine	*NG*	*4.1*	*3.87*	*2.4*
Glutamic acid	*NG*	*12.1*	*11.0*	*8.7*
Proline	*NG*	*4.0*	*3.92*	*2.9*
Glycine	*5.10*	*6.1*	*5.56*	*3.7*
Alanine	*NG*	*8.9*	*8.34*	*7.3*
Tyrosine	*3.80*	*3.6*	*2.41*	*0.8*
Aspartic acid	*NG*	*10.1*	*8.31*	*7.3*

a) Methane-grown, g/16 g N; b) Propane-grown; c) Butane-grown; NG: Data not given

V. APPLICATIONS

A. Single Cell Protein

The possibility of using single-cell protein to alleviate
the world-wide protein shortage has received considerable
attention during the past two decades. Single-cell protein by
hydrocarbon-utilizing microbes has also been explored. Natural
gas is abundant in nature. Various gaseous hydrocarbons are
often a waste product in petroleum refining due to their low
concentration. The conversion of this relatively low-cost
substrate to utilizable food would have considerable potential
from both an economic and a humanitarian viewpoint.

The crude protein content, in percent dry cell weight,
were 55% for methane-grown *Mycobacterium cuneatum* (39), 56.3%
for propane-grown *Mycobacterium petroleophilum* (39), and 73%
for butane-grown *Pseudomonas butanovora* (45).

Amino acid analyses of SCP produced from propane, butane,and
methane are compared in Table 5. The quality of SCP produced
from gaseous alkanes was reported to be comparable with those
of high quality proteins and was also highly palatable.

B. Biocatalysis

Leadbetter and Foster (2) demonstrated the production of
methyl ketones from gaseous alkanes with methane-grown
Pseudomonas methanica. This strain was able to oxidize but not
assimilate propane and butane, in the presence of growth substrate (methane). Products of this co-oxidation of propane
were n-propanol, propionic acid, and acetone; butane yielded
n-butanol, butyric acid, and 2-butanone. Subsequently, Lukins
and Foster (54) reported that propane-grown *Mycobacterium
smegmatis* produced much more methylketones and less n-propanol.

Hou et al. (57-59) found that resting cell suspensions of
methane-grown cells oxidized gaseous alkenes to their corresponding 1,2-epoxides, which accumulated. They also found that
these biocatalysts (cells) oxidized secondary alcohols to the
homologous methyl ketones (38, 60-65). Subsequently, they also
observed the hydroxylation of gaseous alkanes by these resting
cell suspensions of methanotrophs (66-72). Independently, the
ability of methanotrophic microorganisms to oxidize a wide range
of hydrocarbons was reported from two research groups in
England. Dalton compared 12 different methane-utilizing strains
for their ability in epoxidation of propylene (73). *Methylomonas
methanica* was found to produce propylene oxide at a much higher

*Table 6. Products from the epoxidation of alkenes
and other compounds by cells of methanotrophs*

Substrate	Product
Ethylene	*ethylene oxide*
propylene	*propylene oxide*
but-1-ene	*1,2-epoxybutane*
butadiene	*1,2-epoxybutene*
isobutylene	*1,2-epoxyisobutane*
cis-but-2-3n3	*cis-2,3-epoxybutane*
	cis-2-buten-1-01
trans-but-2-ene	*trans-2,3-epoxybutane*
	trans-2-buten-1-01
isoprene	*1,2-epoxyisoprene*
3-methyl-1-butene	*3-methyl-1,2-epoxybutane*
2-methyl-1-butene	*2-methyl-1,2-epoxybutane*
carbon monoxide	*carbion dioxide*
dimethyl ether	*methanol and formaldehyde*
toluene	*benzyl alcohol + cresol*
benzene	*phenol*
benzyl alcohol	*benzaldehyde + ρ-hydroxybenzyl alcohol*
o-cresol	*5-methyl,1,3-benzene diol*
ethyl benzene	*benzoic acid + 2-phenyl ethanol + phenyl acetate + ρ-hydroxyethyl benzene*
phenol	*catechol + 1,4-dihydroxy benzene*
pyridene	*pyridene-N-oxide*
styrene	*styrene epoxide*
naphthalene	*1-naphthol*
isopropyl benzene	*ρ-hydroxy isopropyl benzene*
ρ-xylene	*4-methylbenzoic acid*

rate. Dalton's group was the first to report a soluble methane monooxygenase activity from *Methylococcus capsulatus* (Bath) which catalyzed the oxygenation of alkanes, alkenes, ethers, cyclic, alicyclic, and aromatic compounds (74). Stirling and Dalton (75) demonstrated that the properties of methane monooxygenase from extracts of *Methylosinus trichosporium* OB3b and from *Methylococcus capsulatus* (Bath) are similar. Higgins' group also found that whole organism suspensions of methane-grown *Methylosinus trichosporium* OB3b catalyzed the biotransformation of various hydrocarbons (76). Several patents covering these topics were also filed by these two groups (77-80). Furthermore, cells of methanotrophic microorganisms were found to epoxidize and hydroxylate C_4 and C_5 branch-chain alkenes and alkanes (81). The products of the epoxidation of alkenes and the hydroxylation of alkanes and other compounds are summarized in Tables 6 and 7.

Cells of C_2-C_4 gaseous alkane-grown microorganisms were also found to epoxidize short-chain alkenes to their corresponding 1,2-epoxides (82). Among the substrate alkenes, propylene was oxidized at the highest rate. In contrast to the case with methanotrophic bacteria, the product epoxides are further metabolized. Recently, we reported the production of alcohols (primary and secondary) and methyl ketones from n-alkanes by cell suspensions of C_2-C_4 gaseous alkane grown bacteria (83,84). Among the n-alkanes, propane and n-butane were oxidized at the highest rate. Recently, NAD-linked 1,2-propanediol dehydrogenase activity was detected in cell-free crude extracts of various propane-grown bacteria (85). Data obtained indicated that this enzyme may be inducible by metabolites of propane subterminal oxidation. 1,2-Propanediol dehydrogenase was purified from propane-grown *Pseudomonas fluorescens* NRRL B-1244. The enzyme properties including immunological and catalytic properties were studied in detail (85). Hou et al. (86) also identified and purified a thermally stable secondary alcohol dehydrogenase from propane-grown *Pseudomonas fluorescens* NRRL B-1244. This enzyme is NAD-dependent, and oxidizes secondary alcohols, notably 2-propanol, 2-butanol, and 2-pentanol. The pH and temperature optima for secondary alcohol dehydrogenase activity were 8-9, and 60-70°C, respectively. It shows good thermal stability and the ability to catalyze reactions at a temperature as high as 85°C.

C. Production of Biopolymers

Davis (87) reported the production of a copolymer of β-hydroxybutenoic acids by *Nocardia sp.* during growth on propane or n-butane. The polymer represents about 13-14% of the dry weight of these gaseous alkane-grown cells. Poly-β-hydroxybutyrate (PHB) has been considered as a storage polymer

*Table 7. Products from the hydroxylation of alkanes by
methanotroph cells and their enzyme*

Substrate	Product
methane	methanol
ethane	ethanol
propane	1-propanol + 2-propanol
butane	1-butanol + 2-butanol
pentane	1-pentanol + 2-pentanol
hexane	1-hexanol + 2-hexanol
heptane	1-heptanol + 2-heptanol
octane	1-octanol + 2-octanol
chloromethane	formaldehyde
bromomethane	formaldehyde
fluoromethane	formaldehyde
isobutane	isobutanol + ter. butanol
hexadecane	hexadecane-1-01
cyclohexane	cyclohexanol + 3-hydroxycyclohexanone
cyclohexanol	3-hydroxycyclohexanone
isopentane	methyl isopropyl ketone
2,2-dimethyl propane	neopentyl alcohol

for methanotrophic microorganisms. Accumulation of PHB in
Pseudomonas methanica in large quantities (25% of the dry
weight of four-day-old cultures) was first observed by Kallio
and Harrington (88,89). Many of the methane-utilizing strains
identified by Whittenbury et al. were packed with lipid
inclusions, PHB (10). PHB was also found in vibrio shaped
methane oxidizing bacteria (90) and in thermophilic and thermotolerant methane utilizers (87). Methane-grown type II
obligate methanotroph, *Methylosinus trichosporium* OB3b, was
found to possess PHB up to 30% of total dry weight under
various conditions (92-94). PHB is an aliphatic thermoplastic
polyester with several properties similar to polypropylene.
It is biodegradable, nontoxic, and binignant to tissue. It is
a strong candidate as a surgical material. In addition, its
unusual electrical properties give it potential to become
commercially viable.

VI. *CONCLUSION*

The potential applications for gaseous alkane-utilizing
microorganisms, as described above, are single cell protein,
biocatalysis for the production of chemicals, and the production of useful biopolymers.

Gaseous alkanes have been known to be more advantageous
than liquid alkanes as substrates for the production of biomass.
The monooxygenases that catalyze the initial oxidative attack
on hydrocarbons are either inducible or present at higher
activity in gaseous alkane-grown cells. Therefore, for the
application of cells or enzymes in biocatalysis, it is preferred
to have the cells grown on gaseous n-alkanes rather than grown
on oxygenated hydrocarbons (liquid). Among gaseous alkanes,
ethane, propane, and n-butane are more advantageous than
methane in terms of mass transfer rate. The transfer rates
of these alkanes into water are 1.5-2 times higher than that
of methane under the same conditions (95,96). This translates
into smaller fermentor for these alkane substrates other than
methane for the same mass transfer efficiency. However,
methane is the cheapest and most abundant among the gaseous
alkanes.

Other areas, such as the production of metabolic intermediates by gaseous alkane-utilizers have not been explored.
Undoubtedly, there are drawbacks for using gaseous alkanes
as substrates in fermentation processes. The gaseous
hydrocarbon-utilizers are generally slow growers and yield
less economically feasible amounts of metabolic intermediates.
Improvements can be expected through mutation or recombinant
DNA techniques. The explosive nature of gaseous hydrocarbon
substrates in fermentation processes presents a definite
challenge to process engineers.

REFERENCES

1. Leadbetter, E.R. and J.W. Foster, *Arch. Biochem. Biophys.*
      82, 491 (1959).
2. Leadbetter, E.R. and J.W. Foster, *Arch. Mikrobiol. 35*,
      92 (1960).
3. Higgins, I.J., D.J. Best, R.C. Hammond, and D. Scott,
      Microbiol. Reviews 45, 556 (1981).
4. Perry, J.J., *Adv. Appl. Microbiol. 26*, 89 (1980).
5. Hou, C.T., *Adv. Appl. Microbiol. 26, 1* (1980).
6. Colby, J., H. Dalton, and R. Whittenbury, *Ann. Rev.*
      Microbiol. 33, 481 (1979).
7. Hanson, R.S., *Adv. Appl. Microbiol. 26*, 3 (1980).
8. Dworkin, M. and J.W. Foster, *J. Bacteriol. 72*, 646 (1956).
9. Quayle, J.R., *Adv. Microb. Physiol. 7*, 119 (1972).
10. Whittenbury, R., K.C. Phillips, and J.F. Wilkinson,
      J. Gen. Microbiol. 61, 205 (1970).
11. Davis, J.B., V.G. Coty, and J.P. Stanley, *J. Bacteriol.*
      90, 102 (1964).
12. Foster, J.W. and R.H. Davis, *J. Bacteriol. 91*, 1924 (1966).
13. Galchenko, V.F., *Appl. Biochem. Microbiol. 11*, 447 (1975).
14. Patt, T.E., G.C. Cole, J. Bland, and R.S. Hanson,
      J. Bacteriol. 120, 955 (1974).
15. Hou, C.T., R.N. Patel, A.I. Laskin, and N. Barnabe,
      Appl. Environ. Microbiol. 38, 127 (1979).
16. Davis, J.B., H.H. Chase, and R.L. Raymond, *Appl.*
      Microbiol. 4, 310 (1956).
17. Bokova, E.N., *Mikrobiologiya 23*, 15 (1954).
18. Kuznetsov, S.I. and Z.P. Telegina, *Mikrobiologiya 26*,
      513 (1957).
19. Blevins, W.T. and J.J. Perry, *Z. Allg. Mikrobiol. 11*,
      181 (1971).
20. Leadbetter, E.R. and J.W. Foster, *Arch. Mikrobiol. 31*,
      91 (1958).
21. McLee, A.G., A.C. Kormendy, and M. Wayman, *Can. J.*
      Microbiol. 18, 1191 (1972).
22. Smirnova, Z.S. and S.D. Taptykova, *Mikrobiologiya 36*,
      311 (1967).
23. Cerniglia, C.E. and J.J. Perry, *Z. Allg. Mikrobiol. 13*,
      299 (1973).
24. Davies, J.S., A.M. Wellman, and J.E. Zajic, *Can. J.*
      Microbiol. 19, 81 (1973).
25. Dunlap, K.R. and J.J. Perry, *J. Bacteriol. 96*, 318 (1968).
26. Blevins, W.T. and J.J. Perry, *J. Bacteriol. 112*, 513 (1972).
27. Cerniglia, C.E. and J.J. Perry, *J. Bacteriol. 124*, 285 (1975).
28. Perry, J.J., *Antonie van Leeuwenhoek 34*, 27 (1968).
29. Kormendy, A.C. and M. Wayman, *Can. J. Microbiol. 20*,
      225 (1974).

30. Hou, C.T., R.N. Patel, A.I. Laskin, I. Barist, and
        N. Barnabe, Proc. Amer. Soc. Microbiol. *035* (1983).
31. Hou, C.T., R.N. Patel, A.I. Laskin, N. Barnabe, and
        I. Barist,*Arch. Biochem. Biophys. 223,*297 (1983).
32. Hamer, G., D.E.F. Harrison, J.H. Harwood, and H.H. Topiwala,
        in "Single-Cell Protein II," (Tannenbaum and Wang, eds.)
        357 (1975).
33. Vary, P.S. and M.J. Johnson, *Appl. Microbiol. 15,* 1473
        (1967).
34. Smirnova Z.S. and S.D. Taptykova, *Mikrobiologiya 36,* 311
        (1967).
35. Sheehan, B.T. and M.J. Johnson, *Appl. Microbiol. 21,* 511
        (1971).
36. Orgel, G., E.W. Pietrusza, and C.G. Joris, U.S. Patent
        3,622,465 (1971).
37. Sugimoto, M., S. Yokoo, and O. Imada, *Ferment. Technol.
        Today, Proc. Int. Ferment. Symp. (4)* p. 503 (1972).
38. Imada, O., K. Hoshiai, and M. Tanaka, U.S. Patent
        3,635,796 (1972).
39. Ilizuka, H., N. Seto, and S. Sakayanagi, U.S. Patent
        3,888,736 (1975).
40. Ajinomoto, K.K., DT. Patent 2461189 (1975).
41. Gaz France-Serv. Nat. French Patent 2273870 (1976).
42. British Petroleum Ltd. Belgium Patent 835381 (1976).
43. British Petroleum Ltd. Belgium Patent 841867 (1976).
44. Max Planck Ges Wissensch English Patent 2043686 (1979).
45. Takahashi, J., Y. Ichikawa, H. Sagae, I. Komura, H. Kanou,
        and K. Yamada, *Agric. Biol. Chem. 44,* 1835 (1980).
46. Volesky, B. and J.E. Zajic, *Appl. Microbiol. 21,* 614 (1971).
47. Colby, J. and H. Dalton, *Biochem. J. 171,* 461 (1978).
48. Colby, J. and H. Dalton, *Biochem. J. 177,* 903 (1979).
49. Tonge, G.M., D.E.F. Harrison, and I.J. Higgins,
        Biochem. J. 161, 333 (1977).
50. Patel, R.N., C.T. Hou, A.I. Laskin, and A. Felix, *Appl.
        Environ. Microbiol. 44,* 1130 (1982).
51. Colby, J., H. Dalton, and R. Whittenbury, *Ann. Rev.
        Microbiol. 33,* 481 (1979).
52. Egorov, A.M., T.V. Avilova, M.M. Dikov, V.O. Popov,
        Y.V. Rodionov, and I.V. Berezin, *Eur. J. Biochem. 99,*
        569 (1979).
53. Wolfe, R.S. and I.J. Higgins, *Int. Rev. Biochem. 21,* 267
        (1979).
54. Lukins, H.B. and J.W. Foster, *J. Bacteriol. 85,* 1074 (1963).
55. Vestal, J.R. and J.J. Perry, *J. Bacteriol. 99,* 216 (1969).
56. Takahashi, J., *Adv. Appl. Microbiol. 26,* 117 (1980).

57. Hou, C.T., R.N. Patel, A.I. Laskin, and N. Barnabe,
        Appl. Environ. Microbiol. 38, 127 (1979).
58. Hou, C.T., R.N. Patel, A.I. Laskin, N. Barnabe, and
        I. Marczak, *Appl. Environ. Microbiol. 38,* 135 (1979).
59. Hou, C.T., R.N. Patel, A.I. Laskin, I. Marczak, N. Barnabe,
        FEMS Microbiol. Lett. 9, 267 (1980).
60. Hou, C.T., R.N. Patel, A.I. Laskin, N. Barnabe, and
        I. Marczak, *FEBS Lett. 101,* 179 (1979).
61. Hou, C.T., R.N. Patel, A.I. Laskin, U.S. Patent, 4,241,184
        (1980).
62. Hou, C.T., R.N. Patel, A.I. Laskin, U.S. Patent, 4,250,259
        (1981).
63. Hou, C.T., R.N. Patel, A.I. Laskin, U.S. Patent, 4,347,319
        (1982).
64. Hou, C.T., R.N. Patel, A.I. Laskin, U.S. Patent, 4,348,476
        (1982).
65. Hou, C.T., R.N. Patel, A.I. Laskin, U.S. Patent, 4,368,267
        (1983).
66. Patel, R.N., Hou, C.T., A.I. Laskin, A. Felix, and
        P. Derelanko, *Appl. Environ. Microbiol. 39,* 727 (1980).
67. Patel, R.N., Hou, C.T., A.I. Laskin, A. Felix, and
        P. Derelanko, *Appl. Environ. Microbiol. 39,* 720 (1980).
68. Hou, C.T., R.N. Patel, A.I. Laskin, I. Marczak, and
        N. Barnabe, *Can. J. Microbiol. 27,* 107 (1981).
69. Patel, R.N., C.T, Hou , A.I. Laskin, U.S. Patent 4,266,034
        (1981).
70. Patel, R.N., C.T. Hou, A.I. Laskin, U.S. Patent 4,268,630
        (1981).
71. Patel, R.N., C.T. Hou, A.I. Laskin, U.S. Patent 4,269,940
        (1981).
72. Patel, R.N., C.T. Hou, A.I. Laskin, U.S. Patent 4,375,515
        (1983).
73. Dalton, H., *Adv. Appl. Microbiol. 26,* 71 (1980).
74. Colby, J., D.I. Stirling, and H. Dalton, *Biochem. J. 165,*
        395 (1977).
75. Stirling, D.I. and H. Dalton, *Eur. J. Biochem. 96,* 205
        (1979).
76. Higgins, I.J., R.C. Hammond, F.S. Sariaslani, D. Best,
        M.M. David, S.E. Tryhorn, and F. Taylor, *Biochem.
        Biophys. Res. Commun. 89,* 671 (1979).
77. National Res. Dev. Corp. Japanese Patent J54-017184 (1979).
78. Imperial Chem. Inds. Ltd. English Patent 2,024,205 (1980).
79. Standard Oil Co. (Ind.) Belgium Patent 884004 (1980).
80. Imperial Chem. Inds. Ltd. English Patent 2,081,306 (1982).
81. Hou, C.T., R.N. Patel, A.I. Laskin, N. Barnabe, and
        I. Barist, *Dev. Ind. Microbiol. 23,* 481 (1981).

82. Hou, C.T., R.N. Patel, A.I. Laskin, N. Barnabe, and
        I. Barist. *Appl. Environ. Microbiol. 46,*171 (1983).
83. Hou, C.T., R. N. Patel, A.I. Laskin, N. Barnabe, *Proc.
        Amer. Soc. Microbiol. 054* (1983).
84. Hou, C.T., R.N. Patel, A.I. Laskin, N. Barnabe and
        I. Barist, *Appl. Environ. Microbiol. 46,*178 (1983).
85. Hou, C.T., R.N. Patel, A.I. Laskin, N. Barnabe, and
        I. Barist,*Arch. Biochem. Biophys. 223,* 297 (1983).
86. Hou, C.T., R.N. Patel, A.I. Laskin, I. Barist, and
        N. Barnabe, *Appl. Environ. Microbiol. 46,*98 (1983).
87. Davis, J.B., *Appl. Microbiol. 12,* 301 (1964).
88. Harrington, A.A. and R.E. Kallio, *Can. J. Microbiol. 6,*
        1 (1960).
89. Kallio R.E. and A.A. Harrington, *J. Bacteriol. 80,* 321
        (1960).
90. Hazeu, W. and P.J. Steenis, *Antonie van Leeuwenhoek, J.
        Microbiol. serol. 36,* 67 (1970).
91. Malashenko, Y.R., in "Symposium on Microbial Production
        and Utilization of gases (H_2, CH_4, CO),(H.G.Schlegel,
        et al., ed.) p. 293. Akademic der Wissenschaften,
        (1979).
92. Best, D.J. and I.J. Higgins, *J. Gen. Microbiol. 125,*
        73 (1982).
93. Weaver, T.L., M.A. Patrick, and P.R. Dugan, *J. Bacteriol.
        124,* 602 (1975).
94. Thompson, A.W., J.G.O.O'Neil and J.F. Wilkinson, *Arch.
        Mikrobiol. 109,* 243 (1976).
95. Takahashi, J., *Petrol and Microorgan. 4,* 24 (1970).
96. Matsumura, M., K. Haraya, K. Yoshitome, and J. Kobayashi,
        Abstr. Ann. Meet. Soc. Ferment. Technol., Japan,
        p. 71 (1973).

CHAPTER 3

MUSHROOM CULTIVATION - TECHNOLOGY FOR COMMERCIAL PRODUCTION

Murray C. Tseng

International Minerals & Chemical Corporation
Terre Haute, IN

John H. Luong

John Labatt Ltd.
London, Ontario, Canada

I.   INTRODUCTION

The mushroom has always been considered with skepticism, secrecy and prejudice in the early history of mankind. Early Chinese records indicate that the mushroom was eaten as early as 3000 years ago in the Chow Dynasty (1). In ancient Egypt, people considered mushroom as the plant of immortality and the Pharaohs, respecting its delicious flavor, decreed that mushroom should not be touched by commoners. The Greeks and the Romans were also well aware of the delicious and unique flavor of the mushroom but they did not develop mushroom cultivation methods even though an extraordinarily advanced civilization was developed during their ancient day. In Japan Shii-take *(Lentinus edodes)* was grown on wood logs for perhaps as long as 2000 years ago. However, only until the eighteenth century the first attempt at serious commercial mushroom growing began in France. In 1707 a French botanist, De Tournefort, wrote about the culture of mushroom. The modern mushroom growing technique was developed early this century when a short composting technique and a synthetic compost were developed. The mushroom industry then matured and a significant commercial production was realized.

Mushrooms are heterotrophic, saprophyte, living on dead organic material. The cultivated mushrooms fall into two major groups, *Ascomycetes* and *Basidomycetes*. The fruiting body of these groups of fungi is called the mushroom and the plant body is the mycelium. By the intertwining of many mycelial threads a primordia is formed, commonly called the pinhead, which grows into a full grown fruiting body called a mushroom.

The most common commercial mushrooms are: *Agaricus bisporus,* mostly produced in North American and European countries, and some Asian countries; Straw mushroom, *(Volveriella volvacea)* produced in China, Taiwan, Japan, The Phillippines,

and Indonesia; Oyster mushroom *(Pleurotus ostreatus)* produced
in Asia, Italy, and Hungary; Shii-take *(Lentinus edodes)* produced in Japan and China; and Jew's Ear (known as Wood's Ear by
Chinese), commonly produced in China for centuries.

The mushroom is an important crop in many regions and
countries. The world annual production of various mushrooms is
over 1 million ton (estimated at 1.2×10^6 ton). United States,
France, China (including Taiwan), Japan, The Netherlands, the
United Kingdom, and Italy are the major producing countries.
The production has increased worldwide. The annual compound
rate of production increase in the U.S.A. is about 8.6% for
the last ten years (2). The Netherlands is another country
which has been expanding the production of mushroom in the last
twenty years with an annual compound rate of growth of 13.1%.

In the last twenty years, the per capita consumption of
mushroom also increased at an annual compound rate of 23% in
Spain, 18% in Italy, 15.2% in West Germany, 13.0% in The
Netherlands, 11% in Canada, 10% in Switzerland, and Sweden, 8%
in the United Kingdom, Belgium, Denmark and Austria and 6% in
France. In 1980 the largest per capita consumption was West
Germany, 2.55 kg, while in Switzerland it was 2.25 kg, in
France 2.1 kg, and in Belgium and Canada 2.0 kg (3). In the
U.S.A. the average annual compound rate per capita was 9.5%
for the past ten years and 1.2 kg was consumed per capita in
1980.

The nutritional value of commercially cultivated mushrooms
varies between different varieties of mushroom. Marked differences in protein content are also noted. For example the
protein content in *A. bisporus* is 29% (dry basis), in straw
mushrooms about 30%, in oyster mushrooms up to 30%, and in Jew's
Ear 4.7% (4). The mushroom protein is easily digested and the
quality of protein seems to be better than that of other
vegetables. In general, the most popular commercial mushrooms
are richer in protein than any comparable fruit or vegetable.
The mushroom can be considered as "protective food" since it is
rich in thiamine, riboflavin, niacin, biotin, pantothenic acid,
ascorbic acid and a remarkably high level of folic acid.

The mushroom has always been considered as a flavoring
material in cooking because of its high concentrations of
easily released glutamic acid and other amino acids. Furthermore, due to its low calorie content (about 30 cal per 100 g)
the mushroom could be included in the diet menu.

It has been known for a long time that some strains of
mushrooms can be used as medicinal herbs for therapeutic purposes (5). Experiments with animals and human beings have
demonstrated that the regular consumption of Shii-take (popular
in Japanese and Chinese communities) reduces the cholesterol
content in the blood (5). Its active compounds were isolated
from dried Shii-take mushrooms recently (6).

Mushroom cultivation is a process utilizing waste materials
such as horse manure, chicken manure, even pig manure, wheat
straw, rice straw, corn cobs, wood bark, sawdust, cottonseed
hulls, etc. to produce a delicious, and nutritious food.
Therefore it can be considered as a two-fold beneficial
operation, which may be aptly described by the phrase "killing
two birds with one stone." Furthermore, mushroom technology
has enabled farmers to harvest close to one pound of fresh
mushrooms from one pound of compost (dry weight) spawned. This
compares very favorably with two pounds of dry feed required
for one pound of chicken, three and a half pounds of feed for
one pound of pork, or five pounds of feed for one pound of
beef.

Undoubtedly, in the future, mushroom production can be
predicted to increase at the afore-mentioned rate as it will
be more widely accepted and the per capita consumption will
also increase. In this report the emphasis will be focused on
the technology of the commercial production of *A. bisporus*.

II. *THE GROWING* SYSTEMS

In the past fifty years, the development of mushroom growing technology has become more sophisticated and diversified to
suit the grower's own needs and conditions. Growing systems
are the results of this development. At the present time there
are many different systems in the industry which can be
grouped as follows:

A. *According to the Room Utilization*

1. <u>One-zone system:</u> After the compost is filled, the
subsequent operations of peak heating, spawn running, cropping
and cooking out are done in the same room. It is simple in
operation, scheduling, and there is a less chance of cross
contamination. This system is suitable for smaller farm operations and yet it may be easily expanded. However, it suffers
several disadvantages such as higher building costs, lower
number of growing cycles (3 cycles) per year, greater difficulty
for automatic control, and lower productivity.

2. <u>Two-zone system</u>: There are separate rooms for peak-heating, and mycellium growing and cropping. With the two-zone system, each type of room can be built for a particular operation need; therefore the overall building cost is less. Due to the optimal use of the production rooms, six to ten cropping cycles per year can be achieved in the same cropping room. However, this system also has some disadvantages such as a tighter schedule in operation, a requirement for equipment and a larger hall for moving compost from one type of room to the other, a high risk of infection, uneconomical for small farm operations.

B.   *According to the Compost Containing Device*

1. <u>The bed (shelf) system</u>: This is the most popular growing system developed in the United States. A so-called standard single house has a fixed bed system of 372 m^2 (4000 ft^2) of growing area and 453 m^3 (16000 ft^3) space under one roof with two tiers of beds; each tier has six wooden beds, approximately 1.7 m wide x 20 m long per bed. Compost is filled on beds to about 30 cm at green fill or to 15 cm after pasteurization. The standard single house (7) is shown in Fig. 1.

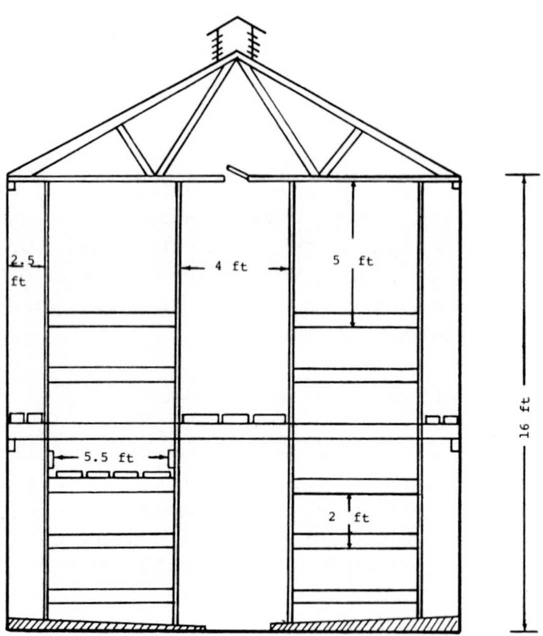

Figure 1. Diagram of a typical mushroom house in Ontario,Canada

A standard double is four tiers of growing beds under one roof with 743 m^2 (8000 ft^2) growing surface. Of course, there are variations of the number of beds in one tier and the number of tiers in one room. The bed system is the most popular growing system in North America, Asia, and Holland. This system, however, suffers from labor intensive requirement for filling, spawning and emptying.

Recently, in Holland, a mechanization of the bed system was developed. A pulling net allows a rapid filling and emptying of the bed. This mechanized bed system offers several advantages such as a labor saving, cost reduction and gives a big competitive edge in the market place in comparison with the other systems.

A so-called Drawer system used in Holland is a bed system with each bed movable in and out of the growing room to a working hall or to a machine for compost filling, tamping, spawning, casing, picking, watering, and emptying. The advantages of the drawer system are the compactness of the beds (10 beds in a tier) in a growing room, more growing sur- face per unit space of room, and easier picking. The disad- vantage is a high demand of air conditioning and power consumption.

2.   <u>The tray system:</u> This system was originated in New York by the Knaust brothers in the early 1930's (8). The compost is filled in trays and moved by truck from one functional room to the other for peak-heating, spawn running, cropping and pasteurization. This is a revolution in the mushroom industry because of its possibility of mechanization, labor saving and large scale production. The usual size of the tray is about 1.2 m x 1.5 m or 1.2 m x 3.0 m and 20 cm in depth. Of course, the size of the trays varies from farm to farm.

Although the tray system was developed in the United States, it was most widely adapted in France, the United Kingdom, Australia, West Germany and by big farms in the United States. It requires a large capital investment for machinery (tray line equipment for tipping, filling, spawning, casing, and emptying). The advantages of this system are that different functional rooms can be built for different particular operational purposes and provide better space utilization. It is more energy efficient, and possibly gives better yields per unit growing surface, and most important, mechanization for large scale producing farm operation. However, due to large capital investment this system is not suitable for small farm operations. It is worth noting that it requires a large amount of power for moving trays around, and there is a high risk of contamination.

    3.   <u>Plastic bag system</u>: A large quantity of mushroom is grown in bags in France, Spain, and Italy because of its low capital investment, simplicity and easy picking. The compost must be pasteurized or spawn run before filling into the bags. Although it was usually filled with compost up to 0.6 - 0.9 m (2-3 ft) high, the system is a low density filling and has less mushroom produced per unit area. It is suitable for low cost buildings such as plastic houses and caves. However, the heat produced during mycelial growth could not be dissipated out of the bag fast enough and the temperature in the bag becomes too high.

    4.   <u>Other minor systems</u>: The deep trough system was pioneered by Flegg and Smith (9) and assessed by Gaze (10) in England. They built a trough of 1 m wide x 1.2 m deep and 5 m long with a wooden slatted (2.5 cm apart) pallet as the bottom. A forced air is circulated through the trough to remove the metabolic heat and to supply oxygen for growth. This is analogous to the bulk treatment of compost in a tunnel. Apparently the yield is reported to be high, 50 kg per m^2 growing surface, or 155 kg per ton of compost at spawning. Some difficult problems are to maintain a uniform temperature and moisture throughout the compost and to prevent condensation of water near the surface layer.

    The other minor but the oldest method is to make a compost bed on the floor for culture growing in caves. This method is practised now on a very small scale.

    A relatively detailed comparative analysis of the various growing systems was discussed in a paper by MacCanna and Dawson (11). They analyzed the capital cost and the production cost based on 1 million kg of fresh mushroom per annum. It was concluded that at a price of 110 pennies per kg, the return on investment from the tray system was 23%, the bed system 18%, the bag system 61%, and the trough system 85%. However, because this estimation was not done on the same basis, a big discrepancy existed. A more recent work of Tschierpe (3) indicated this clearly. Obviously, it is necessary to take the local conditions into account in choosing a mushroom growing system.

III. SUBSTRATE PREPARATION

A.   Basic Requirements

Like other fermentation processes, mushroom cultivation
requires similar basic nutrients such as a carbon source, a
nitrogen source, oxygen, minerals, trace elements, vitamins, etc.
The physical/chemical conditions such as temperature, pH value,
moisture content and humidity in the air, porosity of compost,
ventilation and air velocity, etc. must be optimally controlled.
Since mushroom cultivation is a solid state fermentation process
with a mixed culture system, the complexity of the process and
operation should be appreciated.

Mushrooms preferentially utilize xylose, arabinose, glucose
and fructose. However, in the cultivation of mushrooms the
carbon substrate is not simple sugars but it is derived from
straw, or other cellulosic materials. It is known that mushrooms produces a large quantity of extracellulose hydrolyzing
enzymes such as cellulases, xylanase, lignase, and laccase, to
hydrolyze cellulose, hemicellulose and lignin into simple sugars
for supporting mushroom growth.

It has been demonstrated that during mycellium growth, a
large portion of lignin was utilized while during the growth of
fruiting, pentosan and cellulose were used (12,13).

Mushrooms prefer organic nitrogen such as protein, peptone
and amino acids for their growth. Ammonia and/or its simple
salts are detrimental to the growth of mushroom. This is the
reason why during composting process, ammonia must be expelled
completely. Dried blood, soymeal, cottonseed meal are the best
nitrogen sources. It has been shown that 80% of the nitrogen
is fixed in the humus-lignin complex and 20% in the microbial
cell mass. The browning reaction between sugar and amino acids
during composting at high temperature constitutes a nitrogen
reservoir for mushroom growth (14).

Since all phases of mushroom cultivation are aerobic processes, oxygen supply is another necessity. Air circulation
and ventilation is not just to supply oxygen, but also a means
to control temperature, humidity, and the concentration of
carbon dioxide and other metabolits.

Gypsum and chalk are important ingredients for substrate
preparation. It provides Ca^{++} to neutralize oxalic acid produced during mycellium growth, and improve the physical structure
of the compost. They also prevent greasy compost formation,
maintain a lower pH value, increases water holding capacity,
reduce the nitrogen loss and increase the mushroom yield (15).

Although vitamins are required for yield increases, vitamins are not used in practice. The microflora in the compost usually provide adequate vitamins for the mushroom growth.

B.   *Materials Used for Substrate Preparation*

The mushroom is a saprophyte growing on dead organic materials of vegetable origin. As a result, all agricultural wastes can be used as basic materials. The oldest and most widely used material is horse manure which contains the solid excrement (feces), liquid excrement (urine) and bedding straw. Horse manure is relatively poor in nitrogen content (about 1.2% dry basis).

Straws (wheat, rice, rye, barley) composes the main bulk of the compost. It supplies the carbon source for mycelium and the fruiting body growth and the energy source for biological process use. The straws also provide a physical environment for microflora to live and multiply. The type of straw used depends on location, availability and economics.

In North America a large quantity of corncobs and hay is used for making compost. Corncobs are rich in carbohydrates but very poor in nitrogen. Dry hay is not an agricultural waste but has been used substantially in some locations in the U.S.A. because of its economic feasibility. Hay has 2% nitrogen in dry basis.

Chicken manure is another important ingredient which is used extensively in North America and Europe. It contains about 4% nitrogen in dry basis and is rich in easily degradable carbohydrates and oil. It is usually used as a nitrogen source.

Sawdust and hard wood bark can be used as straw substitutes for making compost. The work of Block et al (16) in 1956 with sawdust and the recent work with wood bark by Schisler (17) provided the background for the utilization of these forest by-products.

Sugarcane bagasse is another useful material to replace horse manure or straw. The work of Kneebone and Mason (18) indicated that an equal yield but larger mushrooms were produced. Tseng et al (19) also reported the utilization of bagasse for mushroom production. The recent work of Peerally (20) showed that sugarcane trash was a good material for compost.

Brewers-spent grains, soybean meals, cottonseed meals, and blood meal are rich in nitrogen and have been used as a nitrogen source for many years. It is anticipated that they continue to be important materials in the mushroom industry.

Pig manure is unpleasant, evil-smelling and not attractive to growers as a substrate for mushroom growing. However, a significant development work on this agricultural waste was performed in Denmark by Rasmussen (21, 22). The conclusion is that pig manure is a first rate cultural medium for mushrooms.

Other materials such as cow manure, waste paper, rice and cottonseed hulls, paper pulp waste, etc. could also be used as substitutes for horse manure in substrate preparation. Further experimental work is required to stimulate the utilization of such materials.

Chemicals such as gypsum, chalk ($CaCO_3$), ammonium sulphate, superphosphate, urea, potash, etc. are also commonly used in substrate preparation as a nitrogen source, or as an activator for synthetic compost manufacturing.

Regardless of the materials used as the carbon source or the nitrogen source in the formulation of composting materials, the difference in production will be insignificant if a proper preparation method is chosen and the correct operation is performed. There are some criteria in the starting materials which must be met, such as a total nitrogen content of 1.5 - 1.8% d.b., a carbon nitrogen ratio of 30:1, a water content of 74-77%, and a N:P:K ratio of 13:4:10 (23). In any composting process gypsum must always be added at about 25 kg which could be reduced to 10 kg per 1000 kg of horse manure (15). The most frequently recommended composting formula can be obtained from the *Penn State Handbook for Commercial Mushroom Growers* (13, p.83) or from the *Vedder's Modern Mushroom Growing* (12, chapter 6) and *Glass Crop Research Institute in the United Kingdom*. In practice, modifications are necessary in order to allow for variables that inevitably exist in the materials at different locations.

C.   *Solid State Fermentation with Undefined Mixed Culture
     System*

Composting could be the largest fermentation found either in nature or in man-made operations. In one location 7000 tons of finished compost was produced each week by CNC at Ottersum in Holland. This is a very complicated ecological and dynamic process, being aerobic, in some portions anaerobic, and thermophilic process involving bacteria, fungi and

Actinomycetes. Hayes (24) described three distinct functional stages in composting and mushroom growing:

1) The establishment stage - In the compost pile where the mesophile flora gradually subsides while the thermophiles and thermotolerants increase, the anaerobic region develops at the center base layer.

2) The maturation stage - The thermophilic microflora dominates while the *Actinomycetes* population increases and the bacteria population decreases.

3) The colonization stage - After spawning the mushroom mycelium gradually colonizes in the compost bulk. A change of the dominant thermophilic flora during composting was summarized by Hayes (25) as shown in Table 1.

Table 1. Summary of changes in the dominant thermophilic and thermotolerant flora during composting

	Establishment Stage	*Maturation Stage*
Bacteria	Bacillus subtilis	B. staerothermophilus
	Flavobacterium sps.	Pseudomonas sps.
Actinomycetes	Streptomyces thermovulgaris	S. thermovulgaris
		S. rectus
		S. sps.
Fungi	Mucor pusillus	Humicola griseus
	Aspergillus fumigatus	Torula thermophila
	Humicola languinosa	

Fermor et al. (26) studied the microflora in a rapidly prepared compost in which a simple sugar was supplemented. It was found that the thermophilic bacteria, *Bacillus* sp. and *Actinomycetes* were dominated at the expense of cellulose utilizing fungi. It should be understood that because an easily degradable sugar was present for bacteria growth, there was little need for cellulose utilizing fungi to produce simple sugar for bacteria use.

In the process described by Sinden and Hauser (27, 28), composts are prepared in two stages. This is the most distinct achievement in mushroom cultivation history. The substrate (compost) can be prepared in a much shorter time with a better yield and a better quality under controlled process conditions.

This two-stage composting process became well known and it is
being practised world-wide in the mushroom cultivation industry.

1.   Phase I composting: This is an outdoor (or an
enclosed yard with a roof) free heating composting process.
The purpose of composting is to prepare a uniform substrate
suitable for mushroom growth but not suitable for growth of
other organisms. Therefore the materials used in a chosen
formulation must be properly mixed. The fresh horse manure
and dry straw must be pre-wetted for about 5-10 days so that
the moisture content in the straw is about 50 to 70%. In this
prewetting period, other organic ingredients should be evenly
distributed with the straw or horse manure. The inorganic
ingredients should be dissolved or suspended in water in order
that the water may be sprinkled onto the main straw bulk pile
which is about 1 m high. A tractor is used to run constantly
over the bulk in order to compact the bulk. If no horse manure
is used, i.e., to make a so-called "synthetic compost," the
straw bales should stand upright soaked in water so that water
can enter the inner hole of the straw. Portions of chicken
manure are also distributed between bales to supply easily
degradable carbohydrates, giving a quick start for self heating
processes. The mesophilic bacteria grow first utilizing some
of those carbohydrates. Some work has been done in Japan to
add culture seed at the beginning to promote the culture population. Since the doubling time of those bacteria is short,
the natural inhabitant of those bacteria in manures is rich
and the bacteria in the return liquid from other composting
bulk are plentiful. Therefore, there is little need to add
culture seeds if the proper condition for bacterial growth is
provided. The returning liquid or fresh water should be
sprayed onto the top of the pile intermittently for several
days. The bacterial growth speeds up and the temperature
rapidly increases to 60-70°C.

After about 10 days of the prewetting period, the bales
are untied and mixed with the rest of the organic ingredients
to make stack piles of about 2 m high and 2 m wide and the
length of the composting yard. Some growers untie the fresh
straw bales, and mix them with other organic ingredients to
make a large loose straw pile for prewetting. They move the
big pile from one location in the composting yard to another
for mixing, aeration, and watering. After stacking the compost
into piles, the solid state fermentation is enhanced. When the
temperature increases rapidly to 60-80°C, the number of thermophilic bacteria rapidly increases and dominates the overall
microbial population. The easily degradable carbohydrates
are gradually utilized for cell mass growth. The free ammonia

is also incorporated into microbial protein but some escapes
into the air. The microbial protein becomes a nitrogen source
for mushroom cell mass production. Aeration is done by natural
convection and air diffusion into the inner part of the
composting pile. To supply air from the bottom for the pile
(compressed air through a pipe and diffuses into the pile)
will improve the efficiency of the fermentation process. After
a few days of fermentation, the pile shrinks and the straw
begins to darken; organic matter is lost due to biological
oxidation to carbon dioxide and water. Gypsum is then applied
onto the top of the pile. A compost turner is used on the
pile to loosen, to mix, to water, and to turn it in order to
make a new compost pile, somewhat smaller in size. If ammonium
sulphate was used as a nitrogen source, addition of $CaCO_3$ is
necessary at this stage. The compost pile will be repeatedly
mixed, turned and a new pile made three more times at two to
three day intervals. At the end of the last turn, the compost
should be ready for pasteurization. The compost should have
the following characteristics: a blackish-green color, a
moisture content of about 69-73% (29), a little water can be
squeezed out when a handful of compost is pressed tightly in
one hand, nitrogen content of 1.8 ±0.2%, and a pH value of 8.2.
Similar methods with small modifications should be applied if
rice straw or other composting materials are used.

    2.   <u>Phase II composting</u>: Sometimes it is called pasteur-
izing, peak-heating, sweat-out, or cook out. The purposes of
this process are to kill the undesirable organisms such as
insects or harmful fungi, to correct and to adjust the compost
so that it satisfied the specific requirements of the mushroom,
and to condition the compost so as to eliminate the easily
degradable carbohydrates and to convert ammonia into microbial
protein or to eliminate ammonia by ventilation.

    Overstizns (30) reported the thermal death of undesirable
pests and pathogens and confirmed that at 60°C for 6 hours or
less they all were killed. The investigator quoted the work
of Fergus (31) that the optimal temperature for growth of
important thermophilic mold and *Actinomycetes* (*Aspergillus
fumigatus*, *Streptomyces*, *Thermomonospora*, *Thermoactinomyces*,
Humicola, *Chaetomium thermophile*, and *Torula thermophila*) was
in the range of 45 to 53°C. However, most of those fungi were
killed at the temperature above 65°C and they all survived for
a long time at 59°C. Therefore the Phase II pasteurization
operation should take place at 57°C (measured in the compost)
for 5 hours, then the compost should be cooled down for
conditioning.

After the completion of the Phase I composting, the compost
is placed either on trays, beds, or in a room (tunnel) for
the Phase II process according to the one-zone or two-zone
growing system. There are two distinct methods for the Phase
II process, namely a thin layer (15-18 cm) of compost method
(on a tray or bed), and a thick layer (about 2 m) of compost
method (in bulk in a room or tunnel).

     a. <u>Thin layer method</u>. This is a conventional method.
The tray or the bed is filled with compost mechanically or
manually to a thickness of 20-25 cm and then pressed to 15-18
cm. If it is filled manually, a uniform thickness must be
maintained. If it is filled on a tray line, the trays must be
transported to a peak heating room. At the start of pasteurization, air circulation is needed to equalize the temperature
in the compost. Then the steam is injected initially into the
air system to heat up the circulating air (no fresh air) which
in turn heats the compost. The thermophiles will produce heat
at 5.54 kcal per kg (10 BTU per lb) dry compost at peak which
would heat the compost. The circulation fan has a capacity of
250-300 m^3 per hr per ton of compost. When the temperature of
the compost reaches 58 to 60°C for a minimum of two hours, the
air temperature must maintain above 57°C for about three hours.
The harmful insects will be killed during the pasteurization
period. When pasteurization is complete, the steam is turned
off and fresh air is breathed in to cool the compost
gradually at a rate of 2-3°C per day for conditioning.

     During peak heating, excess ammonia is liberated into the
air and eventually vented outside the room. Ammonia is also
produced through a biological reaction such as de-amination
process, especially under unfavourable growing conditions.
The ammonia generation can be seen in Fig. 2 (32). During the

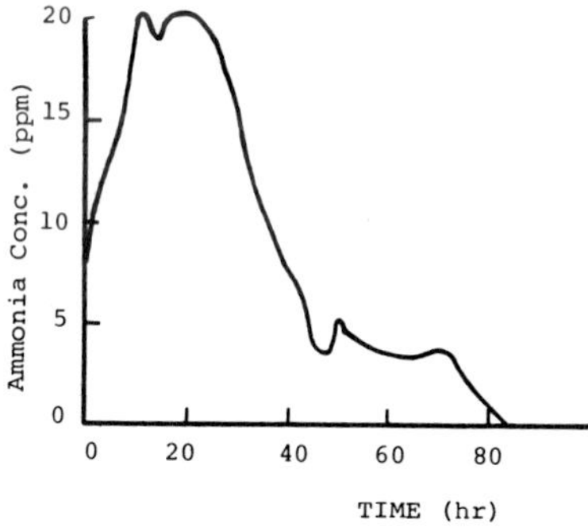

Figure 2. Ammonia production from compost during Phase II
*   composting*

conditioning period, fresh air supplies oxygen for aerobic
process and the thermophiles grow rapidly and convert the
easily degradable carbohydrates and later cellulose and ammonia
into microbial cell mass. The fatty acid compound originally
present in the compost reduces to about 50% while linoleic acid
increases. The Phase II process is complete when the ammonia
present in the air reaches 5-10 ppm. At this time the steam
is turned off and maximum fresh air (filtered) is supplied to
cool the temperature of the compost to 25°C for spawning. The
compost should be dark brown, with a moisture content of 65-67%,
a pH of 7.5 or less, a sweetish smell, a blueish-white mycelium
of *Actinomycetes* and *Humicola* appears on and in the compost. A
typical Phase II composting temperature program is shown in
Fig. 3 (33).

    b. <u>Thick layer pasteurization</u> (pasteurization in bulk or
in tunnel). This method was first developed in Italy and
France and it was later perfected in the Netherlands. The CNC
(Cooperative Dutch Mushroom Growers' Association) has a composting yard of 4.05 hectares (10 acres) and produces 7000 tons
of compost per week using tunnels for the Phase II process and
spawn running. In general, a tunnel has a capacity to process
75 tons or one tunnel with twin compartments to process 100
tons of compost. The dimension is about 4 m x 20 m x 3.5 m
(34) with a grated floor having a total opening of 1/4 to 1/3
of the total floor area. The wall and the ceiling are insulated
to R-value of 20 or so. The circulation fan which has a
capacity of 200 m^3 per ton compost per hr is installed to supply
air from below. The air flows from the grated floor and through
the compost mass and then it returns to the circulation fan
after having mixed with filtered fresh air. Some of the
returned air is vented to the outside of the tunnel. A steam
pipe is installed at the air inlet or under the grated floor
to supply steam for the initial use during pasteurization.
The amount of fresh air depends on the stage of the process.

    At the beginning of the operation the tunnel is loosely
filled with compost using a compost swiveling spreader. Compost is filled up to 2 meters high and to about 900 kg per m^2.
In the worst case, the pressure drop of air flow through the
compost mass is about 150 mm water gage.

    The advantages of pasteurization in bulk are: 1) it
requires much less steam; 2) the temperature difference
between the air and the compost or in various parts of the
compost is small (3°C); 3) it is easier to control and manage
and much less space is required; 4) a uniform oxygen supply to
and a uniform metabolites removal from all parts of the compost;
5) it is easy to maintain at the optimal temperature and

achieve a high carbon dioxide concentration for thermophiles
growth; 6) it is easier to remove ammonia (1000 ppm to below
10 ppm in 5 to 7 days). A typical temperature program (33) is
shown in Fig. 4).

The recent work of Gerrit (35) verified the air circulation
requirement at 150-200 m^3 per ton per hr was the optimal condition. Therefore for energy saving the best tunnel operation
should be at 150 m^3 per ton per hr and the fan capacity should
be designed for 225 m^3 per ton per hr.

When the Phase II process is complete and the compost is
cooled to 25°C for spawning, the compost is taken out with a
front loader and put into another bulk spawn running tunnel to
about 1.5 m thick layer using the same equipment for tunnel
filling. Using a pulling net to pull the whole tunnel compost
out slowly is another means. A compost beater tears the compost which falls on an elevator and then a conveyor transports
the compost to another tunnel filling machine.

Inoculation is done by either spreading the grain spawn
on top of the compost before pulling out or adding the grain
spawn on the compost while being transported by the conveyor.
Inoculum is about 5.5 kg grain spawn per ton of compost.

A recommendation was proposed to re-pasteurize the compost
at the end of conditioning before cooling. This is a very
risky operation since the increase in the temperature for
second pasteurization will upset the ecological balance and
de-amination could occur.

The overall change during Phase I and Phase II composting
processes can be summarized (33, p. 30) in Fig. 5. Figure 5
clearly indicates that a composting process must provide such
proper quality of finished compost at spawning in order to
obtain a good substrate for mushroom production.

3.   <u>Other composing methods</u>: The loss of organic matter
during the composting process is about 50% (loss in Phase I is
30%, in Phase II 20%). This loss is due to microbial oxidation
of organic matter to produce heat for metabolic use, carbon
dioxide and water. Water is also lost due to evaporation.
Methods are now under development to reduce such a loss.

Express preparation of substrate was initiated by Laborde
and Delmas (36) and Smith (37), then it was further extended
by Laborde (38). The purpose of this method was to eliminate
the Phase I process using either a controlled low temperature

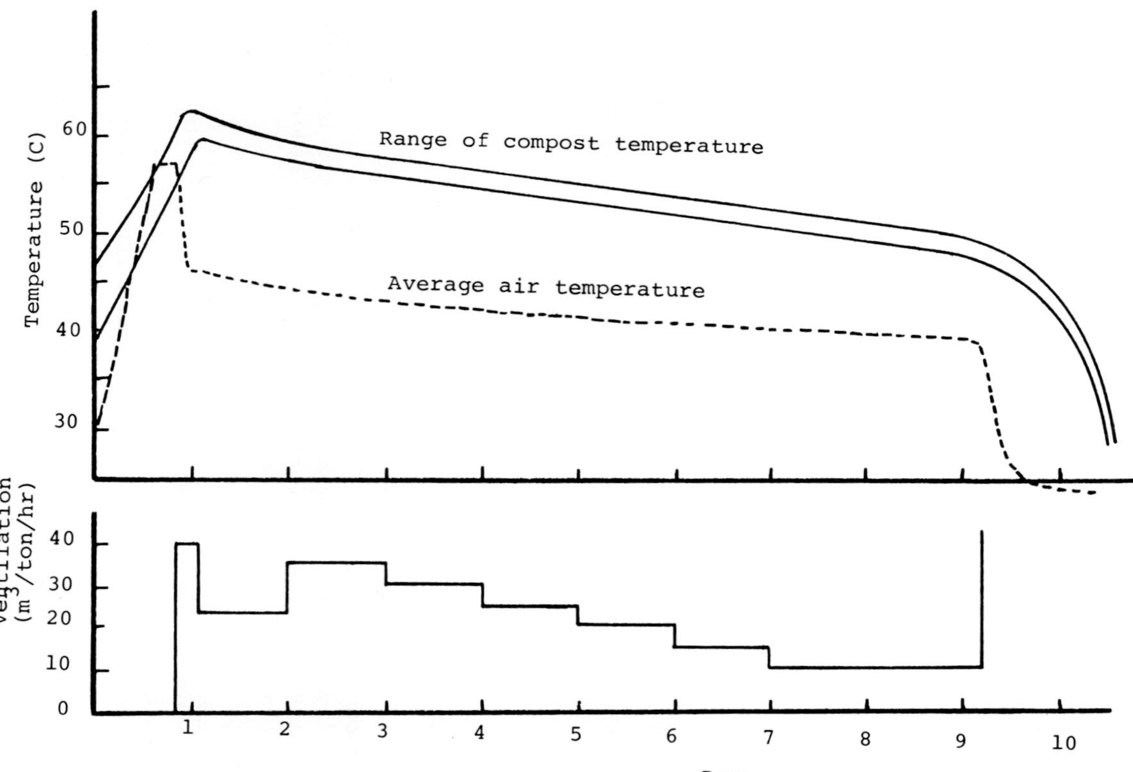

Figure 3. The temperature and fresh air programs of the Phase II process of composting on trays or beds

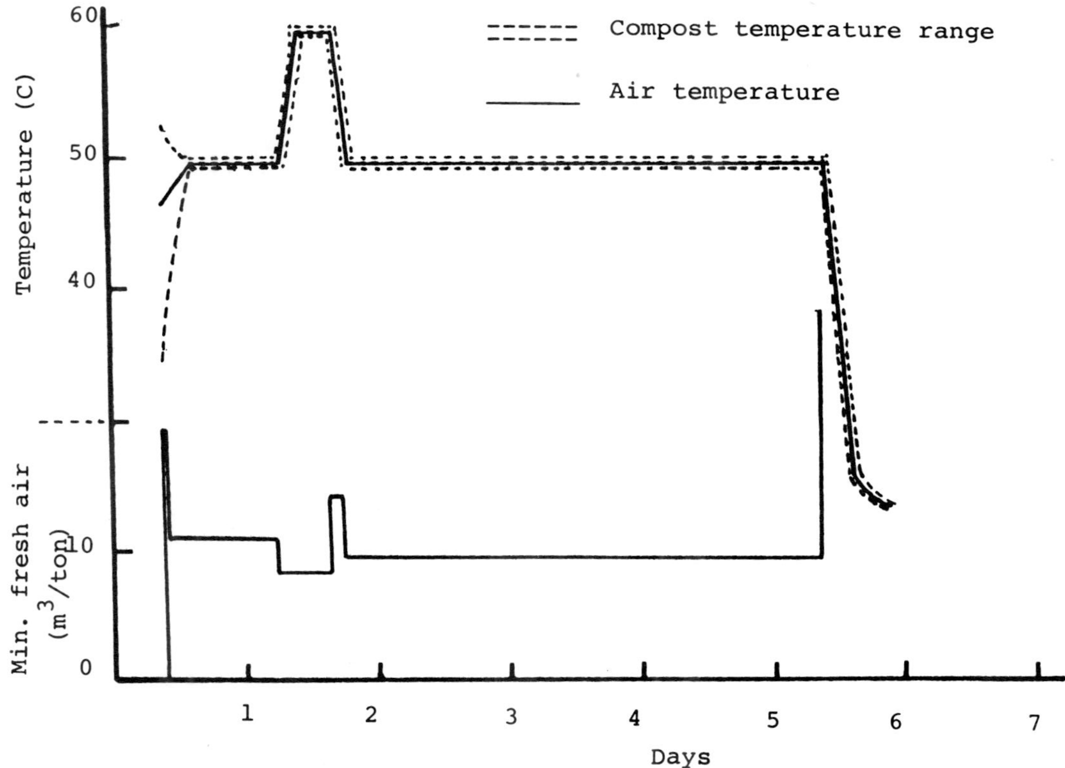

Figure 4. *The temperature and fresh air program of the Phase II process of composting in tunnels*

61

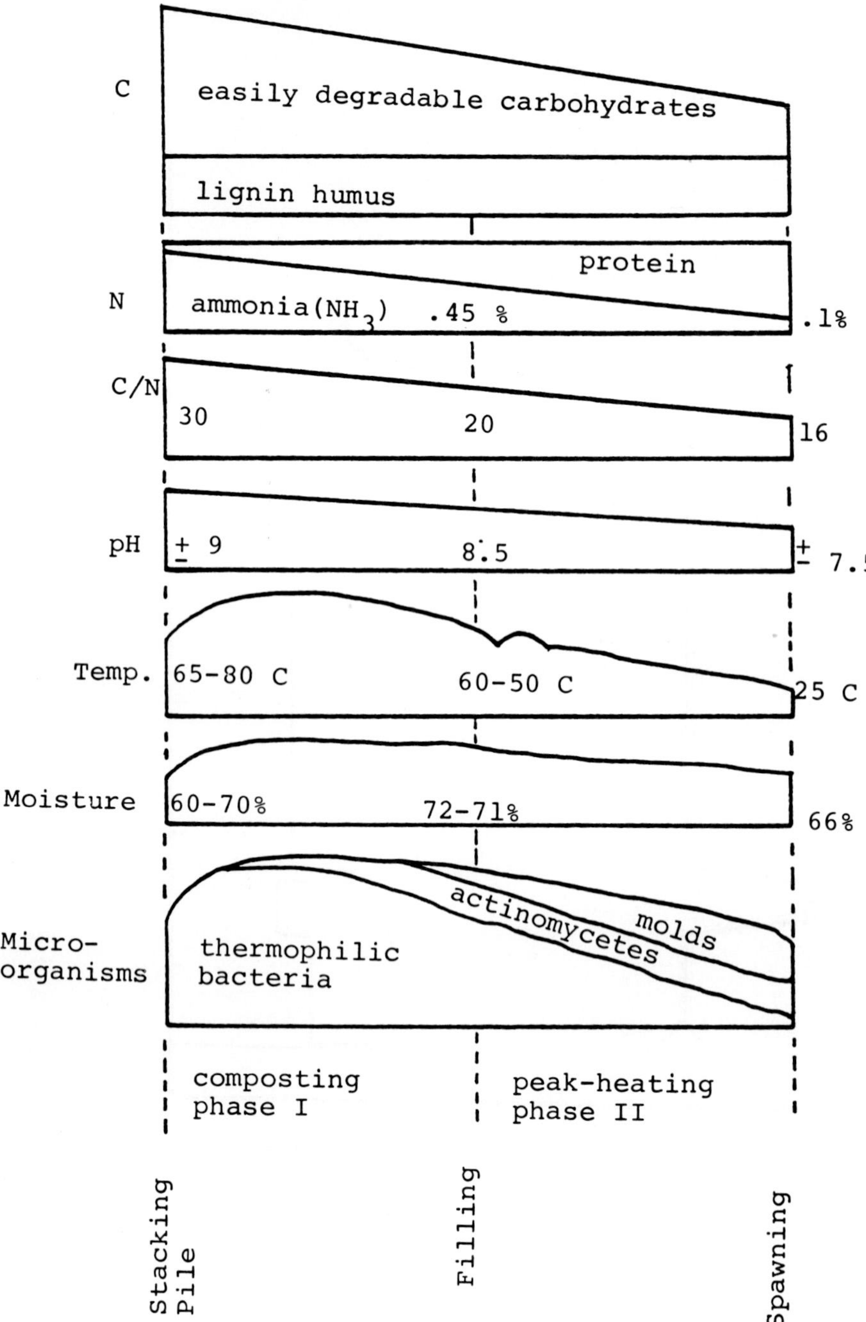

Figure 5. Diagram of the changes during solid-state fermentation of compost

(below 60°C) microbial process or a controlled high temperature
(80-85°C) chemical process to produce black substance and the
lignin-humus nitrogen complex for mushroom growth. The
investigators have achieved a 40 to 50% saving of dry matter
and a reasonable yield of mushroom comparable to normal composting methods. This method, however, has some drawbacks,
e.g., the activator should be added and energy is required for
thermal-chemical reactions.

The Barrel composting method was developed by Baker and
Baker (39) in Australia. An octagon barrel was constructed
of 16 m long and 3.8 m across with a capacity of 40 tons
driven by two 7.5 HP motors for agitation (1 revolution every
eight minutes). Water spray and air supply was also installed.
The activator and the straw was used for compost making.
Although it looks more mechanized and with a controlled environment for composting, it did shorten the time needed but it did
not eliminate the Phase I process. Considering the energy used
for the rotation of the barrel (15 HP total), this method will
not be economically attractive for commercial application.

D.   *Quantity of Compost Prepared for Mushroom Growing*

Experience indicates that before the mushroom fruiting
body appears the substrate loss is about 60% or more of the
starting materials. In general, the substrate loss in the
Phase I process is 30-35%, in the Phase II process about 20%,
and in the mycellium growing process about 10%. The most
important quantity of compost is the amount filled onto the
tray, bed or into the bag for mushroom growing and picking.
Generally, the more compost fill per unit picking area, the
more mushroom will be picked. So called "Biological Efficiency"
in the mushroom industry is the quantity of mushroom picked per
unit dry weight of compost used at filling for a period of 5
picking weeks. Biological efficiency of 1 is the target in
the mushroom industry today. With a standard filling of
9.3 m^2 per ton (100 ft^2 per ton) of compost, 22 to 24 kg of
mushrooms can be harvested per m^2 and with a thicker fill of
7.0 m^2 (75 ft^2 per ton), 29 kg of mushroom can be picked per
m^2. However, the biological efficiency is reduced. Nevertheless, comparing the cost of the compost and the price of
mushrooms, growers tend to fill thicker layers of compost on
their growing surface until the limitation of aeration and
heat dissipation comes into play. Figure 6 shows the relation
of the quantity of compost filled per m^2 area and the mushroom
harvested (12, p. 171). Figure 6 provides a guide to growers,
the optimum of compost filled per unit surface required under
their own conditions.

IV. THE GROWTH OF MYCELIUM AND THE MUSHROOM

After the Phase II composting process is complete, the compost should be cooled to 25°C for inoculation (called spawning in the mushroom industry). Mushroom mycelium will grow through the compost mass in two to three weeks.

A. Preparation of the Inoculum and Inoculation

The mushroom seed which is called a spawn is prepared aseptically by transferring mycelium onto steam sterilized grain contained in a bottle with a cotton plug. Subsequently the bottle is incubated at 25°C until the grain is overgrown with mycelium.

The method of inoculation differs from traditional fermentation in liquid culture. The grain spawn is sowed over the surface of the compost on a fixed bed at 307 g per m^2 (1 oz or more per ft^2) and the mycelium is allowed to grow into the compost.

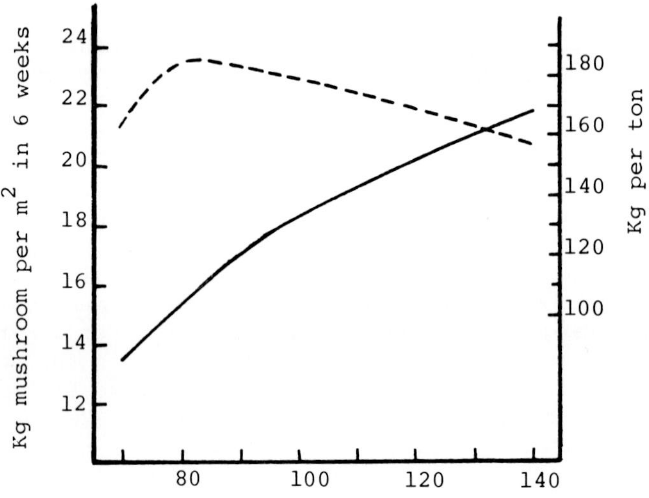

Figure 6. The relation between the quantity of compost filled
and the mushroom harvested
———— yield in kg/m^2
---- yield in kg/metric ton of compost

A better method is after sowing the grain spawn on the surface of the compost, a ruffling machine runs over the bed to mix the spawn and compost so that it is easier for mycelium to grow through. In general, after spawning the bed is tramped for compactness and it is covered with paper or polyfilm to prevent drying, escape of carbon dioxide and contamination. The bed is then sprayed with a disinfectant solution. The best method is to mix the spawn with the compost thoroughly either on a tray line or in a tunnel system as mentioned in a previous section.

B.   Mycelium Growth

The growth of mushroom mycelium must be faster than the growth of other competitive fungi or bacteria so that the compost can be colonized by the mushroom and the antagonism will hamper the growth of other micro-organisms. In order to achieve this the grower must create and maintain the best conditions for mycelium growth.

1.   Optimal temperature: The ideal temperature for mycelium growth is 24-25°C (40) in the compost (Fig. 7).

Since the mycelium growth is an exothermal reaction, with a rate of heat generation about 27 kcal per m^2 per hr (10 Btu per ft^2 per hr), a cooling system is required to cool the circulating air to about 16°C. This could be more critical for mycelium growth in bulk compost. Without a proper cooling, the temperature could rise to 30°C or more and the mycelium could be severely damaged resulting in a crop failure.

2.   Air circulation: There is a need to have air flow through the spawn running beds or trays to remove the reaction heat, to supply oxygen, and to sweep away metabolites. An air

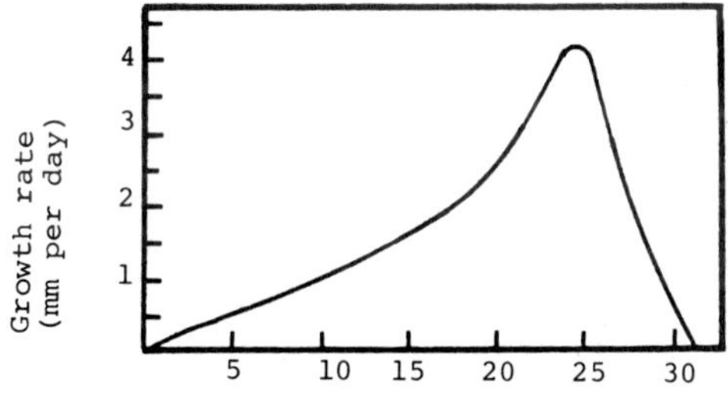

Figure 7. The effect of the compost temperature on the growth rate of mycelium

flow of 13.7 m^3 per hr per m^2 (0.75 cfm per ft^2) of bed (tray) area seems to be adequate. In the case of spawn running in tunnels, a circulation of air at 200 m^3 per ton per hr through the mycelium growing compost is adequate. The temperature of the circulating air is about 2 to 3°C below the temperature in the compost.

   3. <u>Humidity</u>: Humidity in the circulating air should be maintained near saturation (95-100%). This can be achieved by steam injection or light watering or keeping the room wet all the time. Extra precautions must be taken for mycelium growing in bulk, otherwise caking will result due to the dehydration of the compost.

   4. <u>Carbon dioxide concentration</u>: The carbon dioxide concentration in the air has a significant effect on mycelium growth. Although carbon dioxide is a metabolic product, it promotes the mycelial growth. San Antonio and Thomas (41) suggested the optimum carbon dioxide concentration for hyphal growth was 0.1 to 0.5%. Long and Jacobs (42) reported that mycelial growth rate increases with increasing carbon dioxide level up to about 10,000 ppm. Schisler (13, p. 103) suggested that carbon dioxide level should be maintained at between 5000 or 15000 ppm during spawn running. Experience indicates that the carbon dioxide concentration of 2% (by vol. or 30,000 ppm) will give a good mycelial growth.

   5. <u>Duration of mycelial growth</u>: In all practices the first stage of growth takes about 14-15 days to grow through the compost. Then a layer of casing soil is applied on top of the compost and the mycelial growth continues for another two weeks. A ruffling machine is applied in some farms to mix the mycelium with the compost. By doing this a much heavier and more uniform flush will be achieved.

C. Casing Soil Application

   It is known that if casing soil is not used, very few or no mushroom will be produced. The layer of casing soil pro-vides an environment for other micro-organisms such as *Pseudomonas* which would stimulate mushroom fruiting. It provides a water reservoir for mushroom growth. It also provides Ca^{++} for neutralization of organic acids produced during mycelial and mushroom metabolic activities and a physical environment for organisms and pinhead formation.

   1. <u>Casing materials</u>: Clay loam soil, peat with chalk, peat moss with weathered spent compost and clay, sugar plant

waste cake and peat are the most common materials used for
casing the compost. They have a high water holding capacity,
good stable physical texture, pH about 8, and they are free
from diseases, toxic metal and undecomposed vegetable materials.

   2. <u>Treatment of casing materials</u>: To prevent biological
contamination, the casing soil is always disinfected by either
physical or chemical treatments. The harmful molds causing
dry and wet bubbles, bacteria causing spots, pits, eelworms,
and mites must be killed.

   The steaming method--a moist casing is heated with low
pressure steam in various equipment like a cement mixer or a
vault, to 60°C and held at that temperature for a few hours.

   The aerated steam method--this is the most widely used
method in the industry. It was developed by the staff at the
Penn State University (43). Steam is injected into a filtered
air stream to produce aerated steam. It passes through the
casing layer to heat it up to 60°C in a rather short time and
is held in this level for a half hour.

   The chemical method--formaldehyde, chloropicrin, or vapam
are used to mix with the casing soil. After a vaporization
step disinfection is achieved.

   3. <u>Casing application</u>: When the mycelium grows through
the compost (two weeks after inoculation) a casing layer
should be applied on it. For a fixed bed farm a simple device
like an endless belt is used to drop the casing on the surface
of the bed. For a tray plant, it is accomplished on a tray
line machine. In the case where mycelium grows in bulk,
casing is done simultaneously when filling the bed with spawn
run compost.

   The thickness of the casing layer, in general, is about
2.5 to 5.1 cm. It is a common practice to disinfect the
room and the beds with a formaldehyde solution over night after
the casing operation is complete.

D. *Mycelial Growth After Casing and Cropping*

   In the tray plant, this is called "set-back". The conditions for mycelial growth are the same as those mentioned
previously, i.e., at 25°C, near saturation humidity, a carbon
dioxide concentration as high as possible, mild air circulation
with very little or no fresh air.

The casing layer must be moistened to saturation by daily
watering. It is done by a water spray with a Rose face. The
more efficient water spray like a Christmas tree with five
spray arms, each arm with multiple spray nozzles, was developed
in Holland. It can water five layers of bed simultaneously
with adjustable control for flow and pressure. The water content of casing should be brought to its maximum water holding
capacity with 4-5 days of watering.

After one week, mycelium will grow through the casing
layer. A ruffling operation can be carried out to mix the
mycelium and the casing. This will give a more uniform and
a heavier flush of mushroom. The mycelium will re-grow to
the surface of the casing in about one week time. At this
time fructification should be promoted.

Pinhead initiation is done by decreasing the air temperature to about 15°C with 100% cooler and drier fresh air. The
temperature in the compost is also decreased gradually to
about 18°C. The CO_2 concentration in the air is also lowered
from 1 to 2% to 0.03 to 0.1% over night. The air circulation
is at about 13.7 m^3 per hr per m^2 bed area, or 150 m^3 per ton
compost per hr. Mycelial growth is then shocked and changed
to the productive stage and tiny clumps are formed. At this
time the fresh air can be reduced to 4-5 m^3 per m^2 bed area
per hr, while the humidity should be maintained at 90-95%.
When the pinheads grow to the size of a pea, watering should
be started to achieve the soil's maximum water holding capacity.
The mushroom will then be ready for harvest in a few days. In
the crop growing period, the temperature program, the fresh
air requirement, and required water supply can be seen in
Fig. 8 (12, p. 283). The humidity should be controlled at
85-90%, and CO_2 concentration at 0.08%. The CO_2 production
rate (by mushroom) was measured by Tschierpe and Sinden (44)
at 1 g CO_2 per hr per kg mushroom on bed. Therefore the ventilation should be varied according to the quantity of mushroom
growing on the beds. The heat production during cropping is
about 10.9 kcal per m^2 per hr (4 Btu/ft^2/hr). A mild cooling
should be provided to maintain the temperature in the bed.
The air circulation should be about 10-15 m^3 per m^2 of bed
area per hr and the air velocity across the bed surface should
be less than 1 m per sec.

One week is normally required for the pins to mature for
harvest. In general practice, one crop of mushroom with four
flushes is most economical. Any flush after the fourth produces a very small quantity of mushroom but some growers pick
up to the sixth flush. On average, a crop should yield 20 kg
cut mushroom per m^2 growing area in four flushes picking
(Fig. 9).

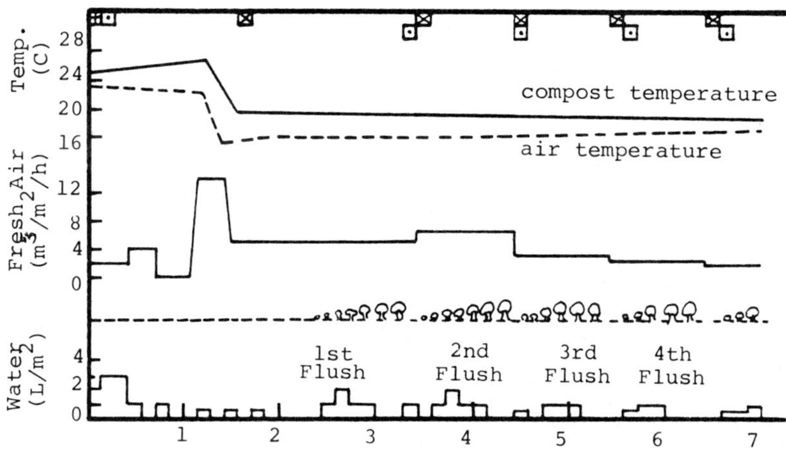

Figure 8.   The temperature, fresh air and water program during
            the entire crop growing period
               insecticides with long time effectiveness
               insecticides or fungicides
               insecticides with short time effectiveness

Figure 9. Mature mushrooms ready for harvest

E.   Supplementation

To increase the yield and the profitability is the desire
of every grower. To provide a better quality substrate is one
important mean to achieve this goal. Supplementation with
materials containing a higher concentration of lipid or protein to the composting material has already been practiced for
some time. Soybean meal, cottonseed meal, vegetable oil,
blood powder, etc. are the materials commonly used for this
purpose. Later on, the attention was not focused on the
supplementation at composting but on supplementation at
spawning or supplementation at casing (SACing) which gave
even better yields. Currently the supplementation is mostly
done either at spawning or at casing.

Schisler (45), Schisler and Sinden (46), and Schisler
and Patton (47) noted that the supplementation of compost with
vegetable oil increased the yield of mushroom up to 12.2 kg
per m^2 (2.5 lb per ft^2) growing area. In general the yield
increases about 15 to 20% in comparison with non-supplementa-
tion. The supplementation prior to the Phase II process is
at the rate of 416 liter (110 gal) of oil per single house
of 372 m^2 (4000 ft^2). The reason for this yield increase
is a thermogenesis activity increase, earlier ammonia disappearance which enhances the mycelial growth.

Carrol and Schisler (48) studied the delayed release
nutrient supplementation by encapsulating the oil droplet
within a protein denatured with formaldehyde. An increase of
yield of 60% was obtained. Similar processes were later
commercialized by Spawn Mate Company.

Gerrits (49) investigated various new products used as
supplement at casing. The author found that the higher
protein content materials gave better yield increases. The
steam or formaldehyde treated products resulted in even
better yield. Among all materials used the following are the
best: Sojax, Milko, HP-meal, Soycomil-p, Soymeal treated
with stream or formalin. These materials applied at 1 kg
per m^2 will give a yield of cut mushroom of 28-32 kg per m^2.
This is almost a 27-45% increase in yield compared with non-
supplemented control which already had a yield of 22 kg of
mushroom per m^2 of growing area.

Similar works were done at Penn State University by
Schroeder et al (50,51). They used ground soybean as a
supplement at casing. They achieved a yield increase such
that the biological efficiency was 1.09 kg cut mushroom per
kg of dry compost.

V.   SOME ENGINEERING AND TECHNOLOGY CONSIDERATIONS

The difficulty of mushroom cultivation is due to the
complexity of solid state fermentation with an undefined mixed
culture system. The scientific knowledge of both the fermentation and the physiology of the culture are not readily
understood. Although in the past fifty years a significant
progress has been achieved through scientific work performed
by various researchers in the world, many natural phenomena
still cannot be explained. Many works have been performed
without knowing the reason why. A trial and error method was
used frequently. Nevertheless, mushroom cultivation is
increasing and improving.

The North American mushroom industry faces another
difficulty, that is "the import challenge" (2). The competitive advantage of imports over domestically canned mushroom
has forced the industry to evaluate the technology used and to
search for efficient, low cost production methods.

A.   Material Handling

Mushroom cultivation differs greatly from other fermentation industries. It cannot be operated in a liquid culture.
The handling of a solid state substrate is much more difficult
than a liquid substrate. Therefore, mixing, aeration, heat and
metabolite removing are entirely different from traditional
means. A mechanized handling method must be applied especially
for bed or tray filling, emptying, inoculating, casing, etc.
At the present time many of these operations are performed
manually in many production plants. The cultivation method
and existing set-up should be modified so that more efficient
handling equipment can be used, for example one equipment
used to fill and to case the bed simultaneously.

Production plant design should take material handling
into consideration so that a streamlined production flow can
be achieved, especially in the Phase I and Phase II substrate
preparation. The dirty (before pasteurization) and the clean
end (after pasteurization) must be separated, without inter-
mixing with each other in order to prevent contamination.
Production rooms and mycelium growing rooms should also be
isolated to prevent contamination.

B.   Air Handling

Mushroom cultivation requires an extensive air handling
system for temperature control (heating and cooling), humidity

control (humidifying and de-humidifying), oxygen supply, and
carbon dioxide control (accumulation and dilution) in various
stages of mushroom cultivation. Tschierpe (52) and Edwards
(53) studied the effect of air in mushroom production plants
in detail. Tschierpe especially emphasized the carbon dioxide
concentration and temperature while Edwards paid more attention
to the fundamental theory.

Although the total quantity of air required in the growing
room can be calculated according to experience or theory, the
criteria which provide a more or less uniform flow of air to
every part of the growing surface is another difficult
engineering problem. In general one air duct is installed in
a single house at the center between two tiers of growing
beds, while two or three ducts are installed for a double
house with four tiers of beds.

Air flows through holes on the duct into the growing
rooms and dissipates to the surrounding space. Some design
provides rough holes at the bottom side of the duct with or
without installation of diffusers; others provide the holes on
the side wall of the duct. Regardless of the design chosen
uniform air flow across the beds must be maintained and the
velocity of air above the mushroom should not be more than
1 m per sec.

Incoming air and discharged air must be filtered. A
micron filter should be used to remove bacteria and virus. A
heating and cooling device should be installed in the air
handling system. Humidification and de-humidification
apparatus should also be installed.

For energy conservation, an energy recovery device, such
as a heat exchanger, an energy wheel, etc., should be
equipped. In the wintertime, the exhaust air is used to pre-
heat the incoming cold air. On the contrary in the summer it
is used to pre-cool the hot air. This method is especially
beneficial for a larger production plant which should install
a central air handling system to supply fresh air to all
growing rooms (54). For smaller plants, application of heat
pump could be another way to save energy.

C.   Carbon Dioxide Control

It has been pointed out previously the importance of the
concentration of CO_2 at the various stages of mushroom cultivation. To achieve the optimal concentration of CO_2 at
various times is not only a means to increase the yield and
productivity, but also to save energy of operation. At the
present time automatic control of the concentration of CO_2

is not widely accepted in the industry because of the lack of
understanding and the lack of reasonably priced equipment.
Engineering and technological developments in this area
should be beneficial to the industry.

D.   *Synchronized Growth and Mechanical Harvest*

Mushroom picking labor cost is the largest single cost
contributing factor. It has been estimated to be about 40-50%
of the total production cost. Therefore any method improving
the efficiency of harvesting will significantly reduce the
production cost. Invention of mechanical harvesters have been
patented and reported (55,56), but it has not yet been applied
in the industry. In Holland two different cutters have been
developed and applied commercially. The drawback of those
harvesters is the damage to the mushroom surface, consequently
the quality of the product is reduced and the shelf life is
also greatly reduced.

The following are the technologies which should be
developed in order to reduce the harvesting cost:

1.   Synchronized growth: Mushrooms grow in periods
called a flush. Within one flush, some grow earlier than the
others. It is required that all mushrooms should be at more
or less the same stage of maturity so that they can be harvested at the same time. This technique should be developed
biologically or environmentally.

2.   Mushrooms with longer stems: This is extremely
crucial for successful mechanical harvesting. Strains with
long stems will be a good candidate. Work should be done in
carrying out strain selection for long stem strains with
high yields.

Tschierpe (57) suggested that a CO_2 concentration of
3-4% will lengthen the stem rapidly. This, although not yet
verified technically, is an interesting recommendation.

3.   Development of better harvesting machines: The
existing machines will cut but also damage the mushroom.
Engineering research work has been conducted recently to aim
at improving or developing entirely new machines so that a
better mechanical cutter can be developed.

E.   *Shelf Life Improvement*

The mushroom is a perishable vegetable having a limited
shelf life, especially in a warm climate. In general,

mushrooms that have been picked are kept in a cold storage
room for static cooling down. The metabolic activity of the
mushroom is very high and heat and moisture are produced.
Although it is a rule of thumb that the reaction rate doubles
with a 10°C temperature increase, for the mushroom it is 3.5
times. Hence the deterioration is accelerated with temperature
increases. For a better shelf life of the produce, the mushroom must be rapidly cooled down using a forced air cooling
method (58) then kept in a cold storage room prior to
delivery.

Langerak (59) studied the effects of irradiation with
γ-ray of 200-300 krad. It was found that the combination of
irradiation and pre-packaged closed mushroom had the optimum
quality without the necessity of extreme cooling. The
organoleptic test and wholesomeness studies showed that this
method is well accepted by consumers as well as the government health authorities.

VI. CONCLUSION

Mushroom cultivation has been developed as an
important fresh produce or processed food in various regions
of the world. It requires only limited land which does not
have to be fertile or suitable for traditional crop cultivation. It is worth noting that the mushrooms do not perform
photosynthesis; therefore it requires no sunshine for growth.

Mushroom cultivation is an intensive production under
a completely controlled growing environment. It has probably
the highest yield of food produced per unit area of land. A
production of 147 kg of cut mushroom per m^2 growing area per
year has already been achieved. It is 4 to 5 times more per
m^2 of land depending on growing installation.

Most of the materials used for mushroom cultivation are
agricultural wastes which are difficult to dispose. The
mushroom technology has been developed with fermentation
techniques to convert these waste materials into delicious,
nutritious food in a large quantity. This is not only just
to increase food production, but also to solve the waste
disposal problem.

Undoubtedly, the mushroom industry will continue to
grow rapidly as it becomes more available, and more widely
accepted as a regular food. The scientific knowledge and
technology will continue to be developed so that the biological
efficiency will exceed "one". Furthermore the mechanization
of cultivation, especially in mushroom harvesting, and the

application of automatic control will increase the production efficiency and minimize production cost. As a result very reasonably low priced mushrooms will become available for consumption. Of course such a technical know-how requires more investigations by biologists and engineers.

The other areas of mushroom technology such as the production of various cellulases, amanitin, and medicinal products, etc. make up the virgin land of technology which remains to be exploited in the future. Further benefit to mankind will be realized if effort is spent to tackle this problem.

REFERENCES

1. Lin, B., Mushroom and Their Cultivation, *Science Publisher,* Peking (1964).

2. Owens, T.R., W.R. Garland, K. Kesecker, and J.L. Runyan, The U.S. Mushroom Industry: The Import Challenge, *USDA Marketing Research Report* 1131 (1982).

3. Tschierpe, H.J., How Far Can Production Method Changes Reduce Cost, *The Mushroom Journal 116,* 259-269, August (1982).

4. Crisan, E.V. and A. Sands, "Nutritional Value, in: *The Biology and Cultivation of Edible Mushrooms,* (S.T. Change and W.A. Hayes, eds. Ch. 6), Academic Press (1978).

5. Lelley, J., The Economic Importance of Macromycetes: The Actual Situation and Future Prospects, *The Mushroom Journal 111,* 77-79, March (1982).

6. Tokita, F., N. Shibukawa, T. Yasumoto, and T. Kaneda, Isolation and Chemical Sturcture of the Plasma- Cholesterol Reducing Substance From Shiitake Mushroom, *Mushroom Science 8,* 783-788 (1972).

7. Blom, T. and F. Ingratta, "Mushroom House" *Ministry of Agriculture and Food, Province of Ontario* (1978).

8. Kligman, A. M., "Handbook of Mushroom Culture" 2nd ed., J. B. Swayne, ed., Kennett Square. Pennsylvania (1950).

9. Flegg, P.B. and J.F. Smith, "Pioneering Deep Troughs," *The Mushroom Journal 89,* 171-177 (1980).

10. Gaze, R.H., "Deep Troughs Assessed," *The Mushroom Journal 95*, 385-397 (1980).

11. MacCanna, C. and W.M. Dawson, "An Economic Comparison of Mushroom Growing Systems," *The Mushroom Journal 100*, 117-129 (1981).

12. Vedder, P.J.C., "Modern Mushroom Growing," p. 78 (1978).

13. Wuest, P.J. and G.D. Bengtson, "Penn State Handbook for Commercial Mushroom Growers," p. 8 (1982).

14. Gerrits, J.P.G., H.C. Bels-Koning, and F.M. Muller, "Changes in Compost Constituents During Composting, Pasteurization and Cropping," *Mushroom Science 6*, 225-243 (1967).

15. Gerrits, J.P.G., "The Significance of Gypsum Applied To Mushroom Compost, In Particular In Relation to The Ammonia Content," *Journal Agricultural Science 25*, 288-302 (1977).

16. Block, S.S., G. Tsao, and L. Han, "Production of Mushrooms From Sawdust," *Journal Agricultural Food Chemistry 6*, 923-927 (1958).

17. Schisler, L.C., "Possible Substitutes for Corn Cobs In Mushroom Composts," *Mushroom News 22 (5)*, 6-15 (1974).

18. Kneebone, L.R. and E.C. Mason, "Sugar Cane Bagasse as a Bulk Ingredient in Mushroom Compost, *Mushroom Science 8*, 321-329 (1972).

19. Tseng, M.C., C.R. Phillips, and J.W. Smith, "Utilization of Bagasse," Annual Conference of the Jamaican Association of Sugar Technologists, November (1972).

20. Peerally, A., "Sugar Cane By-products as Bulk Ingredients in Mushroom Compost," *Mushroom Science 11*, 293 (1981).

21. Rasmussen, C.R., "The Improvement of Composts," *Annual Conference of the Mushroom Growers' Association* (1956).

22. Rasmussen, C.R., "Pig Manure as a Substitute for Horse Manure in Mushroom Growing," *The Mushroom Growers' Association Bulletin 103* (1959).

23. Stoller, B.B., "Principles and Practice of Mushroom Culture," *Economic Botany 8*, 48 (1954).

24. Hayes, W.A., "Mushroom Nutrition and the Role of Micro-
     organisms in Composting," Proceedings of the Second Seminar
     in Mushroom Science. Published for the Mushroom Growers'
     Association, England (1977).

25. Hayes, W.A., "Microbiological Changes in Composting
     Wheat Straw/Horse Manure Mixtures," *Mushroom Science 7*,
     173 (1969).

26. Fermor, T.R., J.F. Smith, and D.M. Spencer, "The Micro-
     flora of Experimental Mushroom Composts," *Journal of
     Horticultural Science 54(2)*, 137-147 (1979).

27. Sinden, J.W. and E. Hauser, "The Short Method of Mush-
     room Composting," *Mushroom Science 1*, 52 (1950).

28. Sinden, J.W. and E. Huaser, "The Nature of the Composting
     Process and Its Relation to Short Compost," *Mushroom
     Science 2*,]23 (1953).

29. Gerrits, J.P.G., "The Influence of Water in Mushroom
     Compost," *Mushroom Science 8*, 43-57 (1972).

30. Overstizns, A., "The Conventional Phase 2 in Trays or
     Shelves," *The Mushroom Journal 97*, 5-17 (1981).

31. Fergus, C.L., "Thermophilic and Thermotolerant Molds and
     Actinomycetes of Mushroom Compost During Peak-Heating,"
     Mycologia 56, 267-284 (1964).

32. Sager, J.C., "A System Analysis of Environmental Variables
     for Control Phase II of Mushroom Composting," Ph.D.
     Thesis, Pennsylvania State University (1973).

33. "Manual of Practical Training School for Mushroom
     Growing," issued by Centrum voor Champig nonteelt
     Onderwijs (1979).

34. Vedder, P. and J. Smith, "Tunnels and Bulk Treatment of
     Compost," *The Mushroom Journal 118*, 333-345, October
     (1982).

35. Gerrits, J.P.G., "Factors in Bulk Pasteurization and
     Spawn Running," *Mushroom Science 11*, 351-365 (1981).

36. Laborde, J. and J. Delmas, "La Preparation Express des
     Substrats," *Bull. F.M.S.A.C.C.* 2093-2109 (1969).

37. Smith, J.F., "Selective Substrate and Rapid Methods of
     Preparation," *The Mushroom Journal 23*, 424-426 (1974).

38. Laborde, J., "Rapid Substrate Making," *The Mushroom Journal 94*, 349-361 (1980).

39. Baker, J.D. and J.W. Baker, "Barrel Composting," *Mushroom Science 11*, 201-209 (1981).

40. Treschow, C., "Nutrition of Cultivated Mushroom," *Dansk Botanical Arkives 11(6)*, 1-180 (1940).

41. San Antonio, J.P. and R.L. Thomas, "Carbon Dioxide Stimulation of Hyphal Growth of the Cultivated Mushroom *A. Bisporus* (Lange) Sing," *Mushroom Science 8*, 623-629 (1972).

42. Long, P.E. and L. Jacobs, "Some Observations of CO_2 and Sporophore Initiation in the Cultivated Mushroom," *Mushroom Science 8*, 373-384 (1968).

43. Aldrich, R.A., P.J. Wuest, and J.A. McCurdy, "Treating Solid, Solid Mixes, or Soil Substitutes with Aerated Steam," Special Circulation 187, *Pennsylvania Cooperative Extension Service* (1974).

44. Tschierpe, H.J. and J.W. Sinden, "Studies on the Composition of Horse Manure Compost from Beginning of Phase II Through Mushroom Cropping as Related to CO_2 Evolution," *Mushroom Science 5*, 61-80 (1962).

45. Schisler, L.C., "Stimulation of Yield in the Cultivated Mushroom by Vegetable Oil," *Applied Microbiology 15*, 844-850 (1967).

46. Schisler, L.C. and J.W. Sinden, "Nutrient Supplementation of Mushroom Compost at Casing--Vegetable Oil," *Canadian Journal Botany 44*, 1063-1069 (1966).

47. Schisler, L.C. and T.G. Patton, Jr., "Yield Response of Selected Mushroom Strains to Vegetable Oil Supplementation," *Mushroom Science 8*, 707-712 (1982).

48. Carroll, A.D., Jr. and L.C. Schisler, "Delayed Release Nutrient Supplement for Mushroom Culture," *Applied and Environmental Microbiology 31*, 499-503 (1976).

49. Gerrits, J.P.G., "Some New Products Used for Supplementation," *de Champignon Cultuur 26*, 463-469 (1982).

50. Shroeder, M., L. Schisler, R. Snetsinger, and V. Crowley, "Yield and Sacing Experiments in the Penn State Mushroom Test Demonstration Facility--10 Year Summary," *Mushroom Science 11,* 405-410 (1981).

51. Shroeder, G.M., L.C. Schisler, and H.R. Muthersbaugh, "Effect of Fragmenting Colonized Compost and Supplementation at Casing on Mushroom Size, Yield and Dry Weight," *Mushroom Science 11,* 419-431 (1981).

52. Tschierpe, H.J., "Environmental Factors and Mushroom Growing," *The Mushroom Journal 1,* 3-28 (1973).

53. Edwards, R.L., "Mushroom House Ventilation in Theory and Practice," *The Mushroom Journal 3,4* (1973).

54. Rucklidge, R.A., "Farm Design, Machinery and Future Trends," *Mushroom Science 11,* 147-163 (1981).

55. Persson, S.P.E., "Mushroom Harvester," U.S. Patent No. 3,635,005 (1969).

56. Persson, S.P.E., "Mechanical Harvesting of Mushrooms and Its Implications," Paper No. 4205, *Pennsylvania Agricultural Experiment Station, University Park* (1972).

57. Tschierpe, H.H., "Grade 1 Mushrooms by Mechanical Harvesting," *The Mushroom Journal 110,* 41 (1982).

58. Baker, J.D., J.B. Watkins, and M. Lawson, "Post-Harvest Management of Mushroom," *Mushroom Science 11,* 645-653 (1981).

59. Langerak, D.I., "The Influence of Irradiation and Packaging Upon the Keeping Quality of Fresh Mushrooms," *Mushroom Science 8,* 221-229 (1972).

CHAPTER *4*

IMMOBILIZED MICROBIAL CELLS

Harvey W. Blanch

Lawrence Berkeley Laboratory
and
Department of Chemical Engineering
University of California
Berkeley, California

I.   INTRODUCTION

There has been a considerable and continuing interest
in the use of immobilized cells, both procaryotic and eucaryotic,
over the last few years, arising primarily from earlier research
and subsequent industrial application of immobilized enzymes.
As the number of potential industrial applications has expanded,
further attention has been given to methods of immobilization
which overcame some of the earlier problems associated with
excessive cell leakage, loss of activity and degeneration of
the support material.

Immobilized cell systems offer the advantages of higher
volumetric productivity than convention fermentation systems
(although many reports of productivities are misleading), the
possibility of continuous operation, and higher stability in
general than immobilized enzyme systems. In cases where an
immobilized enzyme system is considered which requires a
cofactor, it is often the case that an immobilized cell system
is the only possible approach to take. It also obviates the
need to extract and purify the enzyme to be immobilized. As
compared to conventional fermentation systems, it may be
possible to obtain a higher cell density in an immobilized
system, and this approach may in fact favour the production
of secondary metabolites. In the case of plant cells, it
may be possible to control cellular differentiation. There
are some indications that by restricting the ability to
replicate in genetically modified organisms, it may be
possible to obtain immobilized cells which are not as
susceptible to reversion.

On the other hand, immobilization carries with it some disadvantages which may make it impractical. If high molecular weight products are desired, their size will restrict diffusion through many support materials. A related constraint is the transport of gaseous reactants and products through the support material. In cases of highly aerobic systems, the low solubility of oxygen in the support material (especially in some gels and polymers) may reduce the effectiveness of the immobilization procedure. Similarly, the production of acid as a reaction product or by-product may result in a local pH which may be deterimental to cellular productivity if the acid is unable to diffuse away from the site of production or if base cannot effectively diffuse through the support to the cells for neutralization.

There has been considerable research directed at the immobilization of mammalian cells, as these are in most cases anchorage dependent, and immobilization offers clear advantages over surface cultivation. This review will focus on microbial cells exclusively: the cultivation of mammalian cells has been reviewed from an engineering viewpoint recently by Kattinger (1), and Tolbert and Feder (2). Plant cell culture has been examined by Brodelius (3). We will focus here on more recent advances in immobilized cell systems, primarily since the last review in this series (4). The extent of activity in this area can be judged from the large number of significant reviews on these systems. Of these, the series of Berger (5,6), is particularly comprehensive, as is the recent publication edited by Mattiasson (7). Earlier reviews in this series include those of Chibata et al., (4), and Abbott (8). Other comprehensive reviews include that of Cheetham (9), a DECHEMA monograph (10), an ACS Symposium series (11), Venkat and Vieth (12), Jack and Zajic (13), Abbott (14), Vandamme (15), Fukui and Tanaka (16), Dunnill (17), and Klein and Vorlop (18).

One of the factors which inevitably governs the performance of immobilized cell reactors is the extent to which external and internal mass transfer limitations affect the supply of substrate and removal of product from the organism. While much activity has focussed on these effects in immobilized enzyme systems, immobilized cells are somewhat more complex in that there may be limited cell replication occurring and the presence of cellular debris may offer an increased diffusional barrier. The intrinsic kinetics of cell growth and product formation are poorly understood in immobilized systems and this further complicates interpretation of kinetic analyses. For these reasons this review will also include a discussion of reactor performance.

The use of immobilized cell systems in industry has been increasing and now includes the conversion of ammonium fumarate to L-aspartic acid (19,20) using *E. coli,* the conversion of sodium fumarate to L-malic acid (21,22) with *Brevibacterium ammoniagenes,* conversion of penicillin to 6-APA with *E. coli* (21,23), the conversion of 7-ACDA and phenylglycine to cephalexin with *Achromobacter* (14), and production of fructose from glucose for high fructose corn syrups using a variety of microorganisms (24,25), hydrolysis of lactose using a bacillus species, sucrose hydrolysis using yeast, and raffinose hydrolysis to galactose and sucrose (26).

A variety of pilot scale processes have been reported, and these include the conversion of L-histidine to urocanic acid by *Achromobacter liquidium* (21,27), conversion of L-arginine to L-citrulline using *Pseudomonas putida* (21,23,28). Other commercial processes, using non-viable cells, are tabulated in Gestrelius (26). In addition, a number of processes have been developed for water treatment, including denitrification (29), phenol degradation (30), and ammonia degradation (31). In most of the above examples, there is generally one enzyme which is responsible for the conversion of substrate to product,i.e., the cell is functioning as a source of crude enzyme without making use of multienzyme, cofactor requiring reaction steps.

II. *METHODS OF IMMOBILIZATION*

There is a wide variety of methods which have been examined for immobilization of whole cells, in either viable or non-viable form. Many of these stem directly from their counterparts in enzyme immobilization. Various methods for classifying these approaches have been proposed: Table 1 shows the scheme of Klein and Vorlop (18). General articles describing many of the methods of immobilization include Durand and Navarro (32), Messing (33), and Chibata (34).

Table 1. Methods for whole cell immobilization[*]

Adsorption		
Covalent Binding		
Crosslinking		
Encapsulation		
Entrapment	*-gelation*	*-polycondensation*
	-precipitation	*-polymerization*
	-ionotropic gelation	

[*] *from Klein, Vorlop (18)*

A. *Covalent Binding*

This approach has been widely and successfully used for
enzyme immobilization, however the attachment of whole cells
to surfaces with binding agents of a generally toxic nature has
not been very successful. If viable cells are immobilized in
this manner, any newly formed cells would not be bound, and
substantial cell leakage would result. Most reports have thus
employed non-viable cells, and the reported activities of the
system are low. Examples include production of urocanic acid
(35), gluconic acid (36), and 6-APA (36). Kennedy (37) has
proposed an interesting approach using hydrous titanium (IV)
and zirconium (IV) as support matrices, where it is postulated
that partial covalent bonds are formed with cell wall components
such as protein and carbohydrates. Typical polymer supports
and covalent coupling agents which have been used include
derivatized porous glass and glutaraldehyde; carboxymethyl-
cellulose and carbodiimide; and hydroxyalkyl methacrylate gel
and glutaraldehyde. Typical cell loadings are low using
covalent coupling, of the order 0-10 gm wet cells/liter wet
catalyst. Consequently overall productivities of these systems
are low.

B. *Adsorption*

Adsorption involves the formation of ionic and hydrogen
bonds between the cell surface and the support: primarily
determined by the charge distribution of carboxy and amino
groups on the cell wall. The properties of the support
determine the resulting behaviour of the system; cell loading
is determined by the surface area of the support material and
its charge, and the available pore size. Adsorption is a
reversible process, and cell leakage can thus be expected.
Table 2 lists some of the variety of supports and organisms
which have been examined.

A particularly mild form of adsorption is the use of
affinity supports, such as concavalin-A Sepharose (38). This
approach has also been used for cell separation (39).

C. *Entrapment*

Entrapment is the most commonly used method of immobilizing
both viable and non-viable cells. Substrate and products may
diffuse freely within the support matrix, provided their
molecular weights are small enough relative to the support pore
size. A variety of polymeric materials have been used,
including synthetic and natural polymers. Use of synthetic

polymers often require somewhat harsh polymerization conditions,
and cell damage or loss of viability may result. Among the
synthetic polymers, polyacrylamide has been frequently used.
The method for preparation of entrapped cells using this polymer
is described in Klein and Vorlop (18), Mattiasson (7), and by
analogy with enzyme preparation in Chibata (34). The variety
of organisms which have been entrapped in polyacrylamide is
large and lists can be found in the literature surveys of
Sturgeon and Kennedy (50), Fukui and Tanaka (16), Mattiasson
(7), Klein and Vorlop (18), and Cheetham (9), and Chibata
et al., (4). They will be described in this review in
relation to the products formed by the immobilized cell
system.

Polyurethane has also been used fairly extensively for
entrapment, using a prepolymer to avoid the exposure of the
cells to harsh reaction conditions (7). Polyurethane has a
fairly open structure and can be used as a catalyst bed as
formed, or can be cut into any desired form for packed bed or
slurry reactors (18). Epoxy resin has also been used, in
conjuction with alginate added during the curing process.
Following curing, the alginate is dissolved away to yield a
high porosity, mechanically stable bead (18).

Table 2. Cell immobilization by adsorption

Product	*Cell*	*Support Material*	*Reference*
Succinic scid	E. coli	*Dowex I.E. resin*	*40*
Acetic acid	Acetobacter	*ceramic support*	*41*
Lactic acid	Lactobacillus and yeast	*gelatin*	*42*
Ethanol	S. cerevisiae	*anion exchanger*	*43*
		bricks	*44*
	S. carlsbergenesis	*PVC*	*45*
		porous glass	*46*
	Z. mobilis	*glass fiber pads*	*47*
Fructose	Streptomyces	*modified cellulose*	*48*
Phenol degradation	seudomonas	*anthracite*	*49*

Among the natural polymers, recent developments have focussed extensively on κ-carrageenan, as this provides a particularly mild form of entrapment. A small amount of cells is placed in the κ-carrageenan as an inoculum and the cells are later permitted to grow, thus providing a highly dense cell concentration toward the outside of the bead. Locust bean gum has been added to increase the mechanical strength of the beads and lower the gelling temperature (16). The gel has also been hardened with addition of hexamethylaminediamine and glutaraldehyde, which also increases operational stability (16). Alginate has also been frequently used, as it offers similar advantages to κ-carrageenan in that conditions used to form the polymer beads are very mild and the resulting bead is quite porous. Methods for preparation are described in (18), (16), (7) and (34). One processing constraint is the low mechanical stability (10,18).

Collagen has been proposed as a simple, mild immobilization matrix (11), due to the large number of reactive groups present on the surface and the possibility of increasing the matrix strength with tanning of the protein (51). Much of the experimental data on collagen matrices for steroid conversion, citric acid production, and candicidin production are summarized in Reference (11). A fairly recent development has been the use of cellulose for entrapment (52), however, the solvents used to dissolve cellulose and the subsequent tanning operations are likely to result in large losses in cell viability.

A further approach to entrapment is to immobilize cells between membranes: various geometries are possible, including flat sheets, spheres, hollow fibers, and spiral wound fibers. This method is also a gentle approach to immobilization, as an inoculum is allowed to proliferate until the available space is filled with organisms. This method has been used for plant and mammalian cells extensively (2,3,53,54,55) and with bacteria in hollow fiber geometry (56); the hollow fiber system has been reviewed recently (55,56), and high cell densities have been obtained (57).

The preceding discussion of immobilization methods is by no means complete, and new supports are being developed which may permit higher cell loadings (58), maintenance of viability and mechanical strength under the processing schemes proposed. The quarterly reviews in *Enzyme and Microbial Technology* provide an ongoing summary of immobilization techniques.

D.   *Physical Properties of Immobilized Cell Systems*

Despite the wide body of literature on immobilized cells, there is little information on the physical characterization of the resulting immobilized particles which would be required for process design and scale-up. Typical parameters of interest include mechanical strength (primarily compression), resistance to abrasion, pressure drop characteristics in packed beds, and behaviour in fluidized beds. The size distribution of the catalyst particles is also of importance in determining the kinetic behaviour, and also the degree of swelling of the particles.

Some of the broad issues are described in relation to reactor design in Venkat and Karkare (59). Klein and Washausen (60) describe specific techniques related to the determination of compressibility of polymers used for entrapment, and present typical results for calcium alginate. Abrasion in stirred tanks is described in Eng (61), and the pressure drop in packed and fluidized beds has been found to follow the Ergun equation in cases where compression of the bed is not a significant factor (62,63). Recently the *Working Party on Immobilized Biocatalysts of the European Federation of Biotechnology* has published a set of guidelines for the characterization of immobilized biocatalysts (64).

III. *REACTOR PERFORMANCE AND CHARACTERIZATION*

A.   *General Considerations*

The kinetics of immobilized cells differ from the intrinsic rates found in the absence of mass transfer limitations due to the nature of the immobilization procedure used. By analogy with heterogeneous catalysis, two prime factors are important; the external and the internal diffusional resistances, which influence substrate and product transport and may also affect pH in the local environment of the cell. A further, although less well characterized factor, is local heat generation and heat transfer mechanisms.

The external mass transfer resistance depends on the flow past the particle containing the cells, and correlations are available for a wide variety of geometries, particle densities (often free suspension may occur, resulting in a zero slip velocity), and packing types (suspensions, packed beds, fluidized beds, films, solids etc.) which are basically identical to those commonly used in heterogeneous catalysis (65,66,67).

Intraparticle mass transfer is somewhat more difficult to
assess, due to the often porous nature of the support particles,
and their variable geometry. The estimation of diffusivity
within the support matrix is difficult (18,68) and depends on
pore size, charge interactions and the tortuosity of the pores.
An additional factor is the solubility of substrate or product
in the support. The question of diffusion is addressed in
references 69 and 70.

When the mass transfer resistances are coupled with the
intrinsic kinetics of product formation by the immobilized
cells, the resultant equations are, in general, difficult ·to
solve and numerical methods are used. When the kinetics can
be expressed in either zero or first order form, solutions are
available for obtaining the observable intracellular kinetics
in spherical, cylindrical and rectangular coordinates. Approxi-
mate solutions are available for the case of Michaelis-Menten
kinetics (71). One difficulty not encountered with immobilized
enzymes is the asymmetric distribution of cells in particles
where cell growth is occurring; cells tend to preferentially
multiply close to the surface where the substrate concentration
is highest. This makes interpretation of experimental data
particularly difficult. Prediction of reactor performance
depends on the nature of the reactor; plug flow, stirred tank,
fluidized bed, or stationary bed. Overall material balances
can be combined with Thiele modulus-effectiveness factor plots.
The correlation of experimental and theoretical data for
immobilized biocatalysts has been recently reviewed by
Kasche (72), and the effects of both external and internal
mass transfer resistances on reactor performance reviewed.
Other reviews are those of Buchholz (63), Klein and Vorlop (18),
and Buchholz (10). It is clear that adequate theoretical
approaches to the problem are available, particularly when
cognizance is taken of the related literature in Chemical
Engineering, but that relatively few thorough experimental
studies have been made.

B.   *Deactivation of Immobilized Cells*

Key to operational stability of an immobilized cell
reactor is the stability of the catalyst: usually measured by
its half-life. The economics of a particular process may
provide indications of a desirable minimum half-life; beyond
a certain value however, catalyst replacement costs may not be
a significant factor. Half lives from several days to two
years have been reported (73). Interpreting the operational
stability of a system is complex, but a first order catalyst
deactivation, coupled with intraparticle mass transfer limitation, was adequate to explain the long term behaviour of *E. coli*

immobilized in epoxy for production of 6-APA from penicillin
(18). Here the relative activity of the system showed two
linear regions of decay. The implications of catalyst half-
life are discussed in terms of reactor performance in (59).
If viable cells are used, it may be possible to alter reaction
conditions such that new cell growth may occur and consequent
catalyst regeneration. The effect of the support material on
cell half-life may be significant: Tosa et al., show a variation from 160 to 243 days for L-malic acid production by
B flavum, immobilized in κ-carrageenan, polyethyleneimine and
polyacrylamide (74).

C.   *Reactor Productivities*

While the productivity of an immobilized cell system
clearly depends on the substrate and product being considered,
non-viable cells often have productivities in the range of
500 to 2000 moles per liter reactor volume per two half-lives
(26). Often literature reports present productivities in
terms of moles of product formed per unit time and mass of
immobilizing matrix, with no reference to the solids fraction
of catalyst in the reactor; this has been discussed by Karanth
(75). Thus prediction of reactor performance is difficult, and
comparisons with conventional systems are impossible. It is
also necessary to report the conversions obtained at the
conditions when maximum productivity is specified; reports of
extremely high productivities under conditions of very low
substrate conversion are extremely misleading. The operation
of immobilized reactors in a differential mode is not likely
to be an economically attractive approach.

IV. *PRODUCTS*

A.   *Alcohols*

The utilization of biomass as an energy source has given
considerable impetus to investigations on the conversion of
glucose and xylose sugars to ethanol, for use as a trans-
portation fuel. Ethanol production in batch and continuous
systems has been reviewed (76). Cell recycle systems show
productivities of up to 80 gm/l hr, with nearly complete
conversion of glucose, so that an immobilized cell system may
not necessarily offer any savings in reactor costs, but rather
assist in terms of improved stability and yield. Yeast,
primarily *S. cerevisiae,* and *Z. mobilis* have been immobilized
on a variety of supports, including porous glass, Raschig rings,
ion exchange resins, and immobilized by entrapment in κ-carrageenan, and alginate. In most cases leakage of cells from the
reactor has been a problem, particularly when the cells have
been bound to carriers. The operation of a pilot-plant for

ethanol production has been reported by Fukushima and Hanai (77),
and a fluidized bed reactor containing *S. formosensis* entrapped
in alginate was shown to give a 7.8 wt% ethanol concentration
in the effluent at a dilution rate of 0.36 hr^{-1} with 162 gm/l
sugar feed (molasses). In most reports it is clear that high
productivities can be obtained at the expense of conversion,
and that ethanol inhibition can be a significant factor. The
evolution of CO_2 may lead to the disruption of the immobilizing
particle. The economics of ethanol production indicate that
there may be little to be gained in operating an immobilized
cell system as opposed to even a conventional batch fermentation,
as feedstock costs, product recovery and stillage handling are
the more important factors (76). Ethanol production has been
examined in immobilized cell systems in (4) and (78). Table 3
lists the more recent literature on ethanol production. As
can be seen, its simplicity of operation has been appealing,
but few studies have used real substrates or tried to develop
operating systems for commercial practice.

Other alcohols have been produced in immobilized cell
systems and include 2,3-butanediol, n-butanol and iso-propanol,
and dihydroxyacetone. At present the fermentative route to
these products is not competitive with petrochemical synthesis.
These are also listed in Table 3.

B.   Organic Acids

The production of organic acids anaerobically is perhaps
one of the more simple applications of immobilized cell
technology, as no transfer of oxygen is required. Lactic acid
is an interesting example of this in which no gas evolution
occurs. It has been examined recently by immobilization of
L. delbrueckii in calcium alginate (Stenroos et al., (99)).
While complete conversion of glucose to lactate can be achieved,
this produces acid, lowering the pH and reducing the specific
productivity of the cells. Production in a hollow fiber
reactor has been reported, and very high rates of lactic acid
production obtained, albeit at low conversions (57). In tubular
flow reactors the problem of pH control and product inhibition
have not been satisfactorily resolved.

The question of supply of oxygen to cells producing
organic acids is a difficult one and has been examined
theoretically by Krouwel and Kossen (100), for CO_2 evolution
by yeast. In order to be economically competitive with traditional deep tank processes for citric acid production, it is
clear that oxygen supply requires a film type attachment of

Table 3. Production of alcohols by immobilized cells

ETHANOL

Organism	Method of Immobilization	Comments	Ref.
S. cerevisiae	alginate	substrate inhibition at glucose concn. greater than 8%	79
S. cerevisiae	"natural carrier"	28.6 g/l hr ethanol temperature and physiology examined	80, 81,82
S. cerevisiae	Ca alginate	optimum T, pH determined comparison with free cell system	83
S. cerevisiae	polyacrylamide hydrazide	growth of cells inside gel increase alcohol tolerance	84
P. tannophilus	Ca alginate	xylose as substrate 42% theoretical yield at max specific EtOH productivity	85
F. oxysporum, *Mucor sp.* *S. cerevisiae*	Ca alginate	all organisms compared for pentose fermentation using isomerase column	86
Z. mobilis	glass fiber pads	6.4% EtOH concn. at 152 gm/l hr	87
Z. mobilis	ion exchange resin	cationic macrorecticular resin, 377 gm/l hr at 80% glucose conversion	88
S. cerevisiae *S. uvarum* *Z. mobilis*	Ca alginate	three cultures compared on 1,3 and 5% glucose inhibition at 20%	89
Z. mobilis *S. bayanus*		bacterium showed higher productivity and conversion	90

OTHER ALCOHOLS AND KETONES

Organism	Method of Immobilization	Comments	Ref.
E. aerogenes	κ-carrageenan	2,3-butanediol produced	91
Cl. butyricum	Ca alginate	n-butanol	92
Cl. acetobutyricum	Ca alginate	n-butanol	93
Cl. acetobutyricum	Ca alginate	isopropanol	91
S. cerevisiae	κ-carrageenan	glycerol	94
Gl. oxydans	Ca alginate	glycerol converted to dihydroxyacetone using peroxide	95
Gl. oxydans	Ca alginate	improved oxygenation on dihydroxyacetone from glycerol	96
Gl. oxydans	polyacrylamide	glycerol to dihydroxyacetone	97
Cl. beijerinckii	Ca alginate	isopropanol-butanol-ethanol production in stirred tank	98

cells to a very porous support. This is also the case with
acetic acid produced by oxidation of ethanol. The use of
methods for generation of oxygen at the reaction site using
peroxide and peroxidase or catalase is not an economically
feasible approach for large volume products, due to the high
cost of peroxide. Similarly, the supply of oxygen using
carriers such as perfluorocarbons does not provide a
sufficiently large increase in the driving force for oxygen
transfer to overcome the decrease in surface area for oxygen
transfer which occurs in most systems where its use would be
attractive; hollow fiber reactors, packed beds, and membrane
reactors. The general questions of oxygen transport are
addressed in Enfors and Mattiasson (101). Recent literature
on the production of organic acids is summarized in Table 4.

C.   *Amino Acids*

The production of several key amino acids is predominantly
via batch fermentation, however aspartic acid is currently
produced by the Tanabe company using immobilized *E. coli,* and
there are several amino acids which show promise for production
by immobilized cells, including glutamic acid, tryptophane,
aspartic acid, lysine, citrulline, and alanine. Other amino
acids can be produced by use of immobilized enzymes, and both
approaches have been reviewed from an industrial perspective
by Brodelius (113). In several cases the immobilized cell
serves principally as a source of one or two enzymes, and
precursors to the amino acids are supplied as substrates. Thus
aspartic acid is produced from malic acid via fumaric acid as
an intermediate; tryptophane from indole and serine or ammonium
pruvate; tyrosine from phenol and serine or ammonium pyruvate;
alanine from aspartic acid. Table 5 lists more recent reports
on amino acid production. In general, amino acids are formed
from rather complex metabolic pathways, so that prospects for
immobilized cell systems which can compete with batch or
continuous culture are not always bright, except in those
situations where the precursor raw material cost is low.

D.   *Antibiotics, Vitamins, and Cofactors*

Various attempts have been made to immobilize antibiotic
producing microorganisms, with the objective of prolonging the
antibiotic production phase and potentially achieving higher
productivities. One early report (120) examined immobilizing
P. chrysogennm mycelia in polyacrylamide gels by direct polymerization of acrylamide monomer. The immobilized mycelium
produced penicillin at a rate of only 15% of free cells,
primarily due to the toxic effects of the monomer during the

Table 4. Production of organic acids by immobilized cells

Organism	Method of Immobilization	Comments	Ref.
A. niger	Ca alginate	air lift fermentor with sucrose as carbon source for citric acid	102
S. lipolytica	wood chips	citric acid production	103
A. niger	adsorption to glass	citric and gluconic acids	104
L. delbruckii	Ca alginate	lactic acid	99
L. delbruckii	hollow fiber	lactic acid	57
L. bulgaricus	dialysis reactor	lactic acid from whey ultrafiltrate	105
Acetobacter aceti	porous ceramic	acetic acid, high O2 transfer reactor	106
Acetobacter	hydrous titanium	acetic acid, tower fermentor	107
Acetobacter	ceramic support	acetic acid, product inhibition examined oxygen transfer	108/109
Achromobacter liquidium	polyacrylamide	urocanic acid from histidine	23
Micrococcus luteus	carbodiimide activated CMC	urocanic acid from histidine	35
Brevibacterium ammoniagenes	polyacrylamide	malate from fumarate	23/110 8
A. niger	Ca alginate	gluconic acid	111/112 113
Ps. dacunhae E. coli	κ-carrageenan	both organisms coimmobilized alanine produced from fumarate, side reactions eliminated	115
Corynebacterium dismutans	polyacrylamide	alanine produced by viable and nonviable cells from glucose	116
Corynebacterium glutamicum	κ-carrageenan	continuous production of glutamic acid in air stirred reactor	117
Brevibacterium flavum	dialysed collagen	glutamic acid formed continuously; spiral wound membrane	118
E. coli	polyacrylamide beads	tryptophane formed from indole and serine productivity 12 g/l hr	119

formation of the polymer gel. Bacitracin production by
Bacillus sp Ky4515 was also lower by 20-25% of washed cells
(121), again confirming the toxic effects of monomers. However, growth of cells within the matrix, especially close to
the gel-liquid interface, resulted in an increase in bacitracin
production.

Freeman and Aharonowitz (122) immobilized *S. clavuligerus*
by using preformed, linear polyacrylamide chains, avoiding the
above mentioned toxicity problems, and showed that the resultant
resting cells had a higher capacity for antibiotic production.
Yields of cephalosporins, over a 96-hour period, were similar
to those of free resting cells. In both cases, approximately
20 mg cephalosporins/gm dry cells were produced, although the
washed resting cells produced cephalosporins at a higher rate.
The effect of culture age on the effectiveness of cell
immobilization was also examined; cells harvested early in the
period of exponential growth were optimal. Other antibiotics
which have been examined include candicidin production by
S. griseus (12), streptomycin production by *S. griseus* (123),
dihydrostreptomycin (123), and the anti-malaria compound
nisin, from *S. lactis* (124).

Propionic acid bacteria have been examined for vitamin B12
production, and *Propionibacterium* ar 1 AKU 1251 was shown to
have activity in entrapped polyurethane (125), and in alginate
and agar. Late exponential phase cells were entrapped with
prepolymer in a ratio of 10 gm wet weight cells to 1 gm prepolymer. Cobalt (II) ion was effective in assisting B12
production. Cells proliferated inside the gel matrix.

The antiviral agent adenine arabinoside has been synthesized
by chemical methods, but a recent approach using gel entrapped
E. aerogenes successfully produced it from uracil arabinoside
and adenine, making use of the transglycosylation activity of
the cells. Prepolymers of urethane were used and 40% dimethyl
sulfoxide solution was used to increase the solubility of
adenine and the reaction product. The initial cellular activity
was maintained for 35 days at 60°C; the yield from 150 mM uracil
arabinoside and 50 nM adenine was 100% based on adenine (126).

NADP, used in several enzymatic test kits, is obtained by
extraction from yeast. Its production was examined from NAD
and ATP using *B. ammoniagenes* immobilized in acrylamide. The
NAD kinase activity of these cells was shown to be fairly
resistant to freezing and irradiation of the cells. Almost
quantitative conversion of NAD to NADP was obtained in 3 hours

in a rotating drum reactor (127). ATP regeneration from ADP
and phosphate has been investigated with *Mastigocladus
laminosus* entrapped in calcium alginate and exposed to light
(128).

E.   Steroids

De novo synthesis of pharmacalogically active steroids
and sterols is difficult, and enzymatic or microbial conversion of naturally occurring steroids is of considerable economic
importance. Raw materials may be of plant origin; diosgenin,
stigmasterol or cholesterol for example. Many steroid transformations require a cofactor, and when immobilized whole cells
are used, no enzyme isolation and purification is required,
operational stability is improved, and most significantly it
is possible to use organic solvents to increase the steroid
solubility if appropriate supports and methods of immobilization are employed.

The main transformations of industrial importance are
Δ'-dehydrogenation, 11α 11β, 16 α-hydroxylation, and side chain
cleavage at the C_{17} position. The effects of various solvents
(ethanol, methanol, ethylene, glycol, glycerol, propylene
glycol) on the support matrix and cells are important in continuous flow processes intended for commercial exploitation.

Δ'-dehydrogenase activity is inhibited at high levels of
product, however, the formation of an intermediate during the
reaction may give stimulation at low substrate conventrations.
Typically, *A. simplex* has been the immobilized organism. The
11 α-hydroxylation step is of importance in the case of
progesterone as this is a route to cortisone and other corticosteroids. The area of steroid transformation by microorganisms
has been very recently thoroughly reviewed in a two-part series
by Kolot (129, 130), and thus will not be discussed further
here. Various types of reactors are discussed by Kolot, as
are the changes in cell properties, e.g., temperature and pH
optima, stability.

Prostaglandin synthesis from arachidonic acid has been of
interest as this would considerably lower the prostaglandin
cost. Single-celled algae (131), and microbial systems (132)
have been examined for this conversion. It has been previously found that soybean lipoxygenase, together with a
reducing agent, would convert arachidonic acid to PGF2, the
major prostaglandin (133). The prostaglandin synthetase
enzyme complex has been examined in ram seminal microsomes
immobilized in a photocrosslinkable gel (134). In this report,

the operational stability of the enzyme system employed was
considerably better than earlier reports on other oxidoreduc-
tase enzymes.

Further reviews of steroid transformations may be found
in references 135, 16, 4, and 23.

F.   Enzyme Production and Activity

The use of immobilized cells for production of extracellular enzymes has been examined as higher cell concentrations
may be maintained and dilution rates in excess of the maximum
growth rate may allow increased enzyme productivities. Many
enzymes are produced as non-growth associated products and the
use of various immobilization methods may additionally assist
in suppressing growth. Shinmyo et al., (136) found α-amylase,
a non-growth associated product from *B. amyloliquefaciens,*
could be produced at the same specific growth rate as free
cells, while the specific growth rate and respiratory activities
were reduced by one-tenth and one-half respectively. Oxygen
diffusion appeared to limit growth to a layer of 50 micron
within the κ-carrageenan beads used for immobilization.
α-amylase production has also been examined in immobilized cells
of *B. subtilis* (137).

Immobilized cells offer an alternative to isolation, purification and immobilization of individual enzymes and there are
an increasing number of reports in this area. Glucose isomerase
activity in immobilized *Arthrobacter* cells has been well
examined, and the kinetics of glucose isomerization determined
(138, 139). β-galactosidase activity in *S. anamensis* has also
been determined in immobilized cells (140), as has the dextran-
sucrase activity of *Leuconostoc mesenteroides* (141). Whole
cells of immobilized *Pichia polymorpha* showed inulinase activity
with a higher optimal temperature and thermal stability (142).
A complete list of the utilization of single and multiple
enzymes in immobilized non-viable cells is available in
Gestrelius (26).

G.   Miscellaneous Products

In addition to those products described above, immobilized
cells have been used in wastewater treatment, recovery of
metal ions from solution, and production of fuels such as
hydrogen and methane. These are summarized in Table 6.

Table 6. Miscellaneous applications of immobilized cells

Product or Activity	Organism	Comments	Ref.
Uranium recovery	*Streptomyces and Chlorella*	fresh and seawater used; adsorption increases with T.	143
Thorium Biosorption	*Rhizopus, Arrhizus*	thorium coordinates with N in chitin cell wall	144
H_2 production from glucose	*Cl. butyricum*	agaracetylcellulose used. Stoichiometric production from glucose	145
H_2, $NADPH_2$	*Cyanobacteria*	polyurethane foams, photoproduction	146
fatty acids	*C. tropicalis*	dicarboxylic acids formed from n-alkanes carrageenan entrapped	147
Nitrification	*Nitrobacter*	immobilized on sand, fluidized bed	148
phenol oxidation	*Ps. putida*	kinetics in chemostat, washed calls	149
BOD reduction and high BTU gas	*mixed*	controlled pore ceramics; two stage anaerobic system	150
Nitrite (ammonia oxidation)	*Nitrosomonas*	cells retained NH_3 oxidizing activity for over 2000 hr	151
O_2 production	*Chlorella pyrenoidosa*	alginate beads	152
Ethylene oxide	*Mycobacterium*	ethylene oxidation, NADH regenerated	153
NH_3 oxidation	*Nitrosomonas europaea*	Ca alginate	154
N_2 and NH_3	*Azotobacter and Cl. pasteurianum*	review on costs and organisms	155

V.   IMMOBILIZED CELLS IN SENSING DEVICES

The use of immobilized cells and enzymes in analytical
sending devices is receiving increased interest, as their
sensitivity and selectivity allow a wide range of compounds to
be determined even in complex mixtures at low levels. Of
particular interest is the case of immobilized cells coupled
with MOSFET devices (156). A complete review of this area is
beyond the scope of the present paper, but earlier reviews
include those of Mattiasson (157), Suzuki and Karube (158), and
Chibata (34). More recent reports are those of Karube and
Suzuki (159) in this series, Fukui and Tanaka (16),
Mattiasson (160), and the abstracts from the Sixth Enzyme
Engineering Conference (161).

REFERENCES

1.   Kattinger,H., *Acta Biotechnologica,* 2(1), 3 (1982).
2.   Tolbert, W.J. and Feder, J., *Ann. Reports Fermentation
        Process, 6,* 35 (1983).
3.   Brodelius, P., in "Immobilized Cells and Organelles,"
        Vol. *1,* 27, Ed. B. Mattiasson, CRC Press (1983).
4.   Chibata, I., Tosa, T., Fujimura, M., *Ann. Reports
        Fermentation Processes, 6,* 1, Ed. G.T. Tsao, Academic
        Press (1983).
5.   Berger, R., *Acta Biotechnologica 1*(1), 73 (1981).
6.   Berger, R., *Acta Biotechnologica, 4*(2) 343 (1982).
7.   Mattiasson, B., "Immobilized Cells and Organelles,"
        Vol. *I* and *II,* CRC Press (1983).
8.   Abbott, B.J., *Ann. Reports Fermentation Processes, 1,*
        205, Ed. G.T. Tsao (1977).
9.   Cheetham, P.S.J., in "Topics in Enzyme and Fermentation
        Biotechnology," *4,* 189, Ed., A. Wiseman (1980).
10. Buckholz, K., (ed) "Characterization of Immobilized
        Biocatalysts," Dechema Monograph No. 1724-1731,
        Vol. *84* (1979).
11. Venkatsubramanian, K., "Immobilized Microbial Cells,"
        ACS Symp. Ser., 106 (1979).
12. Venkatsubramanian, K., Veith, W., *Prog. Industrial Micro.,*
        15, 61, Ed., M.J. Bull, Elsevier Press (1979).
13. Jack, T.R., Zajic, J.E., *Adv. Biochem. Eng., 5,* 125,
        Ed., A. Fiechter, Springer Verlag (1977).
14. Abbott, B.J., *Adv. Appl. Microbiol., 20,* 203, Ed., D.
        Perlman, Academic Press (1976).
15. Vandamme, E.J., *Chem. Ind.,* (London) *24,* 1070 (1976).
16. Fukui, S., Tanaka, A., *Ann. Rev. Microbiol., 36,* 145,
        (1982).

17. Dunnill, P., *Phil. Trans. Roy Soc.* (London) Ser B, 290, 409-420 (1980).

18. Klein, J., Vorlop, K.D., in "Foundation of Biochemical Engineering," *ACS Symp. Ser.*, *207, 377,* Eds. H. Blanch, T. Papoutsakis and G. Stephanopoulos (1982).

19. Tosa, T., Sato, T., Nishida, Y., Chibata, I., *Biochem. Biophys., Acta, 483,* 193 (1977).

20. Skinner, K.J., *Chem. Eng. News, 33*(53), 22-29 and 32-41, (1975).

21. Chibata, I., *Pure Appl. Chem., 50,* 667 (1978).

22. Yamamato, K., Tosa, T., Yamashita, K., Chibata, I., *Biotech. Bioeng., 19,* 1101 (1977).

23. Chibata, I., Tosa, T., *Adv. Appl. Microbiol., 22,* 1, Ed., D. Perlman, Academic Press (1977).

24. Samejima, H., Kimura, K., Ado, Y., *Biochimie., 62,* 299 (1980).

25. Lee, C.K., Long, M.E., U.S. Patent 3,821,086 (1974).

26. Gestrelius, S., in "Immobilized Cells and Organelles," Vol. *II,* 1, Ed., B. Mattiasson, CRC Press (1983).

27. Yamamoto, K., Sato, T., Tosa, T., Chibata, I., *Biotech. Bioeng., 16,* 1601 (1974).

28. Chibata, I., Tosa, T., Sato, T., Mori, T., Yamamoto, K., *Enzyme Eng., 2,* 293, Eds. K. Pye and L. Wingard, Plenum Press (1974).

29. Karube, I., Hirano, K., Suzuki, S., *Biotech. Bioeng., 19,* 1233 (1977).

30. Allen, B.R., Coughlin, R.W., Charles, M., *Ann. N.Y. Acad. Sci., 326,* 105 (1975).

31. Jeris, J.S., Owens, R.W., Hickey, R., Flood, F.J., *J. Water Pollut. Control Fed., 49,* 816 (1977).

32. Durand, G., Navarro, L.M., *Process Biochem., 13*(9), 14-23 (1978).

33. Messing, R., *App. Biochem. Biotech., 6,* 167 (1981).

34. Chibata, I., "Immobilized Enzymes-Research and Development," Halstead Press (1978).

35. Jack, T.R., Zajic, J.E., *Biotech. Bioeng., 19,* 631 (1977).

36. Nelson, R.P., U.S. Patent 3,957,580 (1976).

37. Kennedy, J.F., in "Immobilized Microbial Cells," *ACS Symp. Ser., 106,* 119, Ed., K. Venkat (1979).

38. Mattiasson, B., Borrebaeck, C., *FEBS Lett., 85,* 119 (1978).

39. Truffa-Bachi, P., Wofsy, L., *Proc. Nat'l Acad. Sci., 66,* 685 (1970).

40. Hatton, T., Furusaka, C., *J. Biochem.* (Tokyo) *50,* 312 (1961).

41.   Ghommidh, C., Navarro, J.M., Durand, G., *Arch. Biochem. Biophys.*, *130*, 384 (1969).

42.   Compere, A.L., Griffith, W.L., *Dev. Ind. Microbiol.*, *17*, 247 (1976).

43.   Daugalis, A.J., Brown, N.M., Cluett, W.R., Dunlop, D.B., *Biotech. Lett.*, *3*, 651 (1981).

44.   Navarro, J.M., *Ind. Alin. Agr.*, *93*, 695 (1976).

45.   Corrieu, G., Blachere, A., Ramirez, A. Navarro, J.M., Durand, G., Duteurte, B., Moll, M., 5th Int'l. Ferment. Symp., Berlin (1976).

46.   Navarro, J.M., Durand, G., *Euop. J. Appl. Micro.*, *4*(4), 243 (1977).

47.   Messing, R.A., Opperman, R.A., Kolot, F.B., in "Immobilized Microbial Cells," ACS Symp. *106*, 13, Ed., K. Venkat (1979).

48.   Johnson, D.E., Ciegler, A., *Arch. Biochem. Biophys.*, *130*, 309 (1969).

49.   Scott, C.D., Hancher, C.W., *Biotech. Bioeng.*, *18*, 1393 (1976).

50.   Sturgeon, C.M., Kennedy, J.F., Literature Survey in "Enzyme and Microbial Technology," Butterworth & Company publishers.

51.   Vieth, W.R., Venkat, K., *Adv. Enzymol.*, *34*, 243 (1976).

52.   Linko, Y.Y., Viskari, R., Pohjola, L., Linko, P., *J. Solid Phase Biochem.*, *2*, 203 (1978).

53.   Shuler, M.L., *Ann. New York Acad. Sci.*, *369*, 65 (1981).

54.   Knazek, R., U.S. Patent 3,821,087 (1974).

55.   Hopkinson, J., in "Immobilized Cells Organelles," 89, Ed., B. Mattiasson, CRC Press, Florida (1983).

56.   Vick Roy, T.B., Wilke, C.R., Blanch, H.W., *Trends in Biotech.*, *2*, (1983).

57.   Vick Roy, T.B., Wilke, C.R., Blanch, H.W., *Biotech. Lett.*, *4*(8), 483 (1982).

58.   Wang, H.Y., Lee, S-S, Takech, Y., Cawthon, C., *Biotech. Bioeng. Symp. Ser.*, *12*, (1982).

59.   Venkatasubramanian, K., Kartare, S.B., in "Immobilized Cells and Organelles," Vol. *II*, 133, Ed., B. Mattiasson, CRC Press (1983).

60.   Klein, J., Washausen, P., in "Characterisation of Immobilized Biocatalyst," 277, DECHEMA Monograph No. 1724-1731, Vol. *84*, Ed., K. Buckholz, Verlag Chemie (1979).

61.   Klein, J., Eng. H, Op Cit, p. 292.

62.   Klein, J., Klugge, M., Op Cit, p. 285.

63.   Buckholz, K., *Adv. in Biochem. Eng.*, *24*, 39, Ed., A. Fiechter (1983).

64.   Anon., *Enzyme and Microbial Technol.*, *5*, 304 (1983).

65.   Satterfield, C.N., "Mass Transfer in Heterogeneous Catalysis," MIT Press, Cambridge, MA (1970).

66. Levich, V.G., "Physicochemical Hydrodynamics," Prentice Hall, N.Y. (1962).

67. Sherwood, T.K., Pigford, R.L., Wilke, C.R., "Mass Transfer," Acad. Press (1978).

68. Klein, J., Eng, H., *Biotech. Lett.*, *1*, 171 (1979).

69. Linko, P., Poutanen, K., Weckstrom, L., Linko, Y.Y., *Biochimie, 62,* 387 (1980).

70. Takaviatsu, S., Tamashita, K., Sumi, A., *J. Ferm. Technol.*, *58,* 129 (1980).

71. Aris, R., "The Mathematical Theory of Diffusion and Catalysts," Clarendon Press (1975).

72. Kasche, V., *Eng. Microbial. Technol.*, *5*(1), 2 (1983).

73. Klein, J., ref (10) p. 309.

74. Tosa, T., Takata, I., Chibata, I., in "Enzyme Engineering," *6*, 237, Eds., I. Chibata, S. Fukuri, L. Wingard, Plenum Press (1982).

75. Karanth, G., Patwardhan, V., *Biotech. Bioeng.*, *24*, 2269, (1982).

76. Maiorella, B., Blanch, H.W., Wilke, C.R., *Adv. in Biochem. Eng.*, *20*, 43, Ed., A. Fiechter, (1981).

77. Fukushima, S., Hanai, S., in "Enzyme Engineering," *6*, 347, Eds., Chibata, I., Fukui, S., Wingard, L.M., Plenum Press (1982).

78. Mattiasson, B., in "Immobilized Cells and Organelles," Vol. *II*, 23, Ed., B. Mattiasson, CRC Press (1983).

79. Cho, G.H., Choi, C.Y., Choi, Y.D., Han, M.H., *J. Chem. Technol. Biotech.*, *32*, 959-967 (1982).

80. Tyagi, R.D., Ghose, T.K., *Biotech. Bioeng.*, *24*, 781-795, (1982).

81. Ghose, T.K., Bandyopadhyay, K.K., *Biotech. Bioeng.*, *24*, 797-804 (1982).

82. Bandyopadhyay, K.K., Ghose, T.K., *Biotech. Bioeng.*, *24*, 805-815 (1982).

83. Williams, D., Munnecke, D.M., *Biotech. Bioeng.*, *23*, 1813-1825 (1981).

84. Pines, G., Freeman, A., *Eur. J. Appl. Microbiol. Biotech.*, *16*, 75-80 (1982).

85. Slininger, P.J., Bothast, R.J., Black, L.T., McGhee, J.E., *Biotech. Bioeng.*, *24*, 2241-2251 (1982).

86. Chiang, L.C., Hsiao, H.Y., Flickinger, M.C., Chen, L.F., Tsao, G.T., *Enzyme Microb. Technol.*, *4*, 93 (1982).

87. Arcuri, E.J., *Biotech. Bioeng.*, *24*, 595-604 (1982).

88. Krug, T.A., Daugulis, A.J., *Biotech. Lett.*, *5*(3), 159-164 (1983).

89. McGhee, J.E., St. Julian, G., Detroy, R.W., Bothast, R.J., *Biotech. Bioeng.*, *24*, 1155-1163 (1982).

90. Amin, G., Verachtert, H., *Eur. J. Appl. Microbiol.
     Biotech., 14,* 59-63 (1982).
91. Chua, J.W., Erarslan, S., Kinoshita, S., Taguchi, H.,
     J. Ferm. Technol., 58, 123 (1980).
92. Krouwel, P.E., van dere Laan, W.F., Kossen, N.W.F.,
     Biotech. Lett., 2, 253 (1980).
93. Hoq, M.M., Tanaka, A., Fukui, S., *Ann. Repts, ICME, 4,*
     139 (1981).
94. Bisping, B., Rehm, H.J., *Europ. J. Microb. Biotech., 14,*
     136 (1982).
95. Holst, O., Enfors, S-O, Mattiasson, B., *Eur. J. Appl.
     Microbiol. Biotech., 14,* 64-68 (1982).
96. Holst, O., Enfors, S-O, Mattiasson, B., *Abstr. Comm.
     Sec. Eur. Cong. Biotech.,* Eastbourne, 156 (1981).
97. Makhotkina, T.A., Pomortseva, N.V., Lomova, I.E.,
     Nikolaev, P.I., *Prikl, Biokhim. Mikrobiol., 17,*
     102-106 (1981).
98. Krouwel, P.G., Groot, W.J., Kosson, N.W.F., *Biotech.
     Bioeng., 25*(1), 281-299 (1983).
99. Stenroos, S.L., Linko, Y.Y., Linko, P., *Biotech. Lett.,
     4,* 159 (1982).
100. Krouwel, P.E., Kossen, N.W.F., *Biotech. Bioeng., 23,*
     651 (1981).
101. Enfors, S-O, Mattiasson, B., in "Immobilized Cells and
     Organelles," Vol. *II,* 41, Ed., B. Mattiasson, CRC
     Press (1983).
102. Vaija, J., Linko, Y.Y., Linko, P., *Appl. Biochem.
     Biotech., 7,* 51 (1982).
103. Briffand, J., Engasser, J.M., *Biotech. Bioeng., 21,*
     2039 (1979).
104. Heinrich, M., Rehm, H.J., *Eur. J. Appl. Microbiol.
     Biotech., 15,* 88-92 Springer Verlag (1982).
105. Stieber, R.W., Gerhard, P., *Biotech. Bioeng., 23,* 535-
     549 (1981).
106. Ghommidh, C., Navarro, J.M., Durand, G., *Biotech. Lett.,
     3,* 93 (1981).
107. Kennedy, J.F., Barker, S.A., Humphreys, J.D., *Nature,
     261,* 242 (1976).
108. Ghommidh, C., Navarro, J.M., *Biotech. Bioeng., 24,*
     1991-1999 (1982).
109. Ghomidh, C., Navarro, J.M., Durand, G., *Biotech.
     Bioeng., 24,* 605-617 (1982).
110. Takata, I., Yamamoto, K., Tosa, T., Chibata, I., *Eng.
     Microb. Technol., 2,* 30 (1980).
111. Linko, P., in "Adv. in Biotech.," Vol. *I,* 711, Eds., M.
     Moo-Young, C.W. Robinson, C. Venzina, Pergamon Press,
     N.Y. (1981).

112.   Vorton, K.D., Klein, J., Wagner, F., *Abst. 6th Intn'l
       Ferm. Symp.*, (London, Canada) Fl2.1.12 (1980).

113.   Klein, J., Manecke, E., "Enzyme Engineering," *6*, 188,
       Eds. Chibata, I., Fukui, S., Wingard, L., Plenum
       Press (1982).

114.   Brodelius, P., "Industrial Application of Immobilized
       Biocatalysts," *Adv. Biochem. Eng.*, *10*, 76, Ed., A.
       Fiechter, Springer Verlag (1978).

115.   Takamatsu, S., Umemura, I., Yamamoto, K., Sato, T.,
       Tosa, T., Chibata, I., *Eur. J. Appl. Microbiol.
       Biotech.*, *15*, 147-152 (1982).

116.   Sarkar, J., Mayaudon, J., *Biotech. Lett.*, *5*(3), 201-206
       (1983).

117.   Kim, H.S., Ryu, D.D.Y., *Biotech. Bioeng.*, *24*, 2167-2174
       (1982).

118.   Constantinides, A., Bhatia, D., Wolf, R., *Biotech.
       Bioeng.*, *23*, 899-916 (1981).

119.   Bang, W-G, Behrendt, U., Lang, S., *Biotech. Bioeng.*, *25*,
       1013-1025 (1983).

120.   Morikawa, Y., Karube, I., Suzuki, S., *Biotech. Bioeng.*,
       21, 261 (1979).

121.   Morikawa, Y., Karube, I., Suzuki, S., *Biotech. Bioeng.*,
       22, 1015 (1980).

122.   Freeman, A., Aharnowitz, Y., *Biotech. Bioeng.*, *23*, 2747
       (1981).

123.   Guttag, A., German Patent 2,304,030 (1973).

124.   Egorov, N.S., Baranova, I.P., Kozlova, Y., *Antibiotiki*,
       23, 872 (1978).

125.   Youngsmith, B., Sonomoto, K., Tanaka, A., Fukui, S.,
       Eur. J. Appl. Microbiol. Biotech., *16*, 70 (1982).

126.   Yokozeki, K., Yamanaka, S., Utagawa, T., Takinami, K.,
       Hirose, Y., Tanaka, A., Sonomoto, K., Fukui, S.,
       Eur. J. Appl. Micro. Biotech., *14*, 225 (1982).

127.   Tanaka, T., Hayashi, T., Kawashima, K., Yokoyama, T.,
       Watanabe, T., *Biotech. Bioeng.*, *24*, 857 (1982).

128.   Sawa, Y., Kanayama, K., Ochiai, H., *Agric. Biol. Chem.*,
       44, 1967 (1980).

129.   Kolot, F.B., *Process Biochem.*, Nov./Dec., 22 (1982).

130.   Kolot, F.B., *Process Biochem.*, *18*(1), 19 (1983).

131.   Ahern, T., Katoh, S., Sada, E., submitted to, *Biotech.
       Bioeng.*, (1983).

132.   Iizuka, H., Oomoto, T., Kosaka, Y., *Japan Kokai, Koko*,
       77, *64*, 484 (1977).

133.   Bild, E.S., Bhat, S.G., Ramadoss, C., Axelrod, B.,
       J. Biol. Chem., *253*, 21 (1978).

134.   Ahern, T., Katoh, S., Sada, E., *Biotech. Bioeng.*, *25*,
       881 (1983).

135. Koshckeyenko, K.A., Turkina, M.V., Skryabin, G.K.,
        Eng. Microb. Technol., *5*, 14 (1983).
136. Skinmyo, A., Kimura, H., Okada, H., *Eur. J. Appl.
        Microbiol. Biotech.*, *14*, 7 (1982).
137. Kokubu, T., Karubi, I, Suzuki, S., *Eur. J. Appl.
        Microbiol. Biotech.*, *5*, 223 (1978).
138. Van Keulen, M.A., Vellenga, K., Joosten, G.E., *Biotech.
        Bioeng.*, *23*, 1437 (1981).
139. Kikkert, A., Vellenga, K., de Wilt, H.G., Joosten, G.E.,
        Biotech. Bioeng., *23*, 1087 (1981).
140. Banerjee, M., Chakrabaty, A., Majumdar, S.K., *Biotech.
        Bioeng.*, *24*, 1834 (1982).
141. Chang, H., Gkim, Y.S., Cho, Y.R., Landis, D.A., Reilly,
        P.J., *Biotech. Bioeng.*, *23*, 2647 (1981).
142. Guiraud, J.P., Bajon, A.M., Chautard, P., Galzy, P.,
        Enz. Microb. Technol., *5*, 185 (1983).
143. Nakajima, A., Horikoshi, T., Sakaguchi, T., *Eur. J.
        Appl. Micro. Biotech.*, *16*, 88 (1982).
144. Tsezos, M., Volesky, B., *Biotech. Bioeng.*, *24*, 955 (1982).
145. Karube, I., Urano, N., Matsunaga, T., Suzuki, S.,
        Eur. J. Appl. Micro. Biotech., *16*, 5 (1982).
146. Muallem, A., Bruce, D., Hall, D.O., *Biotech. Lett.*, *5*(6),
        365 (1983).
147. Yi, Z.H., Rehm, H.J., *Eur. J. Appl. Micro. Biotech.*, *16*,
        1 (1982).
148. Tanaka, H., Uzman, S., Dvnn, I.J., *Biotech. Bioeng.*, *23*,
        1683 (1981).
149. Sokol, W., Howell, J.A., *Biotech. Bioeng.*, *23*, 2039
        (1981).
150. Messing, R., *Biotech. Bioeng.*, *24*, 1115 (1982).
151. Kokufuta, E., Matsumoto, W., Nakamura, I., *Biotech.
        Bioeng.*, *24*
152. Adlercreutz, P., Mattiasson, B., *Eng. Microbiol. Technol.*,
        4, 332 (1982).
153. de Bont, J.A., van Ginkel, C.G., Tramper, J., Luyben,
        K.C., *Enz. Microb. Technol.*, *5*, 55 (1983).
154. van Ginkel, C.G., Tramper, J., Luyben, K.C., Klapwijk,
        A., *Enz. Microb. Technol.*, *5*, 297 (1983).
155. Hallenbeck, P.C., *Enz. Microb. Technol.*, *5*, 171 (1983).
156. Winquist, F., Danielsson, B., Lundstrom, I., Mosbach, K.,
        Appl. Biochem. Biotech., *7*, 135 (1982).
157. Mattiasson, B., in "Immobilized Microbial Cells," *ACS
        Symp. Ser.*, *106*, Ed., K. Venkat, 203 (1979).
158. Suzuki, S., Karube, I., in "Immobilized Microbial Cells,"
        ACS Symp. Ser., *106*, Ed., K. Venkat, 221 (1979).

159.   Karube, I., Suzuki, S., *Ann. Reports Ferment. Processes*,
        6, Ed., G.T. Tsao, 203 (1983).
160.   Mattiasson, B., in "Immobilized Cells and Organelles,"
        Vol. *II*, Ed., B. Mattiasson, CRC Press, 95 (1983).
161.   Enzyme Engineering - 6, Eds., Chibata, I., Fukui, S.,
        Wingard, L., Plenum Press, N.Y., 357-430 (1982).

CHAPTER 5

FERMENTATION SUBSTRATES AND ECONOMICS

Bruce E. Dale
James C. Linden

Colorado State University
Fort Collins, Colorado

ECONOMIC ASPECTS OF FERMENTATION PROCESSES: INTRODUCTION

In 1977 Nyiri and Charles wrote an article for this series on the economics of fermentation processes (1). Now, as they observed then, economics of fermentation are seldom discussed in the open literature. We have combined information from a variety of sources on the economics of fermentation processes. The most surprising conclusion from our literature review is the strong dependence of overall fermentation costs on the cost of the carbon substrate. Although any review is necessarily incomplete, we believe the data summarized here should provide some useful perspectives.

Operating costs are particularly sensitive to the carbon substrate cost. We expected to find, based on commonly held views, that the separation costs and mixing/aeration costs were predominant. The chemical industry has had a similar experience. Frequently as chemical technology improves and processes mature, the raw materials costs tend to dominate both the variable and fixed costs (2). It may well be that for fermentations producing higher-value products, the carbon substrate cost is not so important. The available data on such fermentations are limited. Nonetheless, for many important fermentations, economic progress and commercial viability will be less dependent on sophisticated separation processes or genetic engineering advances than on rather mundane substrate costs. For substrate cost-dominated fermen-tation, it appears that only inexpensive sugar syrups derived from efficient conversion of lignocellulosic materials with carefully-integrated by-product recovery, can substantially reduce fermentation product costs below current levels.

I.   COSTS OF FERMENTATION PROCESSES: SINGLE-CELL PROTEIN

The two largest volume commodities which are being produced or have been recently considered for production by
fermentation are ethanol and single-cell protein (SCP). As
"biotechnology" develops, the knowledge base from these
fermentations should be very useful. The economics of these
two processes should also provide useful insights for biotechnology management. Table 1 summarizes data from many sources
on the economics of single cell protein manufacture (3,4,5,6,
7,8). As others have noted (9,10), the cost of the carbon
substrate obviously dominates the overall costs, representing
approximately 40% of total production costs and about 50-70%
of non-capital related costs. Even relatively cheap "waste"
materials such as bagasse and newsprint represent large
portions of the overall costs. One conclusion derived from
this fact is that as SCP processes develop and improve and
capital costs are reduced, the substrate cost will become even
more dominant.

II. ETHANOL

The economics of ethanol fermentation are particularly
important due to its potential fuel/chemical uses. Sufficiently
inexpensive fermentation ethanol could displace a host of products currently derived from petroleum. Table 2 summarizes
some of the compounds that can be made from ethanol.

Costs of fermentation ethanol from more or less conventional
sugar and starch feedstocks are summarized in Table 3. Different accounting conventions were used in these studies and
different rates of return on investment were also required.
Nonetheless, some important trends are clear. First, as with
single-cell protein, the cost of fermentation ethanol is
dominated by the feedstock costs which usually account for 40
to 70% of production costs. By-product credits were subtracted
from actual feedstock costs to give the net feedstock costs.
In some rows of this and other tables, the production costs
are given for the same feedstock without by-product credits to
illustrate the economic effect if markets for by-products are
not available. It is abundantly clear that by-product credits
are very important in reducing the net cost of fermentation
ethanol from starch and sugar feedstocks. For instance, if
corn is at $3.00 per bushel, without sales of the by-product
distiller grains the feedstock cost alone would be $1.20 per
gallon. Without these by-product credits, the fermentation
ethanol alone has no chance of being economically competitive.
More will be said later about the importance of by-product
values for ethanol from lignocellulosics.

Table 1. Costs and feedstocks for SCP manufacture

Study/Source	Date of Study	Substrate	Substrate Cost $/Ton	Substrate Cost $/Ton SCP Produced	Other Operating Costs $/Ton SCP	Percent Non-Capital Related Costs Due to Substrate	Percent of Total Costs Due to Substrate
Ghose (3)	1969	Sucrose	200	400	---	---	---
Ghose (3)	1969	Molasses	35	70	---	---	---
Litchfield (4) (British Petroleum)	1969	Gas Oil	---	---	---	19	13
(British Petroleum)	1969	Alkanes	---	---	---	53	40
Litchfield (4) (Gulf)	1974	Alkanes	---	---	---	60	46
Litchfield (4) (ICI)	1973	Methanol	---	---	---	74	63
Abbott & Clamen (5)	1973	Molasses	40	78	16	84	---
Abbott & Clamen (5)	1973	Paraffins	80	80	48	63	---
Abbott & Clamen (5)	1973	Methanol	40	100	62	62	---
Abbott & Clamen (5)	1973	Methane	20	32	140	19	---
Abbott & Clamen (5)	1973	Ethanol	120	176	41	80	---
Abbott & Clamen (5)	1973	Acetate	120	334	34	91	---
Moo-Young (6)	1977	Paraffins	---	---	---	50	35
Moo-Young (6)	1977	Methanol	---	---	---	52	38
Moo-Young (6)	1977	Molasses	---	---	---	75	62
Moo-Young (6)	1977	Ethanol	---	---	---	50	---
Moo-Young (6)	1977	Bagasse	---	---	---	29	---
Laine (7)	1974	Paraffins	70	70	30	50	35
Laine (7)	1974	Gas Oil	30	30	32	20	15
Cysewski & Wilke (8)	1976	Newsprint	---	---	---	41	39

*Table 2. Some products derivable from ethanol through various processes**

Oxidation or dehydrogenation	Dehydration	Modification
Acetone	Ethylene	Ethyl chloride
Acetaldehyde	Ethylene oxide	Butadiene
Acetic acid	Polyethylene	Ethyl amines
Methyl isobutyl ketone	Polystyrene	Diethyl ether
Methyl methacrylate	Ethylene glycol	Chloroform
Chloral	Polyethers	
Peracetic acid	Surfactants	
Croton aldehyde	Polyvinyl chloride	
Octoic acid		
Butyl alcohol		
Butyric acid		
Pentaerythritol		
Cellulose acetate		
Acetic anhydride		
Various acetates		
Polyvinyl acetate		

*
Adapted from Paturau (11)

Lignocellulosic materials also contain large amounts of potentially fermentable sugars, usually in the form of high molecular weight polymers called cellulose and hemicelluloses. Glucose is the monomeric sugar comprising cellulose while xylose is the primary sugar monomer in the heterogeneous hemicelluloses from grasses and hardwoods. Many studies are available on the costs of fermentation ethanol from lignocellulosic materials. Some of these studies are summarized in Table 4.

It is important to remember that not one of these economic summaries is based on an operating commercial facility. Therefore, the values are much more speculative than those presented in Table 3 for ethanol from starch and sugar crops. However, some important conclusions can be drawn from the data. First, alcohol from lignocellulosics would appear to be available for around $1 to $2/gallon. As an octane enhancer for gasoline, this ethanol should therefore be competitive with other octane

Table 3. Costs of fermentation ethanol from sugar or starch feedstocks

Study/ Source	Date of Study	Feedstock	Feed- Stock Costs $/Ton	Total Production Costs $/Gal	Total Variable Costs* $/Gal	Net Feedstock Costs $/Gal	Feedstock Costs as % of Production Costs	Feedstock Costs as % of Variable Costs
Martin (12)	1982	Sugarcane	10	1.12	0.51	0.48	42	95
Martin (12)	1982	Wheat (Europe)	200	2.13	1.61	1.53	72	96
Martin (12)	1982	Wheat (Europe) without by-products	200	2.54	2.02	1.95	77	97
Martin (12)	1982	Sugarbeet	20	1.84	0.88	0.84	46	95
Martin (12)	1982	Sugarbeet without by-products	20	2.07	1.11	1.07	51	96
Tyner (13)	1980	Wheat	127	1.45	---	1.23	85	---
Hertzmark, et al. (14)	1980	Corn (wet-milling)	91	1.17	---	0.38	32	---
Hertzmark et al. (14)	1980	Corn (wet-milling) without by-products	91	1.79	---	1.00	85	---
Cysewski & Wilke (15)	1978	Molasses (50% sugar)	50	0.82	---	0.60	73	---
Cysewski & Wilke (15)	1978	Molasses (50% sugar) without by-products	50	0.96	---	0.74	77	---
Borglum (16)	1981	Corn	116	1.53	---	0.82	54	---
Friend & Shahani (17)	1981	Cheese whey	144	1.06	0.64	0.25	23	39
Friend & Shahani (17)	1981	Cheese whey without by-products	144	1.91	1.49	1.10	58	74
DeCarvalho, et al. (18)	1977	Sugarcane	11	1.26	0.81	0.56	44	69
Earl & Brown (19)	1980	Fodder beets	15	1.36	0.69	0.44	33	64
Earl & Brown (19)	1980	Fodder beets without by-products	15	1.68	1.01	0.77	46	76
DeCarvalho, et al. (18)	1977	Cassava	29	1.37	1.02	0.69	50	68
Kosaric, et al. (20)	1982	Jerusalem artichokes	10	0.79	-0.34	-1.44	---	---
Kosaric, et al. (20)	1982	Jerusalem artichokes without by-products	10	2.89	1.74	0.64	22	37
Maiorella, et al. (21)	1979	Molasses (50% sugar)	85	1.90	1.38	1.00	53	72
Maiorella, et al. (21)	1979	Molasses (50% sugar) without by-products	85	2.13	1.60	1.22	57	76

*By-product credits are included to reduce feedstock and variable costs except where indicated otherwise

Table 4. Costs of fermentation ethanol from lignocellulosic feedstocks

Study/ Source	Date of Study	Feedstock	Feed-Stock Costs $/Ton	Total Production Costs $/Gal	Total Variable Costs* $/Gal	Net Feedstock Costs* $/Gal	Feedstock as % of Production Costs	Feedstock as % of Variable Costs
Rugg,et al. (22)	1981	Sawdust	30	1.07	0.45	0.35	33	78
Rugg,et al. (22)	1981	Sawdust (acid hydrolysis) without by-products	30	1.19	0.57	0.48	40	84
Wilke, et al (33)	1981	Corn stover (enzyme hydrolysis)	30	2.90	---	0.65	22	---
Wilke, et al (33)	1981	Corn stover (enzyme hydrolysis) without by-products	30	3.50	---	1.25	36	---
Sitton, et al. (24)	1979	Corn stover (acid hydrolysis	15	1.19	0.70	0.24	20	34
Koutinas, et al. (25	1981	Wheat straw (enzyme hydrolysis)	9	0.72	---	0.29	41	---
Wayman, et al. (26)	1979	Aspen wood (enzyme hydrolysis)	30	1.12	---	0.44	39	---
Tyner (13)	1980	Crop residue or forages	30	1.16	---	0.60	52	---
Bungay (27)	1981	Municipal solid waste (enzyme hydrolysis)	30	1.02	0.71	0.40	39	56
Cysewski & Wilke (29)	1976	Newsprint (enzyme hydrolysis)	---	1.05	--- (Sugar Costs)	0.72	69	---
Roberts, et al. (30)	1980	Wood (acid hydrolysis)	30	1.52	---	0.41	27	---
Moreira, et al. (31	1981	Sudan grass (enzyme hydrolysis)	16	1.82	---	0.57	31	---
Moreira, et al. (31)	1981	Sorghum silage (enzyme hydrolysis)	21	2.12	---	0.87	41	---
Jones & Semrau (32)	1982	Wood (acid hydrolysis)	33	2.71	1.21	0.37	14	31
Jones & Semrau (32)	1982	Wood (acid hydrolysis) without by-products	33	3.15	1.65	0.81	26	49
Martin (33)	1982	Wood waste	30	1.11	---	0.29	27	---
Martin (33)	1982	Wood waste (without by-products)	30	1.65	---	0.83	50	---

*
 By-product credits are included to reduce feedstock and variable costs when available except where
 indicated otherwise.

**
 All production costs not included

enhancers. As a chemical feedstock, fermentation ethanol at
$1/gallon is equivalent to an ethylene price of approximately
$0.25 per pound plus conversion costs. Second, although ligno-
cellulosics are generally less expensive than starch and sugar
feedstocks, they still represent roughly 30 to 60% of total
ethanol production costs versus 40 to 70% of total costs for
the starch and sugar materials. Third, by-product credits are
important in reducing the ethanol cost from lignocellulosics
as they are in ethanol from starch or sugar feedstocks. It is
reasonable to believe that further development of such by-
products as well as process improvements will ultimately lead
to commercial-scale production of alcohol from lignocellulosics.
We will return to this subject after considering the economics
of fermentation processes other than those for SCP and ethanol.

III. ECONOMICS OF OTHER FERMENTATION PROCESSES

Much less published information is available on the
economics of fermentations other than those for SCP and
ethanol. Nonetheless, some data for other fermentations are
available and these are summarized in Table 5. Most of these
products are much higher value commodities than ethanol and
SCP. Nonetheless, the carbon source is still a significant
portion of the overall production costs, accounting for at
least 25% of total costs. From the limited data available,
the carbon substrate cost for an "average" higher value fermentation product would appear to be approximately equal to the
separation or capital costs. For the lower value, more easily
separated commodity products such as ethanol and SCP, the carbon
cost will dominate production costs and breakthroughs in separation technology or genetic manipulation will have comparatively
little effect on overall process economics unless they impact
directly on the efficiency of carbon utilization by the
microorganisms.

Thus, for substrate-cost dominated fermentations, we
naturally wish to know what raw materials are available to us
and at what prices. Sucrose and its by-products such as
molasses are the traditional substrates for the fermentation
industries. Table 6 summarizes sucrose and molasses prices
since 1970. Except for the sharp price increase in 1974-75
(which was followed by an equally precipitous price drop) sugar
and molasses prices have generally trended upward with infla-
tion during the past 12 years. Indeed, in 1970 the overall
consumer price index (CPI) for the United States stood at
116.3 while in 1982 the CPI was 287.1, or a rise of about
2.5 fold. When sugar and molasses prices for 1982 are compared
to their 1970 prices it is seen that the price of bulk sugar

Table 5. *Economic data on other fermentations*

Study/Source	Study	Feedstock	Fermentation Products	% of Production Costs Attributable to:		
				Carbon Substrate	Utilities and Labor	Capital Costs
Gibbs (34)	1983	Woodchip	Acetone, butanol,	28	22	36
Schierholt (35)	1977	Molasses	Citric acid (surface fermentation)	33	49	13
Schierholt (35)	1977	Molasses	Citric acid (submerged fermentation)	27	53	10
Paturau (36)	1969	Molasses ($20/Ton)	Lactic acid	12	33	37
Paturau (37)	1969	Molasses ($15/Ton)	Glycerol	29	22	31
Lodha, et al (38)	1973	Molasses	L-lysine "main proportion of the cost of production . . . is the carbohydrate"			
Inskeep (39)	1951	---	Bacitracin	25*	Recovery 40	
Kobayashi (40	1972	---	Itaconic acid	49*	Utilities 13	
Lenz & Moreira (41)	1980	Molasses	Acetone, butanol, ethanol	63	19	10
Lenz & Moreira (41)	1980	Cheese whey (at trucking cost)	Acetone, butanol, ethanol	46	26	19

*
Carbon cost not separated from other raw material costs

Table 6. Sugar and molasses prices since 1970

Year	Average Refined Bulk Sugar Price Duty Paid, New York ¢/lb Sugar	Average Sugarcane/ Molasses price New Orleans (50% Syrup) ¢/lb Sugar
1970	8.1	2.5
1971	8.5	2.5
1972	9.1	2.7
1973	10.1	5.6
1974	29.5	6.8
1975	22.5	4.7
1976	13.3	5.2
1977	11.0	4.1
1978	13.9	5.1
1979	15.6	8.2
1980	30.1	9.7
1981	19.7	8.7
1982	19.9	5.0

(sucrose) has also increased 2.5 fold while the molasses price
has risen two fold (molasses prices have begun to rise again
sharply during 1983). In other words, the selling price of
these fermentable sugars has risen about as fast as everything
else over the same period while the real (uninflated) price
has changed little if at all.

However, as we have shown above, many fermentation processes could benefit greatly by a less costly source of sugar
and all of the fermentation industry could benefit by less
volatile prices for their major raw materials. (The sugar
price rise in 1974-1975 caused a great deal of anguish in the
fermentation industry.) How might this be achieved? It might
be achieved by making use of "waste" carbohydrates such as
cheese whey or pulp mill effluents, by developing new sugar
crops or by converting the complex carbohydrate polymers in
lignocellulosic materials to fermentable sugars. Waste carbohydrates are certainly a possibility and might be available in
some locations at attractive prices. For instance, hydrolysis
of the lactose in cheese whey has been estimated to yield

glucose at a cost of 5¢/pound (42). However, such wastes are
likely to be available only in specific areas and in limited
quantities. Alternate sugar crops such as sweet sorghum or
Jerusalem artichokes offer another means of producing fermentable sugars, particularly in temperate climates. The
economics of such new crops are necessarily speculative but
many analyses conclude that sugars from such crops will not be
much cheaper than those from sugarcane, if at all. That leaves
us with the lignocellulosics. What are the possibilities there?

IV. *LIGNOCELLULOSE CONVERSION*

Lignocellulosic biomass consists of cellulose, hemicelluloses, lignin and a whole host of other components such
as protein, pectin, soluble sugars, vitamins, minerals and
so forth. In terms of sugars, the component of major interest
is cellulose, a glucose polymer which comprises 40-50% of the
plant weight. If we assume that lignocellulosic biomass is
available at \$30/ton, then the cellulose cost is equivalent to
about 3¢/pound cellulose. If only 50% of this cellulose is
converted to glucose (which is the case in many of the acid
and enzyme hydrolysis processes), the equivalent glucose
cost becomes 6¢/pound and we have already approached or
exceeded the cost of sugar in molasses (see Table 6) <u>without</u>
adding in any of the processing costs. Two crucial areas for
research and development are therefore easily identified by
this simple analysis. First, effective use must be made of
other components of the biomass so that the glucose product
does not need to carry the whole cost. Second, yields of
glucose from cellulose are of utmost importance. Therefore
an effective, economical pretreatment of cellulose is essential
to obtain increased glucose yields and reduce the unit cost of
sugar.

A number of estimates are available on the cost of sugar
from lignocellulosic materials. These data are summarized in
Table 7. Again, it is important to remember that these
sugar costs are necessarily speculative. Nonetheless, it is
clear that in order to obtain a sugar cost competitive with
molasses either the feedstock must be very low cost or the
conversion of cellulose to glucose must be very high. Each
of these factors will now be considered.

IV. *BY-PRODUCTS FROM LIGNOCELLULOSIC MATERIALS*

One way of reducing the cost of a lignocellulosic feedstock for glucose production is to recover and utilize other
components of the material. This type of thinking is common
to engineers in the petroleum refining industry where a
multitude of products are obtained from a barrel of crude oil.

Table 7. Sugar costs by hydrolysis of cellulosic materials

Study/Source	Date of Study	Feedstock	Feedstock Cost $/Ton	Acid or Enzyme Hydrolysis?	Sugar Yield lb/lb Feedstock	Yield as % of Theoretical	Assigned Syrup Cost ¢/lb Sugar
Wilke (43)	1981	Corn stover	0	Enzyme	---	---	10
Wilke (43)	1981	Corn stover	30	Enzyme	---	---	15
Lee, et al (44)	1982	Waste newspaper	55	Enzyme	---	---	17
Lutzen, et al (45)	1983	Wood	30	Acid	---	---	27
Wilke, et al (46)	1976	Waste	0	Enzyme	0.34	50	5.2
McKee & Co (47)	1978	Corn stover	30	Acid	---	90	4.5
Grethlein (48)	1978	Newsprint	0	Acid	0.38	55	2.5
Spano, et al (49)	1978	Urban waste	6	Enzyme	0.45	40	6.8
Linden, et al (50)	1983	Forage crop	13	Enzyme	0.41	81	8.0

Much more work and thinking along such "refiner" lines needs
to be done by individuals involved in biomass conversion.
Some of the products other than glucose obtainable from lignocellulosic materials include hemicelluloses, lignin, protein,
latex and so forth.

Hemicellulose can be rather easily hydrolyzed to the monomeric sugar xylose which can also be fermented, although not as
easily as glucose (51,52). Sugar syrups at a few cents per
pound should be obtainable from the hemicelluloses which make
up 20-30% of biomass. Tsao and his colleagues at Purdue
University have pioneered this approach. Lignin is a complex
polymer of phenylpropane units with many potential applications
(53,54). As these applications are developed, lignin should
command a price of about $0.25 per pound. Lignin makes up
10-25% of many cellulosic materials. Protein is also ubiquitous
in biomass since enzymes (composed of protein) are required to
fix carbon and convert it to cellulose, hemicelluloses, lignin
and so forth. Depending on the species and its maturity,
protein can comprise 5-20% of the whole plant weight and should
also have a value of about $0.25 per pound as an animal feed
when priced against soybean meal (55,56). If we assume a hemicellulose value of 3¢ per pound at 25% of the plant weight and
a lignin and protein value of $0.25 per pound at 15% and 10%
of the plant weight, respectively, a simple calculation will
show that these other components could provide a credit of $140
per ton of biomass. This would be more than sufficient to offset the commonly assumed lignocellulose feedstock cost of $30
per ton. It is obvious that a great deal of work needs to be
done in developing such approaches, particularly in developing
appropriate lignin and protein products. Nevertheless, such
multiple-product, "refining," approaches undoubtedly hold the
key to low-cost sugars from lignocellulosic biomass. Such low-
cost sugars will not result from approaches which (in essence)
use part of the plant and throw the rest away.

VI. INCREASING CELLULOSE CONVERSION THROUGH PRETREATMENTS

The impact of percent cellulose conversion on the net
cost of the resulting glucose was illustrated above. A great
variety of pretreatments have been developed to increase
cellulose hydrolysis. These pretreatments can be characterized
as either chemical or physical in nature while some pretreatments have both chemical and physical effects (57,58,59). A
complete review of these treatments, their cost and effectiveness, is beyond the scope of this article but will be attempted
in the next edition of this series. It seems fair to say,
however, that while no pretreatment has yet achieved widespread commercial acceptance as a means of increasing cellulose

hydrolysis, the various sodium hydroxide and steam explosion pretreatments appear to be the best developed and the most cost effective at the present time (60,61,62,63).

VII. A NOTE ON THE ECONOMICS OF THE CELLULASE ENZYME FERMENTATION

It is widely perceived that ethanol production by enzymatic hydrolysis of lignocellulosic materials is dominated by the cost of the cellulase enzymes themselves. This is true enough and would seem to be an exception to the trends we have noted earlier for substrate-cost dominated fermentations. However, it must be remembered that the enzymes are themselves <u>Products of fermentation</u>. Just what are the cost components of this cellulase enzyme fermentation? The Natick group has produced most of the relevant data. Their work shows that of the $0.57 per gallon ethanol production cost attributable to cellulase enzyme production, $0.31 per gallon is raw materials costs (64). These raw materials consist primarily of the carbon and nitrogen for fermentation. The Natick data also show that 8.9 mg carbon and 0.62 mg nitrogen are required to produce one international unit (IU) of enzyme (65). If we assume that carbon is available from sugar costing $0.05/pound (1982 average molasses price from Table 6) and that nitrogen is available from ammonia costing $200/ton, the calculated enzyme price is $2.52/10^6 IU due to carbon requirements alone plus $0.17/10^6 IU due to nitrogen requirements. These calculated values correspond reasonably well to published total cellulase production costs of $2.50-$5.00/10^6 IU (66). From our simplified calculations, the carbon source may represent 94% of the raw materials costs. Indeed, Ryu and Mandels confirm that "the carbon source used is the major cost factor in cellulase process technology (67). If our figure of 94% of raw materials costs is used, then the carbon source for enzyme production represents a cost of $0.29 per gallon of ethanol of 51% of the total cellulase enzyme production costs of $0.57 per gallon. This figure is very much in line with those quoted earlier for other large volume fermentation products such as SCP and ethanol. If this enzyme production cost is combined with realistic lignocellulosic substrate costs of $30/ton, then substrate costs for enzymatic hydrolysis of biomass represent 45% of the total ethanol costs of $1.87 per gallon from the Natick data (68). Therefore, cellulase enzyme production is also governed by carbon substrate costs. It is worth pointing out that the yield of cellulase enzymes produced from lignocellulosic materials will also be increased by appropriate pretreatments. This effectively reduces the feedstock cost per unit of enzyme produced and would have a significant impact on the cost of cellulase enzymes. Therefore, pretreatments are doubly important in enzymatic hydrolysis and ethanol fermentation of

lignocellulosic biomass; first by increasing the yield of
cellulase enzymes and second by increasing the yield of sugars
when these enzymes are used to hydrolyze the cellulosic
material.

VIII.   *FERMENTATIONS CLASSIFIED BY CARBON SUBSTRATE*

Tabulations of fermentations according to product formation
can be found in the literature; those given on the basis of
substrate are infrequently seen. We have provided in Tables
8-14 a partial list of microbial fermentations in which transformation of carbon substrates have been demonstrated. Each
table gives information on substrates with a given number of
carbon atoms. The so-called C-1 fermentations, given in Table 8
can be subdivided into classes based on the oxidation state
of the carbon. The oxidation of carbon monoxide and the
reduction of the resulting carbon dioxide to methane, methanol,
organic acids, and cell carbon have potential where concentrations of the gases are available, for instance from power
plants and fermentation facilities. The cost of the carbon in
such instances will depend heavily on the particular
circumstances.

Compounds containing 2,3 and 4 carbon atoms abound in
nature primarily as metabolic intermediates of living organisms.
Tabulation of these would serve little purpose. The conversions
listed in Table 9 represent instances of potential biotechnological interest.

Interest in C_5 substrates, tabulated in Table 10, on the
other hand is vital. As already emphasized above, the
utilization of all parts of lignocellulosic biomass is important
to economic substrate costs. Separation of the hemicellulose
fraction for utilization independent of the utilization of
glucose may be easily accomplished in most biomass pretreatment
schemes. The cost of pentose substrates derived from biomass
would have nearly the same value as glucose.

Tabulation of C_6 substrate fermentations has not been
presented because glucose is nearly a ubiquitous substrate for
microbial processes. The value of glucose as fermentable substrate has been covered adequately above. Mannose is another
hexose of importance to the biomass refining as the primary
component of the hemicellulose of softwoods. Since mannose is
convertible to ethanol by yeast, the bioprocessing of softwoods
would simplify the problem of the hemicellulose utilization presented when considering grasses and hardwoods. The inability
to hydrolyze structural polymers of softwoods has restricted the

Table 8. C$_1$ Substrates

Substrate	Microorganism	Product	Reference
carbon monoxide	*Methanosarcina barkeri*	carbon dioxide	Fuchs, et al, (69)
	Methanobacterium formicicum	carbon dioxide	Fuchs, et al. (69)
	Desulfovibrio desulfuricans	carbon dioxide	Fuchs, et al. (69)
	Clostridium welchii	carbon dioxide	Fuchs, et al. (69)
	Clostridium pasteurianum	carbon dioxide	Fuchs, et al. (69)
carbon dioxide	*Methanobacterium thermoautotrophicum*	methane	Taylor, et al. (70)
	Clostridium thermoaceticum	acetate	Taylor, et al. (70)
	Clostridium formicoaceticum	acetate	Taylor, et al. (70)
	Clostridium acidi-urrici	acetate	Taylor, et al. (70)
	Clostridium kluveri	formate	Thauer, et al. (71)
	Clostridium pasteurianum	formate	Thayer, et al. (71)
	Nitrogen fixing organism N34	cell carbon	Ooyama (72)
formic acid	*Methanobacterium thermoautotrophicum*	cell carbon	Taylor, et al. (70)
	Pseudomonas methanica	carbon dioxide	Wadzinski & Robbins (73)
formaldehyde	*Methanobacterium thermoautotrophicum*	cell carbon	Taylor, et al. (70)
	Pseudomonas methanica	cell carbon	Wadzinski & Robbins (73)
methanol	*Methanobacterium thermoautotrophicum*	cell carbon	Taylor, et al. (70)
	Pseudomonas methanica	formaldehyde	Wadzinski & Robbins (73)
methanol (con't)	*Methylomonas clara*	SCP	anonymous (74)
	Candida boidinii	yeast	Peppler (75)
methane	*Methylococcus sp (6)*	methanol	Malashenko (76)
	Methylobacter sp (2)	methanol	Malashenko (76)
	Methylomonas sp (4)	methanol	Malashenko (76)
	Methanomonas methanooxidans	methanol	Malashenko (76)
	Methylosinus trichosporium	methanol	Malashenko (76)
	Methylocystis parous	methanol	Malashenko (76)
	Pseudomonas methanica	methanol	Malashenko (76)

Table 9. C$_2$, C$_3$, and C$_4$ Substrates

Substrate	Microorganism	Product	Reference
acetic acid	*Prototheca zopfi*		Koenig and Ward (77)
	Methanobacterium ruminantium	cell carbon	Bryant (78)
	Methanobacterium omelianskii	pyruvate	Knight (79)
	Clostridium kluyveri	hydrogen, cells	Jungermann, et al. (80)
acetaldehyde	*Pseudomonas methanica*	acetic acid	Leadbetter & Foster (81)
ethanol	*Pseudomonas methanica*	acetaldehyde	Leadbetter & Foster (81)
	Candida utilis	yeast	Peppler (75)
	Clostridium kluyveri	hydrogen & cells	Jungermann, et al. (80)
alanine	*Clostridium propionicum*	acrylic acid, propionic acid	Sinskey, et al. (82)
propane	*Pseudomonas fluorescens*	cells	Hou, et al. (83)
glycerol	*Acetobacter suboxydans*	dihydroxyacetone	Perlman (84)
	Gluconobacter melanogenus	dihydroxyacetone	Perlman (84)
1,2-propandediol	*Arthrobacter oxydans*	lactic acid	Chibata, et al. (85)
butane	*Pseudomonas butanovora sp. nov.*	biomass	Ichikawa, et al. (86)

Table 10. C_5 Substrates

Substrate	Microorganism	Product	Reference
D-xylose	*Clostridium acetobutylicum*	acetic acid, butanol and butyric acid	Saddler, et al. (87)
	Klebsiella pneumoniae	ethanol, acetic acid and 2,3-butanediol	Saddler, et al. (87)
	Pachysolen tannophilus	ethanol	Slininger, et al. (88) Gong, et al. (89)
	Candida tropicalis	ethanol	Jefferies (90)
	Candida sp XF-217	ethanol, xylitol, and arabitol	Gong, et al. (89)
	Fusarium oxysporium	ethanol, xylitol, and arabitol	Gong, et al. (89)
	Mucor sp	ethanol, xylitol, and arabitol	Gong, et al. (89)
	Monilia sp	ethanol, xylitol, and arabitol	Gong, et al. (89)
	Aeromonas hydrophilia	ethanol and 2,3-butanediol	Flickinger (91)
	Clostridium thermoaceticum	acetic acid	Ljungdahl, et al. (92)
	Thermoanaerobacter ethanolicus	ethanol, lactic acid and acetic acid	Ljungdahl, et al. (92)
	Clostridium thermohydrosulfuricum	ethanol, acetic acid and lactic acid	Zeikus, et al. (93)
	Candida utilus	single cell protein	Wiley (94)
L-arabinose	*Clostridium acetobutylicum*	acetic acid, butanol and butyric acids	Saddler, et al. (87)
	Klebsiella pneumoniae	ethanol, acetic acid and 2,3-butanediol	Saddler, et al. (87)
	Candida tropicalis	L-arabitol	Gong, et al. (89)
	Pachysolen tannophilus	L-arabitol	Gong, et al. (89)
	Schizosaccharomyces pombe	L-arabitol and ethanol	Gong, et al. (89)
D-xylulose	*Kluyveromyces lactis*	ethanol	Gong, et al. (89)
	Schizosaccharomyces pombe	ethanol	Gong, et al. (89)

amount of research on the widely distributed pine. However, organosolv pretreatment of pinewood has recently been shown to open this source of fermentable sugars to economic biotechnological utilization (94).

Some of the microorganisms which utilize disaccharides derived from common carbohydrate sources are listed in Table 11. Utilization of cellobiose would be considered useful in simultaneous saccharification and fermentation processes, because of the strong inhibition of cellulase enzymic activity by cellobiose. The value for lactose from cheese manufacturing operations has negative value because of pollution abatement costs.

One of the polysaccharide substrates (Table 12) not discussed in depth above, but which has received world-wide attention is inulin. Hydrolysis of the polymer from the tubers of Jerusalem artichokes (Helianthus tuberosus) yields fructose, which generally has the same fermentation capacity as glucose. Some of the yeast listed carry hydrolytic capability for the β-1,2-fructan polymer.

Lignin degrading fungi listed in Table 13 are thought to partially derive energy from carbohydrate fractions of the lignocellulosic substrate. Lignin represents approximately 20% of the biomass on earth, but the slow rate of bioconversion of lignin with known technologies relegates this resource primarily to a valuable by-product status in biomass conversion processes.

Finally, the conversion of petroleum products to biomass by organisms is tabulated in Table 14. By comparison with the value of substrate carbon from the carbohydrate sources discussed above, crude oil at $30 per barrel breaks down to $0.07-0.10 per pound of carbon, depending on crude oil quality. The utility of paraffin fractions or other refinery by-product streams may have discounted value in localized situations.

IX. *SUMMARY AND CONCLUSIONS*

Depending on the fermentation, it appears that 25% to 70% of the total fermentation costs will be due to the carbon source. As Tong has noted (127), "Raw material cost is now the most significant cost factor for both synthetic and fermentation process routes." At current prices of fermentable sugars and of other carbon sources derived from coal and petroleum, most ethanol and SCP fermentations cannot compete

Table 11. C$_{12}$ substrates

Substrate	Microorganism	Product	Reference
cellobiose	*Clostridium acetobutylicum*	acetone and butanol	Saddler, et al. (87)
	Candida lusitaniae	yeast	Daniell & Raymond (96)
	Brettanomyces intermedius	yeast	Daniell & Raymond (96)
	Thermoanaerobacter ethanolicus	ethanol, acetic acid and lactic acid	Ljungdahl, et al. (92)
	Clostridium thermohydrosulfuricum	ethanol, acetic acid, and lactic acid	Zeikus, et al (93)
	Clostridium thermosaccharolyticum	ethanol, acetic acid and lactic acid	Wang, et al. (97)
	Alcaligenes sp.	single cell protein	Callihan & Dunlap (98)
	Candida wickerhamii	ethanol	Freer (99)
	Candida lusitaniae	ethanol	Freer (99)
lactose whey	*Kluyveromyces fragilis*	ethanol	Blackebrough & Mores (100)
	Aspergillus niger	citric acid	Somkuti & Bencivengo (101)
	Candida pseudotropicalis	ethanol	Vienne & von Stockar (102)
	Kluyveromyces lactis	ethanol	Vienne & von Stockar (102)
	Clostridium acetobutylicum	butanol and acetone	Compere, et al. (103)
sucrose (molasses)	*Saccharomyces cerevisiae*	ethanol	Haraldson & Rosen (104)
	Schizosaccharomyces pombe	ethanol	Haraldson & Rosen (104)
	Clostridium thermocellum	ethanol, acetic acid and lactic acid	Ng, et al. (105)
	Zymomonas mobilis	ethanol	Rogers, et al. (106)
	Clostridium acetobutylicum	acetone, butanol and ethanol	Beech (107)
maltose (starch)	*Saccharomyces cerevisiae*	ethanol	Peppler (75)

Table 12. Polysaccharides

Substrate	Microorganism	Product	Reference
cellulose	*Cellulomonas sp.*	cellobiose, protein	Callihan & Dunlap (98)
	Trichoderma reesei	cellulase, protein	Peitersen (108)
	Aspergillus terreus	cellulase, protein	Garg & Neelakan (109)
	Acetivibrio cellulolyticus	ethanol and acetic acid	Khan & Murray (110)
	Aspergillus fumigtus	cellulase, protein	Vandamme, et al. (111)
	Taralomyces sp.	cellulase, protein	Nishio, et al. (112)
	Thermoactinomyces sp.	cellulase, protein	Haegerdal,et al. (113)
	Penicillium verruculosum	cellulase, protein	Szakacs, et al. (114)
	Chaetomium cellulolyticum	cellulase, protein	Chahal, et al. (62)
	Clostridium thermocellum	cellobiose, protein	Ng, et al. (105)
	Monilia sp.	ethanol	Gong, et al. (89)
inulin (Jerusalem artichokes)	*Klebsiella oxytoca*	acids	Ochuba & von Riesen (115)
	Kluveromyces fragilis	ethanol	Guiraud, et al. (116)
	Kluveromyces marxianus	ethanol	Guiraud, et al. *(116)*
	Torulopsis colliculosa	ethanol	Guiraud, et al. (116)
	Saccharomyces cerevisiae	ethanol	Duvnjak, et al. (117)
	Saccharomyces diastaticus	ethanol	Duvnjak, et al. (117)
	Kluveromyces cicerisporus	ethanol	Duvnjak, et al. (117)
	Clostridium acetobutylicum	acetone, butanol and ethanol	Wendland, et al. (118)
amylopectin	*Klebsiella oxytoca*	acids	Ochuba & von Riesen (115)
	Klebsiella pneumoniae	acids	Ochuba & von Riesen (115)
	Clostridium acetobutylicum	acetone, butanol and ethanol	Beech (119)
chitin	*Mucro lusitanicus*	cell mass	Bemmann, et al. (120)
carrageenan	*Klebsiella oxytoca*	acids	Ochuba & von Riesen (115)
	Klebsiella pneumoniae	acids	Ochuba & von Riesen (115)
	Enterobacter aerogenes	acids	Ochuba & von Riesen (115)
	Enterobacter agglomerans	acids	Ochuba & von Riesen (115)
	Enterobacter cloacae	acids	Ochuba & von Riesen (115)
	Pectobacterium sp.	acids	Ochuba & von Riesen (115)
polypectate	*Klebsiella oxytoca*	acids	Ochuba & von Riesen (115)
	Pectobacterium sp.	acids	Ochuba & von Riesen (115)
tragacanth	*Klebsiella oxytoca*	acids	Ochuba & von Riesen (115)
	Klebsiella pneumoniae	acids	Ochuba & von Riesen (115)

Table 13. Liquins

Substrate	Microorganism	Product	Reference
alkali straw and Kraft pine lignins	*Sporotrichum pulverulentum*	lignin fragments	Janshekar, et al. (121)
	Humicola fuscoatra	lignin fragments	Janshekar, et al. (121)
	Aspergillus wentii	lignin fragments	Janshekar, et al. (121)
	Chaetomium cellulolyticum	lignin fragments	Janshekar, et al. (121)
	Phanerochaete chrysosporium	lignin fragments	Fenn & Kirk (122)
grass lignins	*Coriolus versicolor*	lignin fragments	Antai & Crawford (123)
pine lignins	*Coriolus versicolor*	lignin fragments	Antai & Crawford (121)
	Poria placenta	lignin fragments	Antai & Crawford (123)
	Monodictys pelagica	lignin fragments	Sutherland, et al. (;24)
	Thermomonospora fusca	lignin fragments	Crawford, et al. (125)

Table 14. Alkanes

substrate	Microorganism	Product	Reference
parex – paraffin	*Mucor lusitanicus*	biomass	Bemmann, et al. (120)
	Aspergillus fumigatus	biomass	Bemmann, et al. (120)
hexadecane	*Candida lipolytica*	biomass	Dworkin (126)
	Candida tropicalis	biomass	Dworkin (126)
	Candida maltosa	biomass	Dworkin (126)
n-alkanes	*Prototheca zopfi*	cell mass	Koenig & Ward (77)

economically with synthetically or agriculturally-derived
equivalent commodities. Fermentation processes can compete
with higher-value products for which the carbon source is
likely to be an important, but not overwhelming, cost factor.
Effective conversion of lignocellulosic materials holds great
promise in providing low-cost sugar syrups for fermentation.
Without such cellulose-derived sugar syrups, bulk commodity or
chemicals production via fermentation is unlikely in the
near future.

Sophisticated separation and genetic engineering techniques
will greatly assist the progress of biotechnology. However,
the potential of many important fermentations, some of which
have been listed above, will be severely limited in spite of
such sophisticated technology until equally sophisticated
thinking and technology are applied to effective conversion of
lignocellulosic materials.

REFERENCES

1.   Nyiri, L.K. and M. Charles, in "Annual Reports on
        Fermentation Processes," (D. Perlman and G.T. Tsao,
        eds.), Vol. *1*, p. 365, Academic Press, New York, (1977).
2.   Rudd, D.F. and C.C. Watson, "Strategy of Process
        Engineering," John Wiley & Sons, Inc., New York, (1968).
3.   Ghose, T.K., *Proc. Biochem, 4*, 43, December (1969).
4.   Litchfield, J.H., "Advances in Applied Microbiology,"
        (D. Perlman, ed.), Vol. *22*, 291, Academic Press, New
        York, (1977).
5.   Abbott, B.J. and A. Clamen, *Biotech. Bioeng., 15*, 117
        (1973).
6.   Moo-Young, M., *Proc. Biochem., 12*, #4, 6, May (1977).
7.   Laine, B.M., *Hydrocarbon Processing,* 139, November (1974).
8.   Cysewski, G.R. and C.R. Wilke, *Biotech. Bioeng., 18*,
        1297 (1976).
9.   Einsele, A., in "Biotechnology," (H. Dellweg, ed.),
        Vol. *3*, 75, Verlag Chemie, Deerfield Beach, FL (1983).
10. Humphrey, A.E., in "Single-Cell Protein II," (S.R.
        Tannenbaum and D.I.C. Wang, eds.), 16, MIT Press,
        Cambridge, MA (1975).
11. Paturau, J.M., "By-Products of the Cane Sugar Industry,"
        184, Elsevier, Amsterdam (1969).
12. Martin, S.R., *The Chemical Engineer (London)*, 50,
        February (1982).
13. Tyner, W.E., "Agricultural Energy Production Potential,"
        15, Dept. of Agricultural Economics, Purdue
        University (1980).

14. Hertzmark, S. Flaim, D. Ray and G. Parvin, *Amer. J. Agr. Econ.*, *62*, #5, 965 (1980).

15. Cysewski, G.R. and C.R. Wilke, *Biotech. Bioeng.*, *20*, 1421 (1978).

16. Borglum, G.B. in "Fuels from Biomass and Wastes," (D.L. Klass and G.H. Emert, eds.), 297, Ann Arbor Science, Ann Arbor, MI (1981).

17. Friend, B.A. and K.M. Shahani in "Fuels from Biomass and Wastes," (D.L. Klass and G.H. Emert, eds.), 343, Ann Arbor Science, Ann Arbor, MI (1981).

18. DeCarvalho, A.V., Jr., W.N. Milfont, V. Yang and S.C. Trinidade in "International Symposium on Alcohol Fuel Technology - Methanol and Ethanol," Wolfsburg, Germany, November 21-23 (1977).

19. Earl, W.B. and W.A. Brown in "Proceedings of the IV International Symposium on Alcohol Fuels Technology," Sao Paulo, Brazil, October (1980).

20. Kosaric, N., A. Wieczorek, Z. Duvnjak and S. Kliza in "Biotechnology R&D Seminar," Winnipeg, Canada, March 29-31 (1982).

21. Maiorella, B., H.W. Blanch and C.R. Wilke, "Rapid Ethanol Production via Fermentation," p. 5, presented at the American Institute of Chemical Engineers, San Francisco, CA, November 29 (1979).

22. Rugg, B., P. Armstrong and R. Stanton in "Fuels from Biomass and Wastes," (D.L. Klass and G.H. Emert, eds.), 311, Ann Arbor Science, Ann Arbor, MI (1981).

23. Wilke, C.R., R.D. Yang, A.F. Sciamanna and R.P. Freitas, *Biotech. Bioeng.*, *23*, 163 (1981).

24. Sitton, O.C., G.L. Foutch, N.L. Book and J.L. Gaddy, *Chem. Engr. Prog.*, 50, December (1979).

25. Koutinas, A.A., P. Yianoulis, K. Gravalos and K. Koliopoulos, *Energy Conv. and Mgmt.*, *21*, 131 (1981).

26. Wayman, M., J.H. Lora and E. Gulbinas in "ACS Symposium Series #90," 183 (1979).

27. Bungay, H.R., "Energy, The Biomass Options," 274, John Wiley and Sons, New York (1981).

28. Bungay, H.R., op cit., p. 276.

29. Cysewski, G.R. and C.R. Wilke, *Biotech. Bioeng.*, *28*, 1297 (1976).

30. Roberts, R.S., M.K. Bery, A.R. Colcord, D.J. O'Neal and D.K. Sondhi in "Energy from Biomass and Wastes, IV," (D.L. Klass and J.W. Weatherly III, eds.), 671, Institute of Gas Technology, Chicago, IL (1980).

31. Moreira, A.R., J.C. Linden, D.H. Smith and R.H. Villet,
     in "Fuels from Biomass and Wastes," (D.L. Klass and
     G.H. Emert, eds.), 357, Ann Arbor Science, Ann Arbor,
     MI (1981).
32. Jones, J.L. and K.T. Semrau, "Wood Hydrolysis for
     Ethanol Production - Previous Experience and the
     Economics of Selected Processes," 20, Paper 89,
     Annual Meeting of the American Institute of Chemical
     Engineers, Los Angeles, CA (1982).
33. Martin, S.R., op cit., loc cit.
34. Gibbs, D.F., *Trends in Biotech.*, *1*, #1, 12 (1983).
35. Schierholt, J., *Proc. Biochem.*, *12*, #9, 20 (1977).
36. Paturau, J.M., op cit., p. 208.
37. Paturau, J.M., op cit., p. 219.
38. Lodha, M.L., S.L. Mehta and N.B. Das, *Indian J. Exp.
     Biol.*, *11*, 571 November (1973).
39. Inskeep, G.C., *Ind. Eng. Chem.*, *43*, 1488 (1951).
40. Kobayashi, T., I. Nakamura and M. Nakagawa in "Proc.
     4th Inter. Fermentation Symposium: Fermentation
     Technology Today," (G. Terui, ed.), 215 (1972).
41. Lenz, T.G. and A.R. Moreira, *I. & E.C. Prod. Res. Devel.*,
     19, #4 478, December (1980).
42. Olson, N.F., *Dairy and Ice Cream Field*, *162*, #7, 550
     (1979).
43. Wilke, C.R., in "Alcohol Fuels Program Technical Review,"
     (L. Douglas, ed.), 42, Solar Energy Research
     Institute, Winter (1981).
44. Lee, H.Y., M.M. Gharpuray and L.T. Fan, *Biotech. Bioeng.
     Symp.*, *12*, 121 (1982).
45. Lutzen, N.W., M.H. Nielson, K.M. Oxenboell, M. Schulein
     and B. Stentebjerg-Olesen, *Phil. Trans. R. Soc.
     Lond.*, *B300*, 283 (1983).
46. Wilke, C.R., R.D. Yang, U. Von Stockar, *Biotech. Bioeng.
     Symp.*, *#6*, 155 (1976).
47. Arthur G. McKee and Co., Chicago, IL, a report entitled,
     "Preliminary Engineering and Cost Analysis of
     Purdue/Tsao Cellulose Hydrolysis (Solvent) Process,"
     report HCP/T464.01. Prepared for U.S. Dept. of
     Energy, Fuels from Biomass Branch, October (1978).
48. Grethlein, H.E., *Biotech. Bioeng.*, *20*, 503 (1978).
49. Spano, L., A. Allen, T. Tassinari, M. Mandels and
     D.D.Y. Ryu in "Proceedings Second Annual Symposium
     on Fuels from Biomass," Vol. *2* (W.W. Schuster, ed.),
     671 (1978).
50. Linden, J.C., V.G. Murphy and D.H. Smith in "Fuels and
     Organic Chemicals from Biomass. Chapter 5:
     Bioconversion Systems," (D.L. Wise, ed.), CRC Press,
     Inc., Boca Raton, FL, In Press (1983).

51. Slininger, P.J., R.J. Bothast, J.E. Van Cauwenberge and
       C.P. Kurtzman, *Biotech. Bioeng.*, *24*, 371 (1982).
52. Chiang, L.C., C.S. Gong, L.F. Chen and G.T. Tsao, *Appl.
       Environ. Microb.*, *42*, #2, 284, August (1981).
53. Katzen, R., R. Fredrickson and B.F. Brush, *CEP*, 62,
       February (1980).
54. Glasser, W.G. and C.A. Barnett, *Tappi, 62*, #8, 101 (1979).
55. Dale, B.E., "Biomass Refining: Protein and Ethanol from
       Alfalfa," *I. & E.C. Prod. Res. Dev.*, *22*, 466
       September (1983).
56. Buchanan, R.A., F.H. Otey and G.E. Hamerstrand, *I. & E.C.
       Prod. Res. Dev.*, 489 (1980).
57. Millet, M.A., A.J. Baker and L.D. Satter, *Biotech.
       Bioeng. Symp.*, *#6*, 125 (1976).
58. Fan, L.T., M.M. Gharpuray and Y.H. Lee, *Biotech. Bioeng.
       Symp. #11*, 29 (1981).
59. Mandels, M., L. Hontz and J. Nystrom, *Biotech. Bioeng.*,
       16, 1471 (1974).
60. Tanaka, M., M. Taniguchi, T. Morita, R. Matsuno and
       T. Kamikubo, *J. Ferm. Tech. (Japan)*, *57*, 3, 186 (1979).
61. Ben-Ghedalia, D. and J. Miron, *Nutr. Rep. Intn'l, 19*,
       #4, 499 (1979).
62. Chahal, D.S., S. McGuire, H. Pikor and G. Noble in
       "Proceedings of the 2nd World Congress of Chemical
       Engineering," Montreal, Canada, October 4-9, 245 (1981).
63. Bender, R., U.S. Patent 4,136,207 (1979).
64. Spano, L., et al., op cit., p. P-677.
65. Spano, L., T. Tassinari, D.D.Y. Ryu, A. Allen and M.
       Mandels, "Enzymatic Hydrolysis of Cellulose to
       Fermentable Sugars for Production of Ethanol,"
       presented at the 29th Canadian Chemical Engineering
       Conference, Sarnia, Ontario, October (1979).
66. Ryu, D.D.Y. and M. Mandels, *Enzyme Microb. Technol.*, 2,
       91, April (1980).
67. Ryu, D.D.Y. and M. Mandels, op cit., p. 93.
68. Spano, L., et al., op cit., loc cit. (1978).
69. Fuchs, G., G. Andress and R.K. Thauer in "Symposium on
       Microbial Production and Utilization of Gases,"
       (H.G. Schlegel, G. Gottschalk and N. Pfennig, eds.),
       E. Goltze KG, Goettingen, West Germany, 231 (1976).
70. Taylor, G.T., D.P. Kelly and S.J. Pirt in "Symposium on
       Microbial Production and Utilization of Gases,"
       (H.G. Schlegel, G. Gottschalk and N. Pfennig, eds.),
       E. Goltz KG, Goettingen, West Germany, 173 (1979).

71. Thauer, R.K., G. Fuchs and P. Scherer in "Symposium on Microbial Production and Utilization of Gases," (H.G. Schlegel, G. Gottschalk and F. Pfennig, eds), E. Goltze KG, Goettingen, West Germany, 157 (1979).

72. Ooyama, J., in "Symposium on Microbial Production and Utilization of Gases," (H.G. Schlegel, G. Gottschalk and N. Pfennig, eds.), E. Goltze KG, Goettingen, West Germany, 237 (1979).

73. Wadzinski, A.M. and D.W. Robbins, *J. Bacteriol.*, *123*, 380 (1975).

74. Anonymous, *Chem. Eng. News*, 20, May 15, 1978.

75. Peppler, H.J., *Microbiol. Technol.*, *1*, 2nd Edition, (H.P. Peppler and D. Perlman, eds.), Academic Press, New York, NY 157 (1979).

76. Malashenko, Y.R., in "Symposium on Microbial Production and Utilization of Gases," (H.G. Schlegel, G. Gottschalk and N. Pfennig), E. Goltze KG, Goettingen, West Germany, 293 (1979).

77. Koenig, D.W. and H.W. Ward, *Appl. Env. Microbiol.*, *45*, 333 (1983).

78. Bryant, M.P., S.F. Tzeng, I.M. Robinson and A.E. Joyner, Jr., *Adv. Chem. Ser.*, *105*, 23 (1971).

79. Knight, E., Jr. and R.W.F. Hardy, *J. Biol. Chem.*, *24*, 2752 (1966).

80. Jungermann, K., M. Kern, V. Riebeling and R.K. Thauer in "Symposium on Microbial Production and Utilization of Gases," (H.G. Schlegel, G. Gottschalk and N. Pfennig, eds.), E. Goltze KG, Goettingen, West Germany, 85 (1971).

81. Leadbetter, E.R. and J.W. Foster, *Arch. Biochem. Biophys.*, *82*, 491 (1959).

82. Sinskey, A.J., M. Akedo and C.L. Cooney in "Trends in the Biology of Fermentations for Fuels and Chemicals," (A. Hollaender, ed.), Plenum Press, New York, NY, 473 (1981).

83. Hou, C.T., R.N. Patel, A.I. Laskin, I. Barist and N. Barnabe, *Appl. Env. Microbiol.*, *46*, 98 (1983), ibid. 171, 178.

84. Perlman, D., *Microbiol. Technol.*, *2*, 2nd Edition, (H.J. Peppler and D. Perlman, eds.), Academic Press, New York, 173 (1979).

85. Chibata, I., T. Tosa and T. Sato, *Microbiol. Technol.*, *2*, 2nd Edition, (H.J. Peppler and D. Perlman, eds.), Academic Press, New York, NY, 433 (1979).

86. Ichikawa, Y., S. Sato and J. Takahashi, *J. Ferm. Technol.*, *59*, 269 (1981).

87. Saddler, J.N., E. Yu, M. Mes-Hartree, N. Levitin and
        H.H. Brownell, AIChE Meeting, Orlando, FL
        February 28-March 3, 1982.
88. Slininger, P.J., R.J. Bothast, L.T. Black and J.E.
        McGhee, *Biotech. Bioeng., 24*, 2241 (1982).
89. Gong, C.S., T.A. Claypool, L.D. McCracken, C.M. Maun,
        P.P. Ueng and G.T. Tsao, *Biotech. Bioeng., 25*, 85
        (1983).
90. Jeffries, T.W., *Biotech. Lett., 3*, 213 (1981).
91. Flickinger, M.C., *Biotech. Bioeng., 22*, Suppl. 1, 27
        (1980).
92. Ljungdahl, L.G., F. Bryant, L. Carreira, T. Saiki and
        J. Wiegle in "Trends in the Biology of Fermentations
        for Fuels and Chemicals," (A. Hollaender, ed.),
        Plenum Press, New York, NY, 397 (1981).
93. Zeikus, J.G., A. Ben-Bassat and P.W. Hegge, *J. Bacteriol.,
        143*, 432 (1980).
94. Wiley, A.J., in "Industrial Fermentations," (L.A.
        Underkofler and R.J. Hickey, eds.), Chemical
        Publishing Company, New York, NY, 307 (1954).
95. Murphy, V.G., K. Dockrey, J.C. Linden and A.R. Moreira,
        Paper 89E of AIChE Annual Meeting, Los Angeles, CA,
        November 14, 1982.
96. Daniell, S.D. and J.C. Raymond, ACS National Meeting,
        NY, August 23-28, 1981.
97. Wang, D.I.C., C.L. Cooney, S.D. Wang, J. Gordon and
        H.Y. Wang, Second Biomass Energy Systems Conference,
        Troy, NY, 537 (1978).
98. Callihan, C.D. and C.E. Dunlap, U.S. Nat. Tech. Inform.
        Serv., PB Rep. No 221096/1 (1973).
99. Freer, S.N. and R.W. Detroy, *Biotech. Lett., 4*, 453 (1982).
100. Blakebrough, N. and M. Moresi, *Eur. J. Appl. Microbiol.
        Biotech., 12*, 173 (1981).
101. Somkuti, G.A. and M.M. Bencivengo, *Dev. Ind. Microbiol.,
        22*, 557 (1981).
102. Vienne, P. and U. Von Stockar in "Proceedings Fifth
        Symposium on Energy Production from Biomass," (C.
        Scott, ed.), Gatlinburg, TN, May 10-13, 1983.
103. Compere, A.L. and W.L. Griffith, *Dev. Ind. Microbiol.,
        20*, 509 (1979).
104. Haraldson, A. and C.G. Rosen, *Eur. J. Appl. Microbiol.
        Biotech., 14*, 220 (1982).
105. Ng, T.K., P.J. Weimer and J.G. Zeikus, *Arch. Microbiol.,
        114*, 1 (1977).
106. Rogers, P.L., K.J. Lee and D.E. Tribe, *Biotech. Lett.,
        1*, 165 (1979).

107. Beech, S.C., *Eng. Proc. Dev.*, *44*, 1677 (1952).
108. Peitersen, N., *Biotech. Bioeng.*, *17*, 129 (1975).
109. Garg, S.K. and S. Neclakantan, *J. Fd. Technol.*, *17*, 271
       (1982).
110. Khan, A.W. and W.D. Murray, *J. Appl. Bacteriol.*, *53*,
       379 (1982).
111. Vandamme, E.J., J.M. Logghe and H.A.M. Geeraerts,
       J. Chem. Tech. Biotech., *32*, 968 (1982).
112. Nishio, N., H. Kurisu and S. Nagai, *J. Ferment. Technol.*,
       59, 407 (1981).
113. Haegerdal, B.G.R., J.D. Ferchak and E.K. Pye, *Appl. Env.
       Microbiol.*, *36*, 606 (1978).
114. Szakacs, G., K. Reczey, P. Hernadi and M. Dobozi, *Eur.
       J. Appl. Microbiol. Biotech.*, *11*, 120 (1981).
115. Ochuba, G.U. and V.L. von Riesen, *Appl. Env. Microbiol.*,
       39, 988 (1980).
116. Guiraud, J.P., J.M. Caillaud and P. Galzy, *Eur. J.
       Appl. Microbiol. Biotech.*, *14*, 81 (1982).
117. Duvnjak, Z., N. Kosaric and S. Kliza, *Biotech. Bioeng.*,
       24, 2297 (1982).
118. Wendland, R.T., E.I. Fulmer and L.A. Underkofler, *Ind.
       Eng. Chem.*, *33*, 1078 (1941).
119. Beech, S.C., *Appl. Microbiol.*, *2*, 85 (1953).
120. Bemmann, W., A. Voigt and R. Troeger, *Zentralblatt
       Bacteriol.*, *II*, *136*, 550 (1981).
121. Janshekar, H., T. Haltmeier and C. Brown, *J. Appl.
       Microbiol. Biotech.*, *14*, 174 (1982).
122. Fenn, P. and T.K. Kirk, *Arch. Microbiol.*, *123*, 307, (1979).
123. Antai, S.P. and D.L. Crawford, *Eur. J. Appl. Microbiol.
       Biotech.*, *14*, 165 (1982).
124. Sutherland, J.B., D.L. Crawford, and M.K. Speedie,
       Mycologia, *74*, 511 (1982).
125. Crawford, D.L., *Can. J. Microbiol.*, *20*, 1069 (1974).
126. Dworkin, D., *Chem. Technol.*, *2*, 350 (1972).
127. Tong, G.E., *CEP*, 70, April (1978).

CHAPTER 6

THE IMPACT OF BIOTECHNOLOGY ON THE HEALTH CARE INDUSTRY[1]

William E. Brown

The Squibb Institute for Medical Research
Princeton, New Jersey

It is a great privilege to be invited by the Microbial and Biochemical Technology Division to give the 1983 David Perlman Memorial Lecture. My associations with David go back more than 30 years, both in the Division and at the Squibb Institute for Medical Research. Consequently, to be given this opportunity to honor David, whom I greatly respected as a scientist and held in warm regard as a personal friend, is an occasion that I cherish.

The subject of my talk, The Impact of Biotechnology on the Health Care Industry, is one in which I am certain David would be highly interested because of his many years in the pharmaceutical industry. On receiving his Ph.D. degree in fermentation biochemistry from the University of Wisconsin in 1943 David embarked on an illustrious industrial career as a pharmaceutical research scientist. In 1947 he joined the Squibb Institute for Medical Research where he remained for nearly twenty years during which time he and I worked in close association. Moving from industry to academia he remained responsive to the health care industry as he accepted an appointment in the School of Pharmacy, University of Wisconsin-Madison, serving as professor and as Dean until his untimely death in 1980. Obviously David's ties with the pharmaceutical industry were close throughout his professional career.

David would be the first to extol the promise of the new developments in genetic engineering and cell fusion; emphasizing to us their potential impact on the fermentation industry.

---

[1] *The fourth David Perlman Memorial Lecture sponsored by the Microbial and Biochemical Technology Division of the American Chemical Society, and BioChem Technology, Inc., Malvern, Pennsylvania. Presented*

However, there is no doubt that he would quickly agree with us
that biotechnology is not a new technology, contrary to what the
the layman must infer based on the outpourings of the popular
press and the predictions of the financial community. Although
there are those who relate biotechnology only to the present
unfolding applications of recombinant DNA and cell fusion I
prefer, as I believe David would, the definition given by
A.T. Bull et al (1) wherein biotechnology is defined as "the
application of scientific and engineering principles to the
processing of materials by biological agents to provide goods
and services." The authors include in their definition the
actual step in which the biological agent is used as well as
the steps concerned with its preparation and the processing of
biological material resulting from its action. By this definition
biotechnology in the health care field is therefore restricted
to the production of diagnostic, therapeutic and prophylactic
agents.

Using this definition I will examine the present health
care industry determining to what extent the various segments
are based on biotechnology, reviewing their historical development
and finally to project where present activities in the field will
take us. I particularly want to address the problems faced by
the new biotechnology companies in becoming viable units in the
pharmaceutical economy. Because of the limitations of time I
will not discuss diagnostic products.

I.   THE HEALTH CARE INDUSTRY

Currently we biotechnologists are practicing our skills
and applying our talents in an exciting period. The opportunities
of genetic engineering have burst upon a health care industry that
is already huge and growing and firmly founded on biotechnology.
However, it was not ever thus. The health care industry is a
modern phenomenon. Prior to the 1930's with a few exceptions
such as the treatment of diabetes with insulin and the use of
corrective surgery, the role of the physician was to diagnose
the diseased state, predict the outcome and provide the patient
and his family the solace and strength they required while the
body either triumphed or succumbed (2). However, starting in
the 1920's and 1930's the first medications became available
that truly allowed the physical to affect the outcome of disease.
Like the 1980's, the 1940's and 1950's were exciting periods for
biotechnologists for this was the time when the foundations of
the antibiotic and corticosteroid industries were laid and the
major advances were made in vitamin and amino acid production.
These eras were quickly followed by periods of major emphasis on
vaccines and anticancer agents. This is a record of which we

biotechnologists can be justly proud. Our collective efforts
have helped build a health care industry that has contributed
immensely to the well being of our fellow man. It is difficult
to put values on the numbers of individuals who have lived
longer and more rewarding lives because they were vaccinated
against polio, the dreaded paralyzer of my youth, because they
were given penicillin for their deadly *Streptococcus* infection
or because they received insulin for their dehabilitating
diabetes. We can, however, measure the increase in the average
life expectancy in the United States which grew five years to
73.3 years from 1950 to 1978 (3) and take a large measure of
satisfaction because products of biotechnology helped achieve
this result.

Although we take for granted the immense size of the modern
health care industry it is difficult to get firm figures on its
components because of varying exchange rates and reporting systems.
It is estimated that annual sales of world pharmaceuticals in
1980 approximated $76 billion, with sales predicted to grow 6%
per annum in the next two decades to about $245 billion (4).
The major pharmaceutical market in the world is the United States,
representing 19% of the total, followed by Japan at about 12%.
The United States, Japan and Western Europe combined consume
54% of the total value of pharmaceutical products marketed. The
developing countries' share of the world market is expected to
double by the year 2000 to about 26%.

Forty years ago pharmaceuticals were produced by companies
primarily operating strictly on a regional basis. Since World
War II the industry has come to be dominated by international
class companies with origins in the United States and Europe.
However, the dominant role of these companies is being challenged
on a variety of fronts. The Japanese pharmaceutical companies
are now vigorously seeking markets outside Japan and Southeast
Asia. Although they are moving cautiously because of their
unfamiliarity with western registration, marketing and sales
practices they will, nevertheless, constitute a major force in
the United States and Europe in the years to come. In addition
a number of European based companies although already international in scope, are strengthening their positions in the
United States. The increasing competition is coming from other
sources as well. Worldwide, companies in other less profitable
lines of endeavor such as chemicals, textiles and food products
are lured by the pharmaceutical market. Finally, with the
present explosion in biotechnology we see still another group
of companies, the genetic engineering companies, seeking their
niche in the pharmaceutical market. The roles to be played by
the individual companies and groups of companies in the years
to come remain very much in doubt.

Sales of pharmaceuticals are commonly broken down into a variety of categories and subcategories. Frequently one finds that products of different types based on therapeutic criteria are grouped together. In spite of this there is no doubt that of the dozen or more categories commonly used the major therapeutic categories represented include antiinfectives, 19.0%; cardiovasculars, 16%; gastrointestinal products, 15.0%; musculoskeletal agents, 14.0%; neurological agents, 11.0%; respiratory products, 6.0%; dermatologicals, 5.0%, genito-urinary products, 5.0% and blood products, 3.9% (4).

Most of these categories owe a considerable amount to biotechnology and the perseverance and persistance of its practitioners. The largest category, that of antiinfectives, is clearly founded on biotechnology and could not have developed without it being involved. To a somewhat lesser extent many other categories are dependent on biotechnology, for example:

antiinfectives	–	penicillins, cephalosporins, tetracyclines, aminoglycosides, macrolides, polyenes
agents for metabolic disease	–	insulin
anticancer antibiotics	–	adriamycin, bleomycin, pepleomycin, mitomycin C
cardiovasculars	–	reserpine, digoxin, streptokinase, urokinase
antiinflammatories	–	many corticosteroids
infusion products	–	amino acids
vitamins	–	B_{12}, B_6, C
vaccines	–	many

The absolute values assignable to each therapeutic product group worldwide affected by biotechnology are difficult to determine. Nevertheless the figures in Table 1 representing a data base approximating 60% of the total have been assembled to give you approximations. It is obvious that the antibiotics represent the major category with corticosteroids a distant second.

The nations that produce pharmaceuticals are for the most part the developed nations. Since the developed nations with the largest populations constitute the major pharmaceutical markets these then are the countries where biotechnology will contribute most to the health care industry. It is not surprising that these countries are also the countries for the

most part that lead the world in developing the new biotechnology.
A probably exception to this statement is Switzerland a small
country that has long been in the forefront of pharmaceutical
research and production.

II. THE GENETIC ENGINEERING COMPANIES

Where will the current involvement in genetic engineering
lead us? On one hand the new biotechnology is expected to
participate substantially in the expansion of the health care
market from the current $76 billion. On the other hand it is
predicted that of the 100 or more newly formed genetic engineering companies currently in the biotechnology race only one
third will be viable in five years (5). Why is this? The
answer is not a simple one. Le us see whether we can find the
answers.

Table 1. Market value of therapeutic categories[*]

Therapeutic class	Worldwide market ($ millions)
Antibiotics	$8,250
Corticosteroids	1,700
Sex hormones	1,450
Anticancer agents	1,200
Antifungals	850
Sera and gamma globulin	460
Insulin	430
Vitamins (C, B_{12}, B_6)	400
Digestives	340
Vaccines	235
Cardiac glycosides	190

[*]Affected by biotechnology

In my opinion there are three essential components of a well
balanced biotechnology company: an innovative research and
development unit, a competent production facility and an enter-
prising marketing operation. The degree to which each of these
requirements is satisfied will spell success or failure to the
new company. Unfortunately many companies were organized on

the initial assumption that the first component, that of R&D
(with emphasis on the R), would ensure commercial success.
The shortcomings of this approach are already becoming apparent.
If one looks at the leaders of this new industry one notes that
they are taking steps to acquire both production and marketing
components.

III. LESSONS FROM THE PAST

1.   Antibiotics

     Let us look at each of these elements but let us first
review what past experience teaches us. There are some interesting lessons to learn from an examination of the development
of antibiotic biotechnology, lessons that can be applied to the
future of genetic engineering. For in the 1940's and 1950's
we were confronted with a somewhat similar explosion of knowledge as we are today.

     A.   <u>Transfer of Technology</u> (6,7,8,9). The first lesson
applies to the protection and transfer of technology. After
Sir Howard Florey came to the United States during World War II
seeking help to develop processes and to manufacture penicillin,
a project was set up second only to the Manhattan project which
ultimately involved twenty-two companies, a number of univer-
sities and institutes, and the Peoria laboratories of the USDA.
Despite the pooling of efforts, progress at first seemed pain-
fully slow. However, in 1943 two years before the war's end
penicillin was being produced and was available to the military
establishment. Much to my surprise I learned years later that
Dutch biotechnologists were secretly producing penicillin while
their country was still occupied by the Nazis. Even more
surprising to me was the fact that the Japanese also were producing penicillin in small quantities in late 1944 before the
war in the Pacific ended. Scientists in both countries had
learned of the miraculous penicillin and its method of preparation from the scientific literature and the lay press and quickly
developed their own technology. These two examples clearly
illustrate one of the cardinal rules of biotechnology--once
it has been shown that something can be done it doesn't take
other parties long to learn how to do it too. In general,
exclusivity on process technology cannot be retained either
through patent coverage or by secrecy. In either case alternate
production methods can be rapidly developed by the technologically
alert. As a consequence, although the new biotechnology company
may have the initial leadership, it is predictable that its
position of leadership will be quickly challenged.

Influenced by the commercial opportunities and the lack of
a product patent on penicillin, penicillin technology was
developed by many companies in the early postwar years and the
new technology spread rapidly. In the United States there were
more than twenty companies producing penicillin while in the
United Kingdom at least four companies were involved (6). We
in the United States were eager to export this new found technology both through government action and through commercial
arrangements. In occupied Japan General MacArthur appointed a
special envoy, Jackson Foster, an academician from the University of Texas, to teach the Japanese (8). From this effort
emerged the Japan Penicillin Research Association an organization
of more than 70 companies that coordinated research on penicillin,
participated in its production and published the Japanese lan-
guare Journal of Penicillin. Some years later the parent
organization was renamed the Japan Antibiotic Research Association, an organization that flourishes to this day and includes
a number of U.S. companies on its roster. In the meantime the
Journal of Penicillin has grown into the highly respected international English language publication Journal of Antibiotics.
Japanese biotechnologists building on their traditional
fermentation activities thus became a dominant factor in the
antibiotic industry.

The newly developed antibiotic technology was exported
throughout the world by American and to a lesser extent British
companies based on their experiences initially with penicillin
and subsequently with streptomycin. United States companies
built antibiotic factories in newly established subsidiaries in
Central and South America and licensed their process technology
to companies in Europe and Asia eager to participate. Thus,
we learn from an accounting by David Perlman (6) that approximately 100 companies were concerned worldwide in the 1950's
with the fermentation of penicillin. By 1977 this number had
decreased to about 35 for a variety of reasons (10). Today
there are probably not more than a dozen major producers of
penicillin worldwide. Needless to say this has not resulted
from a major drop in tonnage of penicillin produced - for
penicillin has become a bulk biochemical. Although it still is
used therapeutically as penicillins G and V a major portion of
production goes into the synthesis of 6-APA, the starting
material for the many semisynthetic penicillins and for the
synthesis of 7-ADCA, the intermediate used in the preparation
of the oral cephalosporins. Finally a large proportion of the
penicillin produced worldwide is used in the animal health
industry as procaine benzylpenicillin.

B. <u>Significance of Patents</u>. It is generally accepted that patents, particularly product patents, play a unique role in the development of new technology (11). They act as an incentive for investment of high risk capital in research and development, promoting innovation and facilitating licensing and technology transfer. Let us review the early antibiotic era and determine the role played by patents.

Penicillin is a classic example of a fermentation product that had no patent protection, a product on which a large number of companies initiated production. However as already pointed out, only a select few concentrated on and succeeded in becoming superior producers. As a consequence these companies, most of whom are situated in the developed countries, are the survivors today. The proposal has been made recently that, in the pursuit of high technology, the developing countries should take over the production of penicillin while the developed countries concentrate on the products of genetic engineering such as interferon (12). In my opinion this approach is not acceptable as the developing countries for the most part do not have the technological base nor the stable economies that are necessary to deliver the continuing improvements in penicillin production that are required to remain competitive. This is obviously influenced by the fact that penicillin is a bulk intermediate. With some other antibiotics such as streptomycin it is different. It should be noted however that many of the international companies operate plants in the developing countries thereby making the latest technology available.

A brief look at the streptomycin and tetracycline fermentations is also in order. Discovered by Selman Waksman at Rutgers University in 1974, streptomycin was patented by the Waksman Foundation and licensed to all comers (9). By 1962 there were only four producers of streptomycin in the U.S. and by 1983 the number has dropped to one. Why is this? Streptomycin is a good antibiotic but it has been replaced by other more useful, less toxic agents. Currently, worldwide, there are probably fewer than ten producers with most the production directed to third world economies that cannot afford the newer more expensive antibiotics and where tuberculosis is still a major problem.

The tetracyclines, chlortetracycline and oxytetracycline, were discovered and developed in the late 1940's and 1950's. In contrast to penicillin G and streptomycin they were both patented and the licensing of these agents closely controlled by the patent holders, Lederle and Pfizer. Possession of patents allowed these two companies to protect their intellectual

property in the United States and in most of Western Europe.
As a consequence both companies prospered and laid the foundation for major participation in the antibiotic market. However
control of the products was not complete for the lack of
product patents in Japan and Italy left both exposed to
competition. This was true particularly in the latter country
where bulk tetracyclines were produced and sold at low cost to
the developing countries of the world where neither product nor
process patents existed.

To illustrate another side of the competitiveness of the
situation I remind you of the notorious case in which cultures
used for the production of chlortetracycline were stolen from
Lederle and sold to organizations in Italy (13). Since Italy
protected neither pharmaceutical products nor processes with
patents the stolen cultures could be and were used industrially
to produce the antibiotic. The United States citizens involved
in the case were indicted and prosecuted.

C. <u>Commercial Continuity</u>. The three foregoing examples
illustrate clearly the importance of product patents to
developing a commercial position in a product. Let us now look
at those 22 companies in the United States that participated in
the penicillin saga in the period 1942-1950 and determine where
this involvement led them. Of the original group, nine continue in 1983 as major antibiotic producers. The companies are
grouped in Table 2 according to the nature of the company
businesses in 1942-1950. Of the 22, 12 were already involved in
pharmaceuticals, 3 in fine chemicals only, 2 in fermentation
chemicals, one in alcoholic beverages and 4 in biologicals. It
is of interest that of those still involved in fermentation of
antibiotics for human use, 8 of the 9 survivors marketed
pharmaceuticals in 1942-50. Only one company, not already in
this category, has emerged as a pharmaceutical force. This of
course is Pfizer who entered the program as a producer of citric
acid.

It is apparent that entry of a company into a new field is
a difficult task at best. Not only must superior process
technology be developed and applied, the company must also
develop a strong marketing and sales capability. In addition
the development of exclusive patented products is highly
desirable. These needs are illustrated further by the information in Table 3. Of the nine major health care companies listed,
eight already had in place a strong pharmaceutical marketing
capability in 1950. As already noted Pfizer alone had to develop
a sales capability. In addition, with the exception of Wyeth,
all companies, Pfizer included, developed patented, fermentation-
derived products of major commercial significance.

Table 2. U.S. Penicillin program 1942-50

Company Category 1942-50	No. of major antibiotic producers, 1983	
	Yes	No
Pharmaceuticals	6	3
Pharmaceuticals/Fine Chemicals	2	1
Fine Chemicals	0	3
Fermentation Chemicals	1	1
Alcoholic Beverages	0	1
Biologicals	0	4
	9	13

Table 3. Factors contributing to company success

Company Category	Company	Patented Products**	Pharmaceutical Marketer*
Pharmaceuticals	Abbott	+	+
	Bristol	+	+
	Lilly	+	+
	Squibb	+	+
	Upjohn	+	+
	Wyeth	+	+
Pharmaceuticals/ Chemicals	Cyanamid	+	+
	Merck	+	+
Fermentation Chemicals	Pfizer	-	+

* 1945

** Developed major patented fermentation products

Other entries to the charmed circle of pharmaceutical biotechnology companies exist in addition to those listed in Table 3. Worthy of mention are Beecham, originally an OTC company which, following 1960 developed a line of patented semi-synthetic penicillins based on the discovery of the penicillin intermediate 6-APA which allowed that company to become a major factor in the ethical pharmaceutical market. Similarly Schering-Plough with the introduction of their patented aminoglycoside gentamicin in 1966 built it into the largest selling brand in the world. Clearly product innovation, patent protection and competent marketing all contribute to commercial success.

Interestingly in Japan the situation took a somewhat different course because process patents but not product patents were granted until 1975. Prior to that time a "different" process would allow a company to market a competitor's anti-biotic. As a consequence, building on their historic fermentation base, the Japanese biotechnologists became superb practitioners of the art of process modification and process improvement. With a limited number of exceptions the Japanese biotechnologists concentrated on process studies rather than on developing new agents. But even in Japan efforts were made to identify new agents for production locally and for licensing abroad where product patents could be attained. As a consequence products such as josamycin, kanamycin, and cefazolin emerged initially, to be joined by an ever increasing number of antibiotics and anticancer agents in the more recent period. At the present time there are more new antibiotics coming out of Japan and being licensed worldwide than from the rest of the world combined. This outpouring of new products was clearly stimulated by the new Japanese patent law.

There are, of course, companies that have made a success of bulk production of pharmaceutical biochemicals and biochemical intermediates. They are, however limited in number. An excellent example of a successful producer is Gist-Brocades of Delft who provide bulk 6-APA and 7-ADCA to the world. Its success obviously stems from superior technology and process patent protection. Still another example is Antibioticos of Spain that owes its existence to the protective attitude of the Spanish government and to, of course, its excellent technology. Still other examples exist in Italy where the inventiveness of the biotechnologist combined with laissez-faire government attitude, for no patent protection has existed, has resulted in the emergence of many antibiotic companies supplying selected products to the world. Nevertheless it must be borne in mind that commercial success in the health care industry is primarily associated with the marketing of patented agents.

2.   *Corticosteroids*

So far I have chosen to draw examples from the antibiotic field. The corticosteroids as a class constitute the second major segment of the health care industry in which the bio-technologists have played a major role. The corticosteroids provide an interesting comparison to the antibiotics because of the major commitment on a large scale to the use of micro-organisms to carry out specific chemical reactions. Following the chemical synthesis of cortisone at Merck in the late 1940's the successful introduction of the use of whole cell cultures for carrying out chemical modifications of sterol and corti-costeroids occurred. The basic studies by Upjohn scientists was followed by other entries in the field. David Perlman was one of those scientists that quickly became involved; he was a superb development man in that he sensed where the opportunities existed.

By this time chemists in several companies were active in the synthesis of even more potent corticosteroids than cortisone. A major breakthrough came when Squibb scientists (14,15) dis-covered that the biological activities of cortisone and hydrocortisone could be enhanced by 9α-halogenation. This finding based on a chemical synthesis, had a biotechnology input. 11-Hydroxylation had been accomplished relatively easily, microbiologically, in the α position. Unfortunately the β-hydroxyl was required for biological activity. On converting the 11α-epimer to the 11β-epimer chemically, 9α-fluorohydro-cortisone was produced as an intermediate and found to be 10-fold more active in both antirheumatic and topical anti-inflammatory activity in man. Unfortunately incorporation of the 9α-fluoro moiety increased salt retention more than 50-fold.

With the recognition that the various properties of hydro-cortisone could be modifed disproportionately the creation of new steroids with increased anti-inflammatory activity and decreased mineral corticoid activity became the new goal of steroid chemists. First it was reported that uniform 2- to 3-fold enhancement of all biological activities could be accomplished by introduction of a double bond into the A ring to produce prednisone and prednisolone, a synthetic step that was easily accomplished microbiologically. However the big breakthrough came when with the synthesis of triamcinolone, 1, by Lederle scientists. They showed that the desired high potency of the 9α-fluorosteroids augmented by 1-dehydrogenation could be maintained while depressing the undesirable salt retention through 16α-hydroxylation. Despite their success Lederle found triamcinolone was dominated by Squibb and Upjohn

CH_2OH
CO
OH
CH_3
HO
OH
CH_3
F
CH_3
O

1

patents. In a conference involving representatives of all
three companies the high point of the meeting came in a dramatic
moment when Squibb announced that the Lederle synthesis of more
than 10 synthetic steps could be replaced by a two-step bioconversion involving 16α-hydroxylation and 1-dehydrogenation. The
biologically based process was adopted, with Upjohn supplying
the starting material and Lederle and Squibb developing the
large scale biosynthetic process. Those days were exciting-
and suspenseful-particularly when we committed steroid intermediate worth 5 to 10 times our annual salaries to a production
batch. The continuing role played by biotechnology in this
class of agent was recently emphasized by the Upjohn announcement
that a new plant has been brought onstream for chemically converting dexamethasone 2, derived from sitosterol by microbiological conversion, to steroid intermediates (16).

3.   Problems of Dogma and Biology

      Let us look at the development of process technology from
another angle. Introduction of new technologies presents
many problems, with a large proportion of them unpredictable.
There is no doubt in my mind that you practitioneers of the
new biotechnology are encountering such problems today and
will continue to cope with them in the years to come. There
is also no doubt in my mind that you will surmount these
problems, solving them with unexpected solutions and infrequently
with some very gratifying results.

      Let me give you examples from my experience and that of
David Perlman. You must have realized by now that problems
with people arise as frequently as with the technology. For
example one problem frequently encountered is that of convincing
management to adopt a new course of action. An early lesson
learned was the value of microorganisms as chemists, a point
frequently emphasized by David Perlman in his later years. At
least one large chemical company while participating in the

wartime penicillin project in the U.S. deliberately chose to
hold back on building the required fermentation plant. Company
chemists, despite the arguments of the microbiologists,
assured their management that given a little time penicillin
could be synthesized more simply and cheaply by traditional
chemistry. How wrong the chemists were proven; sometime later
that same company made a major but belated commitment to
fermentation. Fortunately it was not too late and that company
has emerged as one of the foremost antibiotic companies of
this day.

The chemists were not the only ones proven wrong. Many
of our engineers proved to be too dogmatic in their ways to
accept the new developments. A traditional producer of butanol
and acetone by anaerobic fermentation started aerobic riboflavin
production in the 1940's with *E. ashbyii* by taking advantage of
an abandoned tank rusting in the open factory yard, upending it
and installing an air sparger at the bottom. No provision was
made for agitation or for proper sterilization--indeed the tank
was not sturdy enough to take the stresses of agitation and of
operation under pressure. This same philosophy resulted in
that company's first penicillin plant being equipped with
unagitated fermentors. Indeed, the company engineers argued
successfully for a number of years that fermentors could not
be provided with mechanical agitators. Needless to say some
years later the more aggressive biotechnologists made their
point and the company constructed its first agitated, aerated
fermentors. This same company failed to develop its own
patented products and a pharmaceutical marketing capability.
Today it is not one of the survivors in the human antibiotic
market.

Construction of these new fangled antibiotic plants became
a major experiment. We biologists and engineers knew what
results we wanted but we did not know with certainty how to
achieve them. Nothing daunted, Squibb build in 1949 what, for
that time was the world's largest antibiotic fermentation plant,
it contained 42 fermentors of 65,000 liter capacity. It was a
good plant in many respects, but we quickly found out its
limitations. By way of example the agitators were underpowered,
the process air was not adequately filtered and pipes and
fittings were not properly designed to permit sterile operations.
Not surprisingly, engineering modifications of the plant were
needed and initiated as soon as the lessons became apparent.

Problems encountered in developing new technology are not
confined to people. Frequently mother nature reminds us that
we are not as all knowing as we would like to think we are.
Back in the late 1940's and early 1950's contamination of the
new *Streptomyces* fermentations was a particular hazard because
of the limitations of both equipment and techniques. In this

period major producers of streptomycin, one by one, encountered
a new disastrous phenomenon which struck our plants despite all
precautions. The fermentation batches lysed, the heavy mycelial
suspension turned to a thin watery soup and little or no antibiotic was produced. From these observations ultimately came
the discovery that a bacteriophage was responsible. Knowing
the cause did not readily yield the answer. Like other industrial plants the Squibb plant stumbled along for many months
at about 75% of capacity before the cause of the problem was
recognized and the situation brought under control by the
introduction of laboratory-produced phase-resistant production
cultures. This problem was one of many tackled and solved by
two men we are honoring this week. Not only did David Perlman
and Asger Langlykke produce the replacement phage resistant
culture-they did an additional bit of genetic engineering--
they were able, seemingly by pure serendipidity, to increase
yields in the fermentor, and to eliminate the troublesome
by-product streptomycin B which complicated the recovery
process and led to major losses on isolation.

IV. POTENTIAL OF THE NEW BIOTECHNOLOGY

Where do we go from here in the health care industry? We
are all aware that the field of biotechnology is undergoing one
or more period of transition. This has come about because of
the application of new tools to our field, i.e., the use of
recombinant DNA techniques and cell fusion techniques. These
tools have already had a major impact on the health care
industry. Overnight we have become aware that through
recombinant DNA we can synthesize on demand complex polypeptides
and proteins at a potential cost well within the realm of
practicality. Suddenly we have realized that a new class or
compound, the biologically active polypeptides and proteins
exemplified by human insulin, the interferons and other agents
such as tissue plasminogen activator, human growth hormone,
blood factors VIII and IX and erythropoietin can now be produced
for evaluation and potential use. The element of drama is
high because major projects have been undertaken at great expense
to develop processes to make these agents, in many cases without
prior proof of therapeutic or prophylactic efficacy or superiority
Furthermore these agents cannot be covered by prodct patents and
therefore marketing exclusivity is unattainable. Despite these
potential drawbacks the competition is fierce and intense.

The future relevance of these activities was recently
brought home to me while in Japan. Acquired immune deficiency
syndrome or AIDS is very much on the minds of the Japanese, to
a greater extent than I would have expected. Puzzled I dug
further and learned that Japan imports a high percentage of its
human blood products from the United States. There is a definite

fear that AIDS will become prevalent in Japan through the use
of blood products of U.S. origin. This fear has also been
expressed in Europe. Production of blood components through
the new biotechnology is an obvious answer to the problem.

2. New Products

A. Insulin. Let us take a closer look at some of these
potential projects. Human insulin is a special case, for
diabetics have been successfully treated with beef and pork
insulin isolated from animal glands for close to 60 years.
Over this period of time, worldwide marketing of insulin products has become the purview of a very few companies. As a
consequence when genetic engineering made production of human
insulin possible, close cooperation between the young biotechnology company and the established marketing company was
necessary. Marketing of human insulin soon became a major
objective even though proof of superiority over animal insulins
was not available and would be difficult to demonstrate. However because of the proven utility of pork and beef insulins
there was no doubt that human insulin would be at least as
effective and safe. Consequently a market for this product
existed.

The drama still is unfolding. Eli Lilly, Novo and Hoechst
all have produced and tested human insulin and the products of
these three companies are already competing side-by-side in
world markets. Interestingly, the technology developed by
these companies for its production is as dissimilar as night
and day. Lilly with the help of Genentech biotechnologists
produces human insulin by a technology that is classic to the
modern era of genetic engineering (17). By one published
route, two *E. coli* strains are used into which have been
inserted the human gene for either the A or the B peptide the
insulin molecule. Yields are limited because the peptides are
produced intracellularly. The two peptides must be isolated
absolutely free of bacterial protein--no mean feat in itself--
and then the two peptides must be chemically joined to form the
insulin molecule. A second potential route involves producing
human proinsulin by a modified *E. coli*, isolating the proinsulin and converting it to human insulin.

The Novo biotechnologists elected to synthesize human
insulin by a different route (18). Porcine insulin and human
insulin are identical with the exception of the terminal amino
acid on the B peptide. In the former case the terminal amino
acid is alanine, in the latter threonine. A process was
developed replacing the alanine of the B peptide with threonine,

thereby producing human insulin directly from pork insulin, without introducing new problems of producing product free of bacterial antigens.

Let this be a lesson. The new glamorous methodology may not always be the best--older methodologies may provide a simpler answer. As biotechnologists we must carefully consider the alternative solutions to the problems at hand. There is no doubt that the chemical approach in the short term is the most cost and time effective presenting fewer technological problems. However, the recombinant approach may indeed be the better for the long erm in terms of both production volume and unit cost. As already stated the market for human insulin is assured. We know that it will at least be as effective as animal insulin and patients will use and the medical profession prescribe it in the hope that it will be more favorable to the patient over a lifetime of therapy.

B. <u>Interferons</u>. The interferons represent a completely different situation. This is a family of related compounds whose commercialization has been undertaken by a large number of companies representing established pharmaceutical and chemical companies and the new biotechnology companies working alone and in cooperation. There are undoubtedly in excess of 25 companies in this race worldwide in the expectation that the interferons will be proven efficacious and safe in the treatment and prophylaxis of a variety of viral diseases and cancers. The expenditures ultimately will be in the hundreds of millions of dollars covering development of processes, construction of plants and conduction of safety and efficacy trials for at least three major classes of human interferons, α, β, and γ, in a multiplicity of diseases. Both cell culture and recombinant DNA processes are under development. Despite the fact that human trials have been underway for one or more years the situation is still not clear. Clinical failures in cancer therapy have been reported as have isolated successes. In the prophylaxis of influenza the side effects of the agent mimic the disease. Whether these side effects are due to impurities or to the agent has not been determined. The therapeutic utility and the ultimate commercial success of the interferons consequently remains very much in doubt.

If we assume therapeutic efficacy and safety are demonstrated the commercialization of the interferons will probably follow the pathway of penicillin. The lack of product patents means that the interferons when introduced will be the subject of intense marketing competition. Predictably, many companies initially in the field will fall by the wayside with only a

limited number of survivors. The intense competition experi-
enced by companies involved in the production and marketing of
penicillin will unquestionable be repeated.

    C. <u>Human Growth Hormone.</u> Human growth hormone, another
agent receiving attention, currently is isolated from the
pituitaries of cadavers. The potential market is expected to
remain small even when lower cost growth hormone becomes
available. Hence commercial interest is limited to a few
companies even though no product patent protection will be
available. Again, once production problems are solved and
efficacy and safety have been established, success will go to
the most agressive in the marketplace.

    D. <u>Thrombolytic Agents</u>. Of the many other candidates
under study that class of agent that dissolves blood clots
deserves special mention. About 25 years ago streptokinase,
an enzyme isolated from the lowly *Streptococcus,* was hearalded
as providing the means for removing clots, a condition that
occurs frequently postsurgery and in various cardiovascular
diseases. Despite a real medical need its acceptance by the
medical profession has been slow because of its side effects--
hemorrhaging at sites throughout the vascular system away
from the clot as well as reactions attributable to introduction
of a foreign protein. Despite these problems it, at long last,
is being given more attention by physicians because of the
major medical need for treatment of patients with infarct and
for prophylaxis and treatment after surgery. More recently
urokinase, an enzyme initially isolated from human urine and
subsequently produced by cell culture has been made available
for medical use. Unfortunately its use does not overcome the
problem of hemorrhaging and consequently its acceptance outside
Japan has been limited.

    A most exciting development combining the new biotechnology
and human health care is the biological agent human tissue
plasminogen activator (TPA), an agent that in animal trials
acts specifically at the site of the clot (19). No evidence of
hemorrhaging at sites distant from the clot has been observed.
If this prediction of safety and efficacy holds up an agent
will be made available to the medical profession that will have
both therapeutic and prophylactic potential in a wide variety
of cardiovascular problems, including heart attack and stroke.
With this indication the market could be truly immense. Indeed
projections have been made for sales worldwide ranging from
$200 million to more than $1 billion for this product alone.
As a consequence many companies are involved in developing
production technology and jockeying for eventual market position
has already started.

   E. Vaccines. In terms of providing protection against
infectious diseases vaccine development has been one of the
major success stories of the health care industry. More than
20 vaccine preparations are available today for protection
against a variety of viral and bacterial pathogens. For the
most part these are viral vaccine preparations using live
attenuated viruses or killed virus preparations. In the case
of the bacterial vaccines against the *Pneumococcus* and the
Meningococcus purified polysaccharide preparations are used.
Despite the unqualified medical success of these vaccines they,
for the most part, have not been commercial successes.
Marketing of these agents has been highly competitive, characterized by bidding on government contracts. The infrequent
occurrence of undesirable side effects associated with their
use followed by liability suits against their producers has
further contributed to the commercial unattractiveness of this
product line. As a consequence production is limited to but
a few companies. Despite this situation the new biotechnology
companies are rushing in, for the opportunity exists to
develop a variety of safer and in some cases more effective
preparations involving the use of specific antigens of the
pathogens (20). Predictably only a limited number of producers will survive.

   F. Monoclonal Antibodies. Cell fusion and the production
of hybridomas with the consequential ability to produce highly
purified monoclonal antibodies has opened exciting opportunities.
The applications are broad-ranging from their use in the
isolation of highly purified proteins and peptides to their use
in the isolation of highly purified proteins and peptides to
their use in the identification of microbial pathogens and cancer
cells. One of the more exciting possibilities is their therapeutic potential, particularly in cancer chemotherapy.
Cytotoxic antibiotics such as mitomycin C, bleomycin,
pepleomycin and adriamycin may in time be joined by chemically
modified monoclonal antibodies that have the ability to
preferentially seek out the cancer cell and on reaching its
target deliver a highly potent toxic agent (21,22). An example
of this type of agent is a monoclonal antibody with the
cytostatic antibiotic adriamycin attached to one of the two
arms of the antibody IgE molecule. Whether agents such as this
will live up to their expectations remains to be seen.
Clearly a company developing a product of this type will reap
major commercial benefit for the agent could be both patentable
and therapeutically useful.

G. <u>Other Opportunities</u>. The new biotechnology undoubtedly
gives promise of many other opportunities, still unidentified,
that do not necessarily involve recombinant DNA technology. I
anticipate that biosynthetic modification of chemicals will
continue to play a substantial role as Sih has emphasized (23).
Let me cite some examples to illustrate this point.

A recent example is that of amoxicillin, <u>2</u>, one of the
most widely used of the semisynthetic penicillins worldwide.
An oral agent, it is active against a variety of gram positive
and gram negative pathogens. You will note that the side chain
has an optically active carbon. To make this penicillin one
must have parahydroxyphenylglycine of the appropriate optical
configuration. Kanegafuchi, a Japanese firm, has initiated
in recent years biosynthetic production of this acid on a large
scale. Other examples of similar technological ingenuity of
which you are aware are the use of immobilized enzymatic and
live cell synthesis in the production of key chemicals such as
6APA and aspartic acid. Clearly similar new opportunities
will be identified in the future.

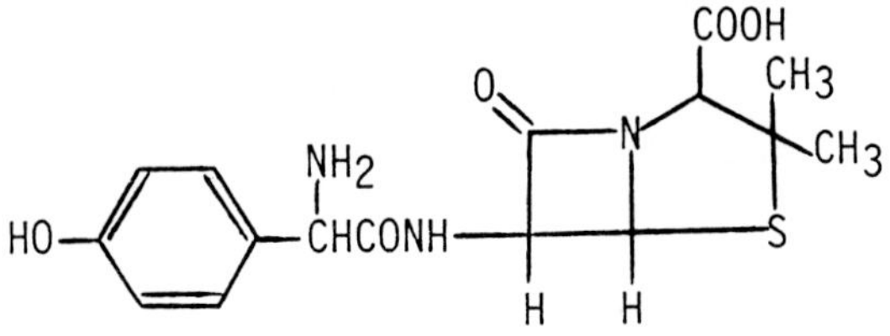

$$\underline{2}$$

One area of biotechnology that has had appeal for decades
comes under the heading of the production of pharmacologically
active substances by microorganisms (24,25). With the development and application of new screening procedures invariably new
and potentially useful agents are identified. One of the more
noteworthy breakthroughs of recent years, is an example in the
animal health field, the antiparasitic agents, the avermectins.
Of particular interest in recent years has been the application
of screening systems utilizing key enzymatic reactions such as
pioneered by Umezawa (26). One objective that is being actively
pursued is the control of body cholesterol in the prevention of
atherosclerosis by inhibition of a key enzyme, 3-hydroxy-3-methyl
glutaryl coenzyme A reductase involved in the synthesis of
cholesterol. At least two companies have agents under development that block this reaction (27,28).

A field of investigation that is burgeoning is that of
immunology. Interferons are of course classic examples of
agents affecting the immune system. In addition to this lead,
many other agents from microbial sources are being examined and
evaluated worldwide as modifiers of the human immune system.
An interesting example is the development of the microbially
derived cyclosporin by Sandoz for depressing the immune system
so that organ transplants will take. The potential application of these agents to human medicine ranges from phophylaxis
and treatment of microbial and viral disease and various types
of cancer to the prevention and treatment of autoimmune
diseases such as rheumatoid arthritis.

These examples serve to illustrate that all phases of
biotechnology undoubtedly will continue to forge ahead making
major contributions to the health care industry with the result
that the world should be a healthier and hopefully happier
place to live in, in the years to come.

2.   *Some Research Considerations*

Early on in my talk I emphasized the three key elements
required in the development of a successful biotechnology
company in the health care field namely R&D, production and
marketing. We have discussed the desirability of developing
unique products for the marketplace and stressed the fact that
commercial success usually is associated with those companies
that develop safe and useful products on which product patents
can be obtained (29).

The weakness as I see it of the newly formed biotechnology
companies oriented toward the health care field is that they
are focussing of necessity on natural biochemicals on which
product patents for the most part cannot be obtained. This
position is acceptable for an interim period and the problems
may be alleviated in part by the issuance of use patents.
Nevertheless it is imperative that these same companies move
forward into the development of patentable products at an early
date. In addition to the motivation for product patents
modification of these molecules also hold out promise for
improved properties. Since these products are mostly peptide
hormones oral administration is not feasible because of enzymatic
destruction in and poor absorption from the gastrointestinal
tract. Despite the example of insulin, routine administration
of a drug by the injectable route, especially if it is used
chronically, is not considered acceptable outside the hospital.
Consequently it is predictable that many of the natural peptides
under development today will ultimately be replaced by patentable
molecules with improved biological properties.

A good illustration of this point is the experience of
Squibb in the development of a new class of drug, the
inhibitors of angiotensin converting enzyme for the treatment
of hypertension and congestive heart failure (30). Key to the
therapeutic problem is control of the production of angiotensin II,
a hormone that acts as a vasoconstrictor thereby increasing blood
pressure. Some years ago Squibb identified the nonapeptide
SQ 20,881, 3, as an inhibitor of the enzyme that converts
angiotensin I to angiotensin II. Unfortunately SQ 20,881 was
not therapeutically useful nor commercially viable for two
reasons--the high cost of synthesis, for this was before the
recombinant DNA revolution and the fact that it had to be
administered by injection. Returning to the laboratories
Squibb scientists identifed the proline analog captopril, 4, as
being a potent inhibitor that can be synthesized economically
and is orally active. This drug the first member of its class
already has annually sales approaching $100 million with
expectations that sales could grow three to four-fold in the
years ahead.

$$< \text{GLU-TRP-PRO-ARG-PRO-GLN-ILE-PRO-PRO}$$

$$\underline{3}$$

$$\text{HS}-\text{CH}_2-\overset{\overset{\text{CH}_3}{|}}{\text{CH}}-\overset{\overset{\text{O}}{\|}}{\text{C}}-\text{N}\diagdown\diagup\text{C}=\text{O}$$

$$\underline{4}$$

This example supports my contention that many of the
injectable peptides proven to have therapeutic activity
through agonist activity or blockage of key receptor sites will
be replaced by smaller, patentable molecules capable of oral,
transdermal or buccal administration. Indeed once therapeutic
efficacy of the various agents currently under investigation is
proven it is predictable that this is the course research will
take.

It is not surprising that drug development is proving to
be a major hurdle to the new biotechnology companies. For
cloning the gene and demonstrating that a human peptide can be
produced by a microorganism is only the beginning. Obtaining
an acceptable yield is a major problem since these peptides
and proteins are not excreted by the cell. Because they are
injectable products, isolation absolutely free of bacterial
protein is a must. Following demonstration of an acceptable

process the development of the product begins. The pharmaceutical industry estimates that it costs $40 to $75 million
to carry a new product through the preclinical pharmacology
and safety, the clinical pharmacology and clinical evaluations
required to get a New Drug Application approved by the FDA.
The financial burden this imposes on the small struggling
genetic engineering company is heavy. It is no wonder that
partnerships are being formed with the major pharmaceutical
companies on most of these projects.

I have stressed that product patent protection is needed
for the ultimate success and continued commercial health of
the health care oriented company for drug research involves
high risk investiments that lead to infrequent commercializable
results. Fortunately product patents are attainable and
enforceable in the major pharmaceutical markets including the
United States, Japan, West Germany and the United Kingdom (29).
Product patents although attainable in Italy, France and
Canada are difficult to enforce in the first two countries and
greatly weakened by the granting of compulsory licenses at
low royalties in the last named. Little or no patent protection
exists in most developing countries.

V.   CONCLUSION

In closing I would like to leave the thought that biotechnology in the health care industry has an exceedingly
bright future. There is no doubt the new biotechnology
companies are leading the industry into new fields of fruitful
endeavor. The successful melding of the new approaches with
other lines of pharmaceutical research will, I am certain,
lead to the development of useful therapeutic agents unidenti-
fiable and unpredicatble as of now. Throughout this development
it is critical for the industry that many of the new companies
grow and prosper.

Once again my thanks to the Division and to each of you
for giving me this opportunity to pay tribute to our former
colleague and friend David Perlman. I wish you success in all
your undertakings. The future of biotechnology is in good hands.

REFERENCES

1. Bull, A.T., Holt, G. and Lilly, M.D., Biotechnology,
       International Trends and Perspectives. Organization for
       Economic Cooperation and Development (1982).

2.   Thomas, L., "The Youngest Science - Notes of a Medicine
       Watcher," The Viking Press, New York (1983).

3.   Vital Statistics of the United States, 1979, Volume II -
       Mortality, U.S. Dept. Health and Human Services, Public
       Health Service, Hyattsville, MD (1982).

4.   Scrip No. 694, 9, May 19, 1982.

5.   Lueck, T.J., "Investing," The New York Times, July 17,
       1983.

6.   Perlman, D., *ASM News 40*(12), 910 (1974).

7.   Helfand, W.H., et al., "The History of Antibiotics,"
       (J. Parascandola, ed), American Institute of the
       History of Pharmacy, Madison, WI (1980).

8.   Yagisawa, Y., ibid.

9.   Lechevalier, H.A., ibid.

10. Perlman, D., *Chemitech.*, 434, July 1977.

11. Patents and Pharmaceuticals, Buchdruckerie Gassei & Cie,
       Basle (1980).

12. Reich, R.B., "The Next American Frontier," Time Books,
       New York (1983).

13. Herald Tribune, June 30, 1962.

14. Fried, J. and E.F. Sabo, *J. Am. Chem. Soc.*, *75*, 2273
       (1953).

15. Borman, A., "Monographs on Therapy," *3*(2), 106 (1958).

16. Scrip No. 803, 6, June 15, 1983.

17. Gentically Engineered Pharmaceuticals, Newswatch,
       McGraw-Hill, Inc., 23 (1983).

18. Ibid, 24.

19. Bergmann, S.R., et al., *Science, 220*, 1181 (1983).

20. Marmion, B.P., *Phil. Trans. R. Soc. Lond. B., 290*, 395
       (1980).

21. F-D-C Reports, 11, August 15, 1983.

22. Genetically Engineered Pharmaceuticals, Newswatch
       McGraw-Hill, Inc., 14 (1983).

23. Sih, C., *Am. J. Pharm. Educ., 41*, 432 (1977).

24. Woodruff, H.G., *Science, 208*, 1225 (1980).

25. Matthews, H.W. and B.F. Wade, *Adv. Appl. Microbiol., 21*,
       (D. Perlman, ed.) Academic Press (1977).

26. Umezawa, H., "Enzyme Inhibitors of Microbial Origin,"
       University of Tokyo Press, Tokyo (1972).

27. Mabuchi, H., et al., *New Eng. Jour. Med., 305*, 4786 (1981).

28. Albers-Shonberg, G., Joshua, H. and Lopez, M.B., U.S.
       Patent 4,294,846, October 13, 1981.

29. Patents - Why, Buchdruckerei Gassei & Cie, Basle (1982).

30. Angiotensin Converting Enzyme Inhibitors (Z.P. Horovitz,
       ed), Urban & Schwarzenberg, Baltimore (1981).

CHAPTER 7

INDUSTRIAL MAMMALIAN CELL CULTURE:
PHYSIOLOGY-TECHNOLOGY-PRODUCTS

Nikos K. Harakas

Monsanto Company
St. Louis, Missouri

Mammalian cell culture is examined in terms of cell physiology, growth characteristics and products. Cell physiology is discussed in terms of structure, composition, and characteristics of normal and cancer cells. The technology of growing such cells in large, industrial quantities is elaborated via the methods of suspension, roller bottles, microcarriers and flat-bed hollow fiber reactors. Products are discussed in terms of production and acquisition with examples such as interferon, etc. The genetic engineering of the mammalian cell is introduced.

I.   INTRODUCTION

The successful culture of the mammalian cell in relatively large industrial scale has only been achieved with increasing confidence and reproducibility during the last fifteen years and especially during the last seven to eight years (1-7c). The present article examines the advances for industrial mammalian cell culture in terms of the physiology of the cell and its culture *in vitro*, the technology to grow large quantities of cells to produce marketable products, and examples of such products derived from mammalian cells.

Cell physiology is discussed in terms of cell components and its environment and the basic characteristics of both normal and transformed or cancer cells. The culture of the mammalian cell is discussed in terms of the Hayflick cell generation diagram for primary culture to: a) cell strains (normal or diploid) or b) cell lines (cancer or aneuploid), the Monod cell growth phase diagram, and the fundamental chemical and physical parameters for its growth in culture. Mammalian cell culture technology is elaborated from the viewpoint of raw

<table>
<tr>
<td>

</td>
<td>

159

</td>
<td>

</td>
</tr>
</table>

materials used to grow cells in large quantities, the starting up of cultures and their propagation for production.
Both laboratory and industrial equipment are described for
propagating normal and cancer cells via suspension, roller
bottles, flat bed hollow fiber reactors and microcarriers.
The cell growth parameters for large-scale culture propagation are discussed in terms of the Monod cell growth phase
diagram and growth factors, and transport phenomena of mass,
heat and momentum relations of Fick, Fourier and Navier-
Stokes, respectively. Products are discussed both in terms
of product acquisition and examples of actual products, e.g.,
interferon and cell surface antigens. The genetic engineering of mammalian cells is introduced briefly.

II. CELL PHYSIOLOGY AND CULTURE

A. Mammalian Cell

The complexity of the structure, chemical and physical
composition of the mammalian cell has hindered its nearly
complete characterization until recently.(8-14) An indication of such difficulties is that it was only recently, in
1956, that the correct number of chromosomes in the human
cell was determined to be 2N=46 instead of forty-eight as
previously thought for a long time.(15)

B. Cell Components

The one-dimensional arrangement of the major components
of the mammalian cell is given in Figure 1.(16) Nearly all
the major components are separated and compartmentalized via
thin ($\sim$4-5 nm) bilayer phospholipid membranes. The nucleus
is enclosed by a series of such membranes nearly connected
but separate. It contains the various components (proteins,
DNA, various types of RNA) which are necessary for the replication of the cell via the four cell cycle phases: pre-
DNA synthesis (G_1); DNA synthesis (S); post-DNA synthesis
(G_2); and mitosis (M) as shown in Figure 2 for a mouse hepa-
toma cell. (9,12,17,19) The plasma or cell membrane which
surrounds the cell is the most important component with respect to its communication with the environment. It is a
phospholipid bilayer about 4-5 nm in thickness and is extensively folded with many invaginations. Thus, it has a
very large surface area to facilitate the transport of several hundred, if not thousand different type molecules that
enter or exit the cell for its various functions including
its replication. The Singer-Nicolson fluid mosaic plasma

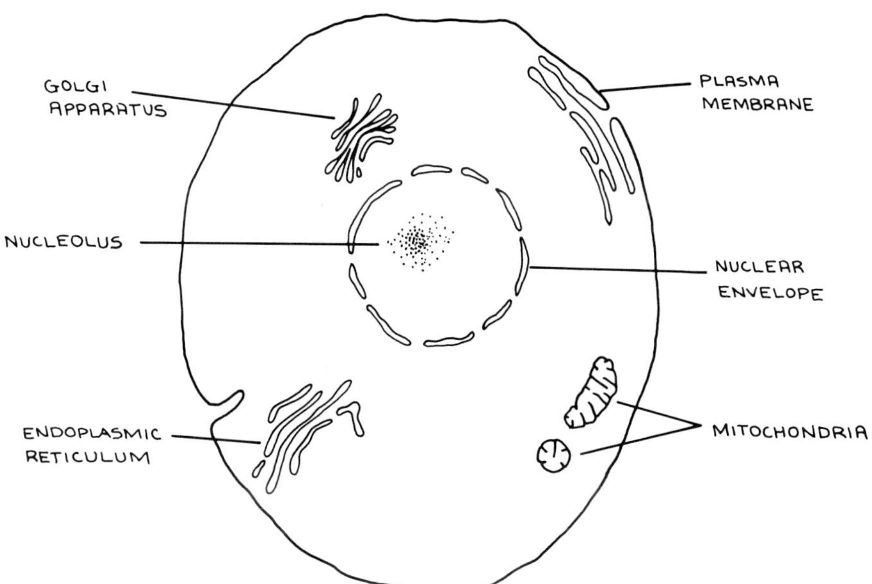

Figure 1. Major Components of Mammalian Cell

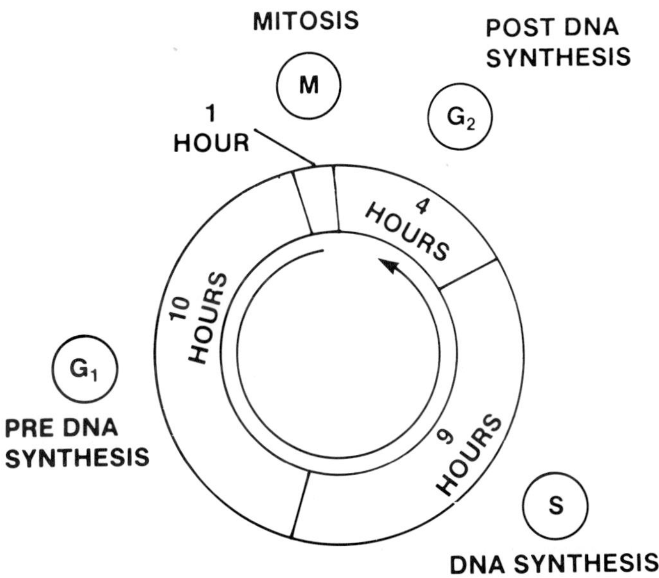

Figure 2. Typical Mammalian Cell Replication Phases

membrane model is illustrated schematically in Figure 3 show-
ing some of the types of molecules found there, but in a
dynamic state.(17) These molecules (proteins, cholesterol,
etc.) play key roles for the various functions of the cell
itself.

C.   Cell Environment

The shape of mammalian cells varies widely from a spherical to a tubular and other shapes depending on its particular differentiated function as shown diagrammatically in Figure 4 for some cells.(9,20-26) Internally, there is a cyto-
skeleton network of microtubules and filaments which support
the various parts of the cell such as the nucleus, mitochon-
dria and plasma membrane (21,27-30) and change the shape of
some cells from nearly flat to nearly spherical during the
mitosis phase in its replication cell cycle.(9) The size and
weight of mammalian cells varies widely as indicated in Table
1 and compared with several other biological specimens. The
cells of interest used for cell culture have a characteristic
dimension of about 10-30 μm and weigh about 2-5x 10^{-10} g.
The chemical composition and physical parameters of the extracellular and intracellular fluid differ considerably with respect to concentration as shown in Figure 5 for only a small
portion of the various molecules for a human cell as, for example, the pH is 7.4 and 7.0 and Na^+ concentration is 142
meq/liter and 10 meq/liter respectively. The normal functional temperature of the cell is 37°C. Control of both chemical and physical parameters of the coenvironment within
physiological limits (8,31a-33) are critical for the various
cell functions including its replication.

All the differentiated cells of a given mammalian
species contain all and the same genetic information regardless of their individual function. About 200 different distinct cell types exist in adult humans.(8) *In vivo* stem
cells differentiate into mature cells which have different
characteristics and functions and thus make up as a collection the individual within a given species, (9) e.g., the
chicken egg which is a very large, single cell (Table 1),
after embryonation differentiates into a whole chicken which
is comprised of a number of mature cells with different
functions. Some of these mature cells can replicate *in vivo*
to produce identical cells, some can secrete certain molecules and others have certain molecules on the plasma membrane as required by the organism for its normal function.
(8,9)

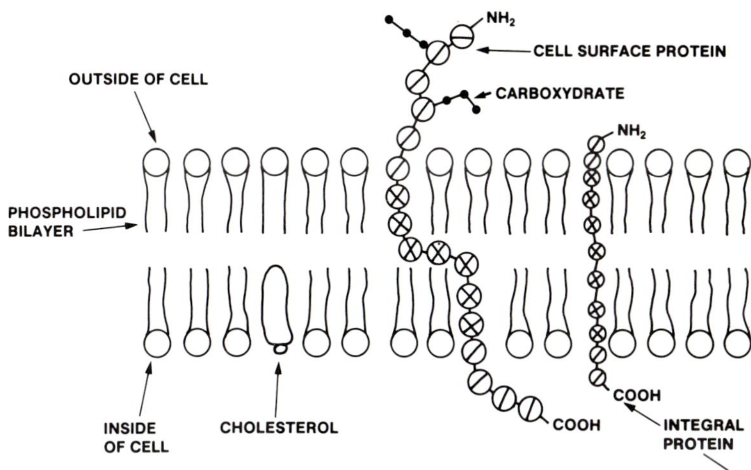

Figure 3. Plasma Membrane of Mammalian Cell with Typical Components

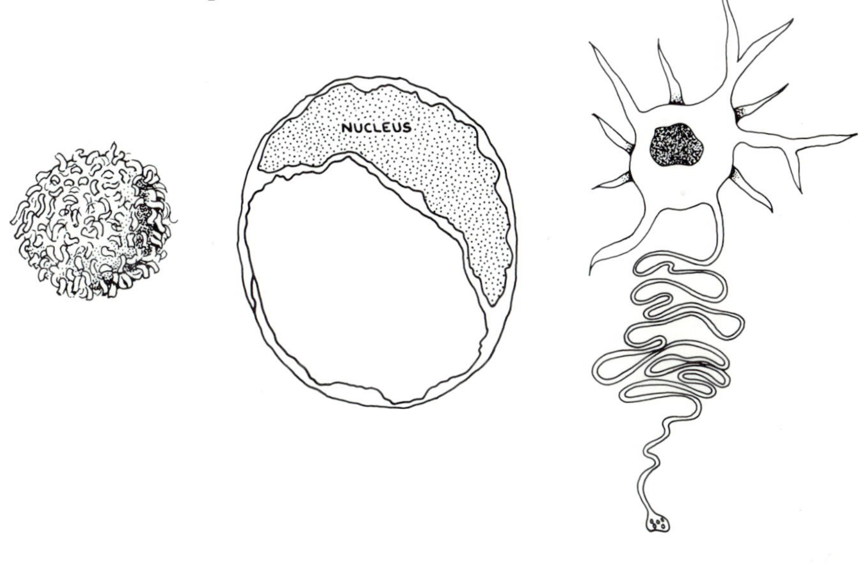

Figure 4. Three Examples of the Two-Hundred Different Types of Differentiated Cells in the Developed Body of Vertebrates: a) Lymphocyte, b) Endothelial Cell, c) Neuron

*Table 1. Mass and dimensions of various biological speci-
          mens (8,10a,10b)*

Specimen	Mass (g)	Greatest Dimension nm	Characteristic Dimension nm
Tobacco necrosis viron	10^{-18}	20	10
Mycoplasma laidlawii	10^{-16}	~50	~50
Mycoplasma gallisepticum	10^{-14}	250	250
Plural pneumonia-like organism (PPLO)	$\sim10^{-14}$	250	100
Mature E. coli bacterium	10^{-12}	2,000	1,000
Red blood cell (human)	10^{-10}	8,000	4,500
Smooth muscle cell	10^{-7}	500,000	45,000
Liver cell	10^{-7}	50,000	50,000
Most mammalian differentiated cells	$\sim10^{-9}$	--	10,000 to 30,000
Large animal eggs	$10-3x10^{3}$	--	2-20 cm

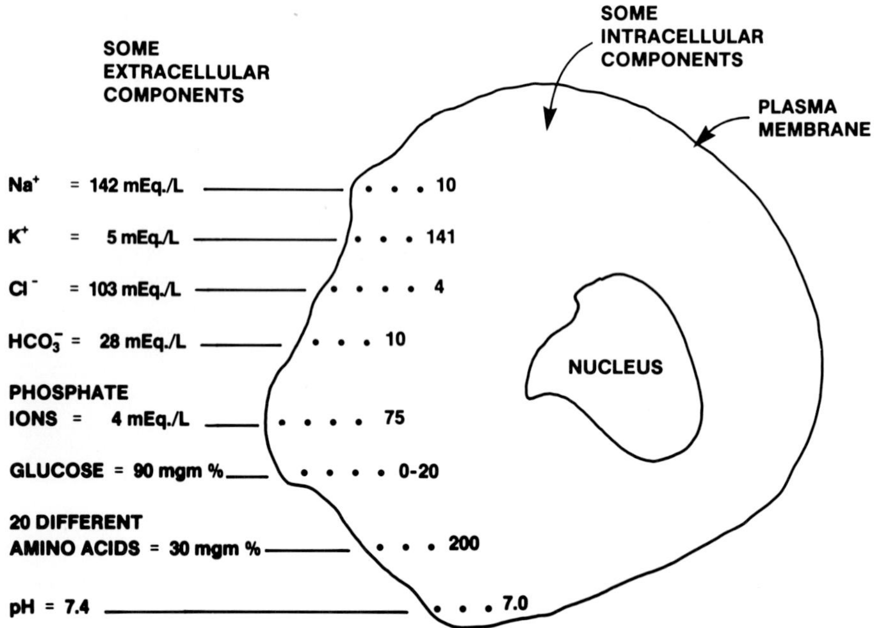

*Figure 5. Only a Small Fraction of the Components of the
           Intracellular and Extracellular Fluids (Modified
           After Guyton (Refs. 31a-32))*

D.   Culture of Mammalian Cell

     Ross G. Harrison is the inventor of mammalian cell cul-
ture in tissue form *in vitro* (34) who while at Johns Hopkins
University in 1907 reported that bits of a frog embryo did
not die when sealed in a closed chamber in lymph fluid from
an adult frog; rather they grew and differentiated similarly as normal embryonic development.(35) In 1912 and
later, Alexis Carrel, Rockefeller Institute, improved the
technique in two important ways: a) fed the cells in the
form of tissue with fresh growth media;(36) b) used a closed
flask with a curved entry port upwards to grow the cells instead of the hanging drop method on a microscope slide.(6,
34,37a) During this same period, Rous and Jones working in
the same institution grew single chick cells in culture as
simple microorganisms. (37b) In the late 1940's and early
1950's Earle, Gey, Eagle and Puck developed more appropriate
techniques and methodology to grow mammalian cells either as
small clumps of several cells or as single cells.(38-44)

E.   Cell Types in Culture

     In the early 1960's Hayflick, Wistar Institute, studied
in painstaking detail the multiplication of mammalian cells
in culture and established the cell strain and cell line
generation diagram as illustrated in Figure 6.(45) A small
amount of normal tissue (<1 gram) excised from various
mammals is dissociated into single cells using the proteolytic enzyme trypsin, (43,44,46) and then such cells can be
placed in culture and are designated as primary culture or
Stage I (Figure 6). When the cells double in number after
a certain time in culture, usually between 10 to 50 hours,
the first generation or first subcultivation is established.
Thus the cells go in Stage II (Figure 6) giving rise into a
cell strain which can be subcultivated up to about 50 generations when senescence occurs (Stage III, Figure 6) for a
continuous culture time of about ten months. Hence, the
given cell strain which comes from a primary culture has a
limited life span, but an enormous total number of cells or
total amount of material can be generated as shown in Figure
6, e.g., cells grown for 50 generations yield a total of
about $2x10^{7}$ metric tons of cell material.(47) The cell
strain or normal cell thus generated has the same morphological, physical and chemical characteristics as the original cell *in vivo*. Such cell strains are also called diploid indicating that for each chromosome there is a duplicate (2N). At any point in the cultivation of a cell strain
in Stage II, there may be an alteration of the cell either

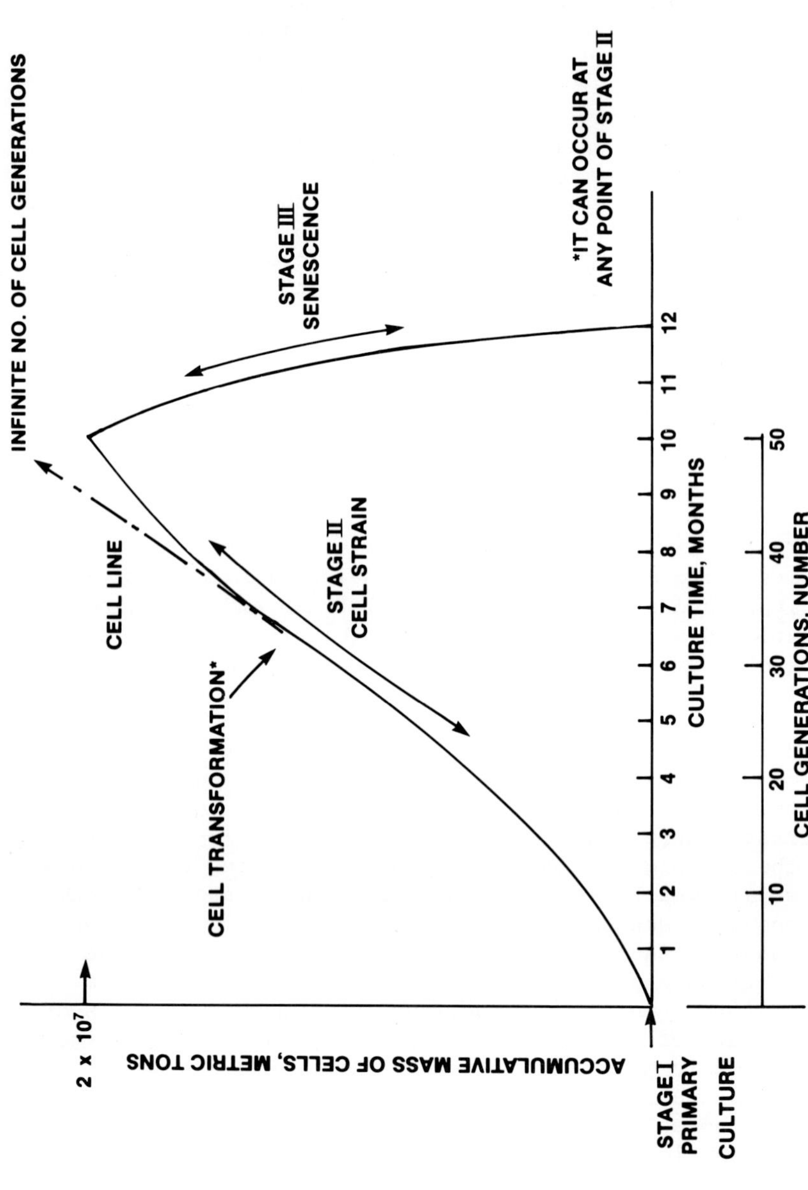

Figure 6. Cell Strain and Cell Line Generation Diagram (Modified after Hayflick (Ref. 45,47))

morphologically, physically or chemically which as a result
of this change gives rise into a cell line. The thus established cell line can be grown in culture for an infinite
number of generations. The cell line generates cells that
are all called either transformed, cancer or aneuploid indicating that all the chromosomes do not appear in duplicate
amongst many other alterations.

   Cell lines also can be generated directly from excised
tumor tissue as was done, e.g., for the HeLa human cell line
obtained from cervical carcinoma tissue.(39) A list of some
commonly used cell strains and cell lines is given in Table
2. Small quantities of cell strains and cell lines are frozen and stored in liquid N_2 in small vials (about one ml of
cell volume per vial) as early as possible with respect to
the number of gnerations after their establishment in culture
via a prescribed protocol.(48) Cell strains and cell lines
can be obtained from the American Type Culture Collection,
Rockville, MD,(49a) and other institutions and investigators. The same institute can re-characterize them for a fee
after culturing them for some time (six months to one year)
to ascertain that they still are axenic and have not been
altered.(44)

   Cell strains or normal cells have a tendency to grow in
culture by adhering to surfaces and as single layers of cells
with ordered orientation and stop to grow when they reach the
sides of the container as shown in Figure 7a. They also are
called anchorage-dependent cells.(8,9,49b) Cell lines or
transformed cells can grow both on surfaces and on top of
each other with random cell orientation (Figure 7b) and can
grow well in suspension either a single cell or as small
clumps of cells.(8,9,49b) Cell strains also can grow on
surfaces of small polymeric spheres such as dextrans which
can be suspended, and the cells grow well in such suspended
microcarrier surfaces.(1,4,50) Many other differences between cell strains and cell lines have been reported extensively elsewhere.(44,49b) Very extensive exploratory research has been done, especially in the last twenty years,
to establish the causes(s) for the alteration of a cell strain
into a transformed or cancer cell and significant progress
has been made but with no real definitive explanation (44,51,55).

Table 2. *Some common cell strains and cell lines (8,9,45, 86,149)*

Name	Species of Origin	Tissue of Origin	Ploidy
3T3	Mouse	Connective Tissue	Aneuploid
CHO	Chinese Hamster	Lung	Diploid
BHK21	Syrian Hamster	Kidney	Diploid
FS4	Human	Foreskin	Diploid
WI-38	Human	Embryonic Lung	Diploid
HeLa	Human	Carcinoma of Cervix	Aneuploid
BW5147	Murine	Thymus	Aneuploid

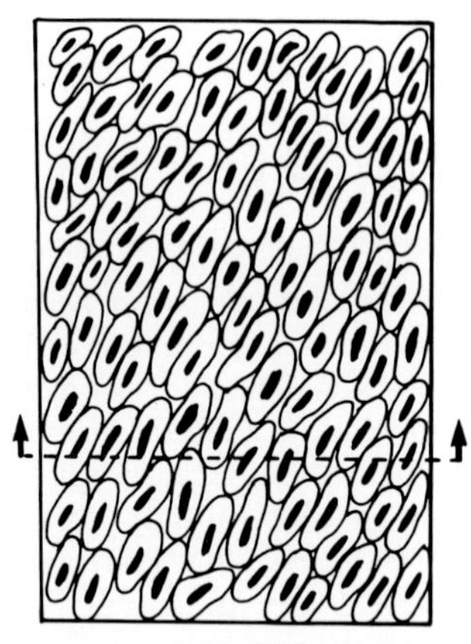

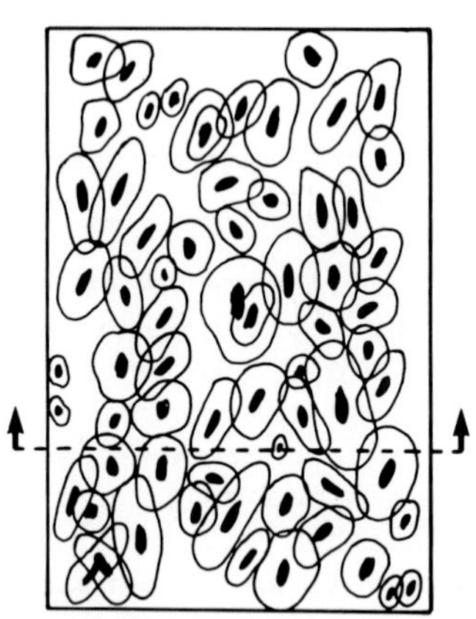

Figure 7. (a) *Normal Cell Growth from a Cell Strain on a T-Flask: It is a Monolayer and It is Oriented.*
(b) *Cancer Cell Growth from a Cell Line on a T-Flask: It is a Multilayer and It is Random.*

F.   Cells in Culture

Cell strains are cultured in plastic T-flasks (manufactured, for example, by Corning Glass Works, Corning, NY) which
are modifications of the Carrel flask as shown in Figure 8
and are available with surface areas from 25 to 175 cm^2. A
frozen vial (about 1 ml cells $\simeq 10^6$ cells) of a given cell
strain is brought to room temperature and quickly thawed in
a 37°C bath and added to a 75 cm^2 T-flask with 15-20 ml of
the appropriate cell growth media whose pH is 7.0-7.4.(56a)
It is placed in an incubator which is kept at 37°C, is 100%
humidified and is purged with air/CO_2 mixture whose composition is 95%/5% respectively.(56b) After two to three days
the 75 cm^2 bottom surface area of the T-flask becomes confluent but with only one layer of ordered cells which have
stopped growing at the edges of the bottom of the flask.(57-
59)

Similarly, cell lines or transformed cells can be grown
on the same T-flasks at the same conditions, but they can
grow on top of each other as discussed previously. Such
cells also can be grown more efficiently and effectively in
suspension in a spinner cell culture reactor first developed
in mid-1950's by McLimans,(60) Wistar Institute, as shown in
Figure 9 (manufactured, e.g., by Bellco Glass, Inc., Vine-
land, NJ). They are available in working capacity volumes
from 25 ml to 9 liters. The 25 to 50 ml reactors can be
inoculated from a frozen vial of cells and the appropriate
volume of fresh growth media added or from one confluent T-
flask after the first subcultivation and trypsin treatment.
Reactors with capacities up to 500 ml are usually inoculated
from several confluent T-flasks. Larger capacity reactors
are inoculated with cells grown in suspension from smaller
capacity reactors with the guiding principle to assure that
the inoculation cell density is more than 0.25×10^6 cells/cm^3
(3,44,61-68) in order to assure a minimum concentration of
cell growth factors which the cells secrete into the medium
and which are indispensable for their replication at their
optimum generation time. The inoculated reactor is either
statically or continuously purged with air/CO_2 mixture of
95%/5% and placed in a warm room which is kept at 37°C. It
is stirred magnetically but gently at a speed so that a
liquid vortex is just starting to be formed on the stirrer
shaft. It can be completely or partially harvested for
further propagation after one to four days depending on the
optimum generation time of the particular cell line. The
cell density of single cells is measured via a hemocytometer
(69) and reported in cells/cm^3. The cells can be separated

*Figure 8.   T-Flask Cell Culture Reactor for Anchorage-
Dependent Cells*

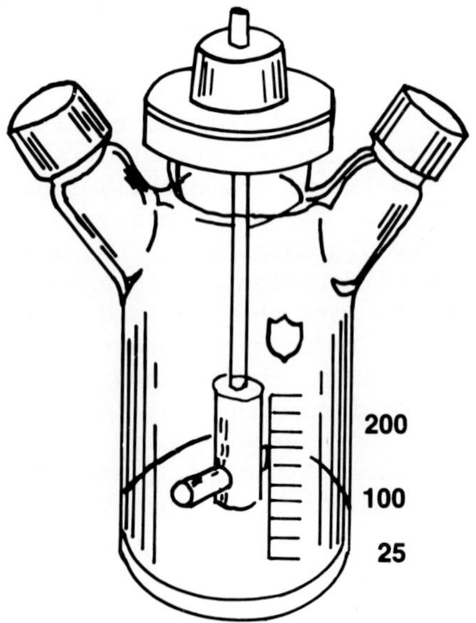

†U.S. Pat. No. 3,622,129

Figure 9.   Suspension Cell Culture Spinner Reactor

from the conditioned or spent growth medium via low speed
centrifugation usually 500-1,000 x g for 10-20 minutes.
For cells grown as small clumps a given volume of the culture is sampled and centrifuged in special pipet tips or
tubes which are calibrated for the given cell line and reported in ml cells/liter.(70,71) Cells in culture can grow
well and can be further propagated provided their viability
is >80% as established by trypan blue dye exclusion in which
dead cells take up the dye.(69)

The growth media for mammalian cells has a very complex
and not completely defined chemical composition because of
the presence of serum obtained either from fetal bovine,
calf, horse, etc. (manufactured, e.g., by K. C. Biological,
Lenexa, KS; and Gibco Laboratories, Grand Island, NY). The
serum should be collected and processed from the respective
animals under very stringent conditions.(72) Serum is used
at a concentration of about 2-10% by volume.(3,33) Besides
serum, a chemically defined supplement is added to the growth
media first established in the 1950's.(40) The actual chemical composition of a very common and useful supplemental
growth media is given in Table 3, comprised of several amino
acids, salts, and vitamins for a total of about forty chemicals, and it is known as Dulbecco's MEM (minimal Eagle
media). In certain cases penicillin and streptomycin antibiotics are added to the medium at 100 units/ml and 100
µg/ml respectively. Dulbecco's MEM is manufactured by the
same suppliers as given previously for serum. $NaHCO_3$ is the
common and inexpensive buffer for mammalian cell culture. It
has two limitations, namely: 1) equilibrium thermodynamics,
partially offset by the 5% CO_2 purging, 2) its pK_a=6.1 but
the operational pH for cell culture is 7-7.4.(57,73) HEPES
(N-2-Hydroxyethylpiperazine-N'-2-ethane sulfonic acid) is an
alternate buffer which eliminates the $NaHCO_3$ problems with
exceptional buffering capacity at pH 7.0-7.4 (pK=7.31), but
it is prohibitively expensive for large-scale cell culture
beyond the T-flask or 500 ml spinner operating capacity.(57,
73) The osmolality of the complete growth medium should be
between about 290 to 320 mOsm/liter which can be checked via
a laboratory osmometer.

Mammalian cells in culture grow according to the bacterial cell growth phase curves established by Monod (74,3,5) as
shown in Figure 10 in six separate phases, φ: φ-1 lag phase;
φ-2 acceleration phase; φ-3 exponential phase during which
the cell density increases logarithmically with a linear increase of time in culture at a constant rate, and the slope
of the straight line curve establishes the generation, or

Table 3. Dulbecco's MEM, high glucose supplement mammalian cell growth media (40)

Components	Concentration mg/liter
INORGANIC SALTS	
$CaCl_2$ (anhydrous)	200.00
$Fe(NO_3)_3 \cdot 9H_2O$	0.10
KCl	400.00
$MgSO_4$ (anhydrous)	97.70
NaCl	6,400.00
$NaHCO_3$[a]	
$NaH_2PO_4 \cdot H_2O$	125.00
OTHER COMPONENTS	
Glucose	4,500.00
Phenol Red	15.00
Sodium Pyruvate	110.00[b]
AMINO ACIDS	
L-Arginine HCl	84.00
L-Cystine $\cdot$ 2HCl	62.59
L-Glutamine	584.00
Glycine	30.00
L-Histidine $HCl \cdot H_2O$	42.00
L-Isoleucine	104.80
L-Leucine	104.80
L-Lysine HCl	146.50
L-Methionine	30.00
L-Phenylalanine	66.00
L-Serine	42.00
L-Threonine	95.20
L-Tryptophan	16.00
L-Tyrosine	72.00
L-Valine	93.60
VITAMINS	
D-Ca Pantothenate	4.00
Choline Chloride	4.00
Folic Acid	4.00
i-inositol	7.00
Nicotinamide	4.00
Pyridoxal HCl	4.00
Riboflavin	0.40
Thiamine HCl	4.00

[a] Added during media mixing at a concentration of 3700 mg/liter

[b] Optional component.

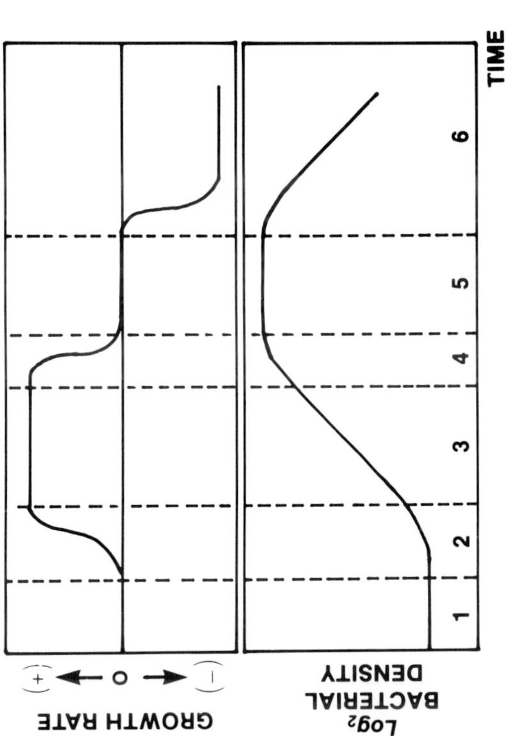

Figure 10. *Monod Cell Growth Phase Diagram (Ref. 74).*

doubling time of the given cell line; ϕ-4 retardation phase;
ϕ-5 stationary phase; ϕ-6 declining or senescence phase. The
length of time for each phase and the level of maximum cell
density that can be achieved are strong functions primarily
of the appropriate physiological environment such as temperature, pH, growth media, dissolved O_2 level in the growth
media, low shear for suspension cells and initial cell density in the culture reactor. The Monod diagram is an indispensable tool for the quantification for mammalian cell
culture and such measurements, i.e., change in cell density
with increasing time in culture at appropriate intervals
should be made for each cell culture experiment. The cell
culture products and other characterization of a cell culture experiment may be a strong function of the growth phase
of the culture at the time of product harvest or measurements. (33,75)

Products from mammalian cell culture are primarily used
for therapeutic purposes. The acquisition of such products
in some cases is via cell secretion into the growth medium
during culture or under other conditions, (76,77) and in
other cases via special processing from the various cell
components such as the plasma membrane. (75,78,79) The
number (8,80a,80b) of molecules per cell for a given desired
product is in the order of 10^4 to 10^6. Hence, the paucity
of the product that can be obtained from mammalian cells requires that both normal and cancer cells be produced in large
quantities via large and technologically efficient systems
which will be discussed in the next section.

III. TECHNOLOGY

The paucity and high cost of a given product acquired
from mammalian cells either in terms of per cell or per unit
volume of culture are a direct function of three inherent
characteristics of the mammalian cell and its replication
requirements in culture: a) limited number of product
molecules of about 10^4 to 10^6 per cell as discussed previously;
b) long replication or generation times in the range of 10 to
50 hours, thus dictating either large anchorage surface area
for normal cell strains or large reactor volume for cancer
cell lines for the respective type of cell culture reactors
requiring special mechanical design both for low shear for
cell agitation and high degree of sterility to keep the
culture axenic for long reaction times; and c) the expensive,
limited supply and quality of serum required for their optimum
growth (2,4,5,33). Some developments in the genetic engineering

of mammalian cells during the last three to four years may
possibly solve some of these limitations. The technology
requirements and developments for the culture of both anchorage
dependent normal cells grown on either fixed or suspended microcarrier surfaces and cancer cells grown in suspension at the
industrial scale will be summarized in the sections that follow.

A. Raw Materials

     1. <u>Water</u>. The most important element for successful
mammalian cell culture is the quality of water used both for
the preparation of equipment and cell growth media. High quality,
distilled water prepared from deionized (D.I.) water (34,72), or
Super-Q^R-Millipore water prepared from deionized water followed
by reverse osmosis quality water (3,33,81) with a resistivity of
18 megaohms must be used for cell growth media preparation.
Also provisions must be made to insure the D.I., reverse osmosis,
stills, and Super-Q^R systems are properly maintained and checked
on a daily basis. Steam used to sterilize various types of
equipment either via autoclaving or in-line must be prepared
with similar quality water as that used for the growth media
and via special steam generators (manufactured, e.g., by AMSCO,
Erie, PA). Common laboratory or plant steam must not be used
since it contains antirustants and antioxidants which are
detrimental for mammalian cells. Similar water quality must be
used in washing and rinsing all process equipment. To keep the
endotoxin or pyrogen level low (<0.1 ng/ml) the final polishing
step for water preparation should be with 10,000 daltons ultrafiltration membranes which retain such polysaccharides, and the
holding tanks must be drained via a bottom drain on a regular
basis.

     2. <u>Serum</u>. Next to water, the quality of the various types
of animal serum is the most important ingredient for successful
mammalian cell culture. Serum is processed by a limited number
of reputable suppliers and special attention should be given
for selecting such suppliers for a given application. Some of
the serum qualities have been discussed previously. Additional
quality control requirements by the suppliers are as follows
for each lot of serum which may be up to about 500 liters
aliquoted in 500 ml containers: a) final sterile filtration
through a double 0.1 μm filter (manufactured by Pall Corporation,
Glen Cove, NY) and with a pressure drop not to exceed about
10 psi; b) minimum sterility testing for bacteria, yeast and
fungus in the growth media, e.g., blood agar, thioglycolate
broth, Sabouraud's dextrose slant (83,84) both at room temperature and 37°C for up to four weeks; c) mucoplasma testing (85);

d) virus screened or tested depending on the particular use of
the product; e) bacteriophage testing; f) efficacy of cell
growth on at least one cell strain on T-flasks and compared to
a control serum lot and a control lot of supplemental medium
usually Dulbecco's MEM; g) total protein content and immumo-
globulin electrophoresis pattern which is especially important
for fetal bovine serum; h) pH after filtration should be 7.2-
7.8; i) osmolality which should be about 290-310 mOsm per
liter; j) endotoxin or pyrogen level should be less than about
2 ng/ml by the LAL (limulus amebocyte lysate) test (manufactured,
e.g., by Mallinckrodt, St. Louis, MO); the measurement indicates
the level of aseptic conditions used to collect the serum from
the animal; k) for some special cases the serum is heat deactivated at 56°C for 30 minutes to eliminate the C3b complement
component before sterile filtration, thus eliminating possible
cell membrane lysis (49b,86); l) the hemoglobin level should
also be measured and should be less than 20 mg%; it indicates
the care given in the collection method and separation of red
cells from the serum. Depending on theparticular application,
donor herd serum is probably the best choice to assure absence
of any organisms in it (72,82). The quality control of a lot
of serum usually takes 30-40 days. Usually serum lots are
pretreated by the user for efficacy for the particular cell
and kept in reserve by the supplier.

   3. <u>Supplemental Media and Buffers</u>. Dulbecco's MEM is an
example of chemically defined supplemental media; $NaHCO_3$ and
HEPES are common buffers used in cell growth media as previously
discussed.

   4. <u>Defined Media</u>. During the last twenty years many
attempts have been made to substitute serum in the growth media
with chemically defined substances such as hormones, proteins,
lipids, etc. Some progress has been made towards that goal for
some cell strains and cell lines grown for research purposes
as discussed elsewhere (87-90).

   5. <u>Process Gases</u>. Air used for cell culture should be
oil free and dry (-70° F dew point). CO_2 and O_2 can be used
from standard cylinders, but old CO_2 cylinders may have some CO
which is detrimental to cells.

   6. <u>Sterility and Endotoxin Tests</u>. The standard sterility
tests for mammalian cell culture are a) blood agar, b)
thioglycolate and c) Sabouraud's slant at room temperature and
37°C (83,84) for up to 7-28 days. The standard endotoxid test
is the LAL (limulus amebocyte lysate).

B.   Sterilization

   1. <u>Process Equipment</u>. All cell culture, growth media
equipment and media filter assemblies must be sterilized either
via autoclaving or in-line steam sterilization in order to
achieve axenic culture. The minimum autoclave steam sterilization conditions are 121°C at 100% saturated steam (20 psig)
for 15-20 minutes. For large industrial equipment for in-line
steam sterilization, the conditions are 129°C at 100% saturated
steam (30 psig) for 30 minutes. All three conditions must be
met simultaneously to achieve sterility (3,91,92a,92b). It is
important that proper provisions be made to remove the air from
the interior of equipment placed especially in autoclaves and
before the sterilization cycle starts; otherwise the minimum
121°C will not be reached (92a). Sterilization via steam is
the only assured method to achieve sterility since under the
above specified conditions the replication enzymes of the
various organisms are deactivated via their hydrolysis (91).

   2. <u>Growth Media</u>. In an appropriate capacity vessel
(Pyrex glass or 316 stainless steel and, if possible, 316L),
high quality water (distilled or Super-Q^R) of the desired volume
is mixed well with Dulbecco's MEM powder, $NaHCO_3$ and desired
volume percent serum and the pH is adjusted usually with 0.1 N
HCl about 0.3 units of pH below the operational level. The
mixture is sterile filtered through a series of prefilters and
then through 0.1 μm porosity membranes (manufactured by Pall
Corporation, Glen Cove, NY) either in flat or cartridge form
(92c,92d). Such prefilters and final filters in their respective housings (manufactured, e.g., by Millipore Corporation,
Bedford, MA for flat filter assemblies and Pall Corporation
for housings of cartridges) are steam sterilized either via
autoclaving or in-line steam at the conditions specified
previously. During filtration, the pressure drop across the
filters should not exceed 10 psi. The sterile medium is
pumped into a sterile vessel and is stored in a room which is
kept at 4°C. Samples are taken from the filled vessel under
appropriate sterile conditions and tested for sterility as
discussed previously.

   3. <u>Aseptic Methods</u>. Mammalian cells starting from
excising tissue from animals for establishment of primary
culture or tumors must be handled under clean and sterile
conditions in order to achieve axenic cultures. Particular
operational problems can be encountered in the initiation of
starting cultures for the establishment of cell strains and
cell lines and their vailing for storage in liquid N_2, and
subsequently in the initial propagation in either T-flasks or

cell culture spinners of small volume (< liters). Such culture
vessels are either sterilized by the manufacturer (T-flasks) or
prepared and sterilized via autoclaving in the laboratory.
Such systems have to be opened to inoculate the appropriate
volume of culture and growth media. Hence, such vessels are
handled in Class II, Type A, laminar flow hoods, as shown in
Figure 11 (manufactured, e.g., by Baker Company, Sanford, ME)
which are equipped with HEPA filters (high efficiency particle
accumulation) which remove all particles and organisms >0.3μ
with a 99.999% efficiency, thus providing a clean environment,
but which is not sterile (93). Such hoods are placed in areas
with no air currents to preserve the front laminar flow air
curtain. The inlet room air is also cleaned with similar HEPA
filters. More stringent conditions can be achieved using
similar but glove-operated hoods (94). In addition, to operate
at as clean as practical conditions, surgeon-type hats, cover
shoes, uniforms and sterile rubber gloves must be used by the
operators. The hood surfaces are rendered sanitized by cleaning
them carefully with 70% ethanol-water solution and the vessel
ports, caps, pipettes, etc. are flamed with blue natural gas
flame for sufficient time until all the condensed moisture is

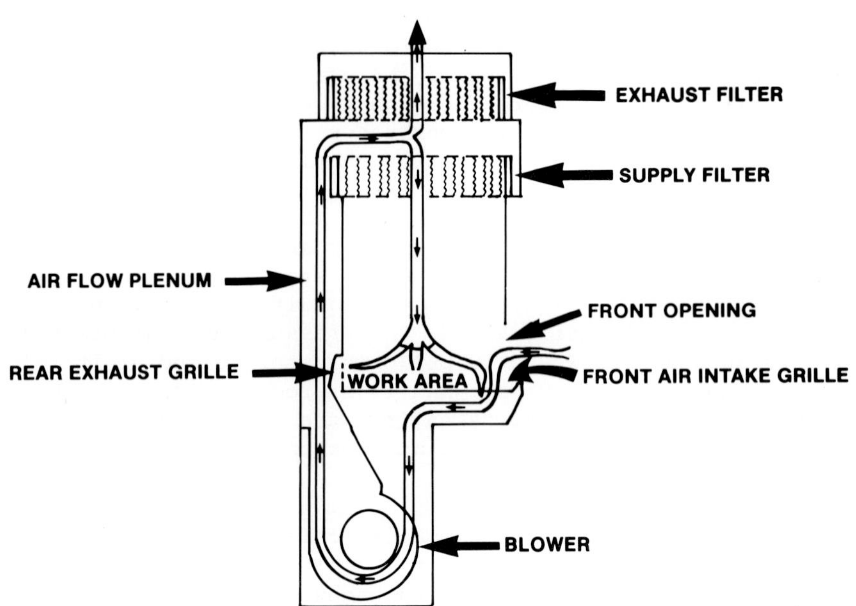

*Figure 11. Sketch of Biological Safety Air Laminar Flow Hood
with Inlet Air and Exhaust Air HEPA [High Efficiency
Particle (and Organism) Accumulation] Filter of
0.3 μm Porosity and 99.999% Efficiency*

completely removed. Recently, some mini-hoods were developed
to make aseptic culture and media vessel connections (95). The
conditions achieved via such open systems are aseptic and not
sterile, but with appropriate care sterile cultures can be
initiated and with good reproducibility (33).

C.   *Industrial-Scale Mammalian Cell Culture*

The gigantic advances in the area of molecular biology of
the cell and the gene (8,9,96-106) since the early 1970's have
provided the feasibility to isolate a number of proteins and
other biomolecules from mammalian cells which can be applied
as very effective therapeutic entities with no adverse effects.
Common examples of such protein products from mammalian cells
are the histocompatibility transplantation cell surface antigens
(107,108), interferons (109-112), and lately neurotransmitters
or neuromodulators (113) and T-cell growth factors of thymocyte
cells (114,115) and other lymphokines (116). To proceed with
in vitro evaluation, characterization and isolation of such
factors from mammalian cells (normal or cancer), large quantities
of cultures of such cells and conditioned media or supernatant
are required (2,6). Larger quantities are required for *in vivo*
testing of essentially pure factors both in animal studies and
clinical evaluation (111,112). Large-scale cell culture systems
for growing both anchorage-dependent cells and suspension cells,
along with major cell growth parameters and transport phenomena
dynamics are described and examined.

D.   *Suspension Culture*

An integrated and fully automated industrial-scale
suspension culture center with a 70-liter and 200-liter working
capacity cell culture vessels is shown schematically in
Figure 12 which is described in detail elsewhere (2,3,5,6,117,
118). Other culture vessels with large volume capacity previous
to this system have been reviewed (3). The system can be
divided into three main areas: a) cell growth medium preparation,
b) cell culture and c) culture separation and harvest into cells
and conditioned media or supernatant as shown in Figure 13.
Other ancillary areas are: cell culture quality water preparation; steam generation; compressed air; and a biohazard containment system for both spent-culture liquid deactivation by high
temperature (kill tanks) and spent culture gas deactivation by
incineration as shown in Figure 14 (3). The complete system
(Figure 12) is *in situ* steam sterilizable (30 minutes/100%
saturated steam - 30 psig, 129°C) with in-line, continuously
flowing steam except for the two centrifuges which are rendered

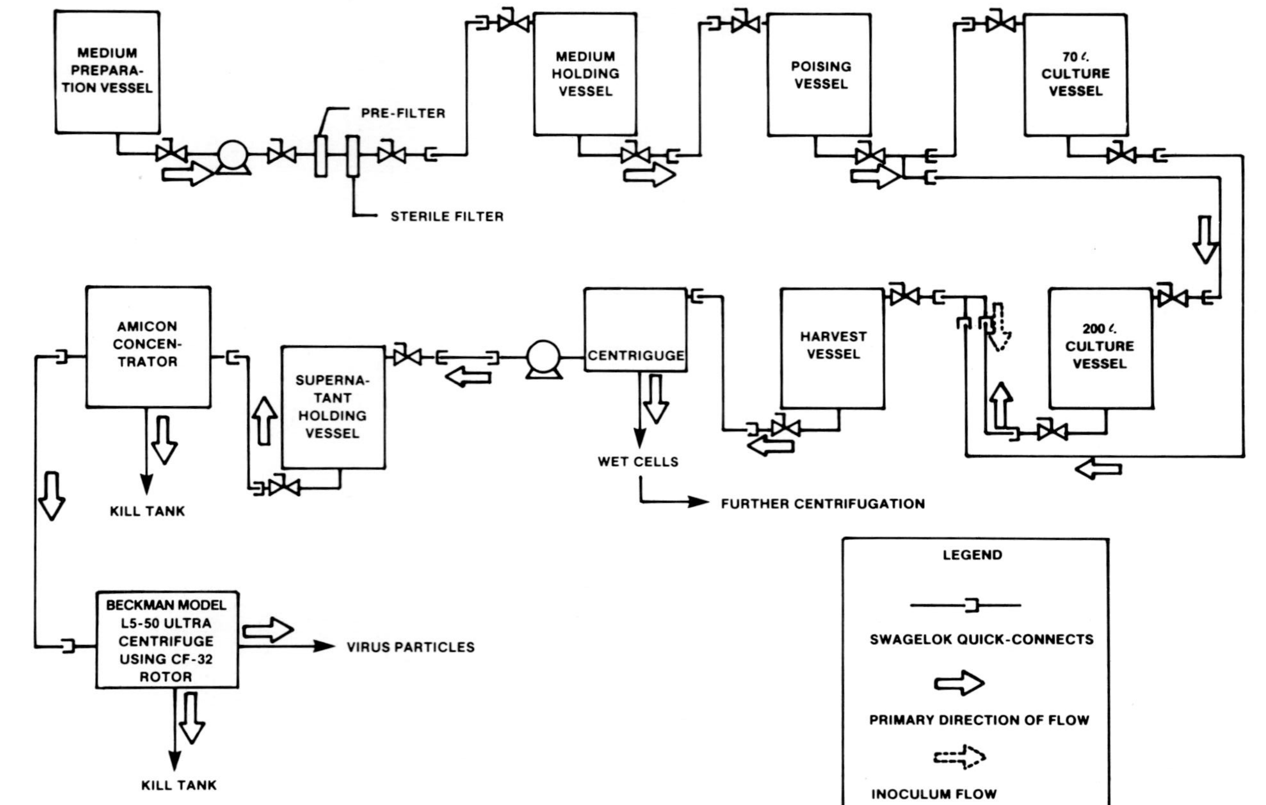

Figure 12.   Schematic Diagram of Integrated Suspension Mammalian Cell Culture Center (Ref. 5).

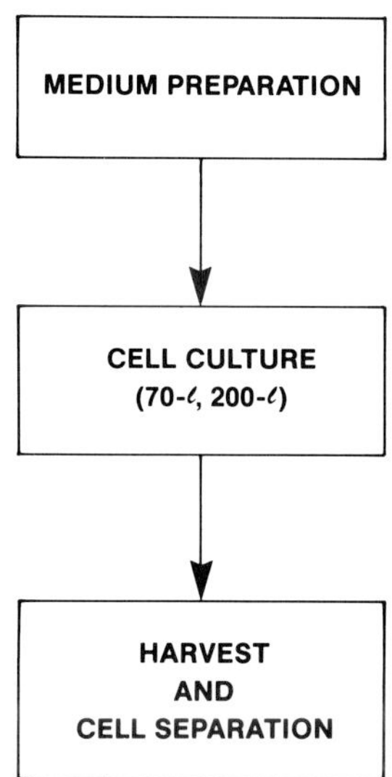

Figure 13. *Major Divisions of Integrated Suspension Mammalian Cell Culture Center (Ref. 5)*

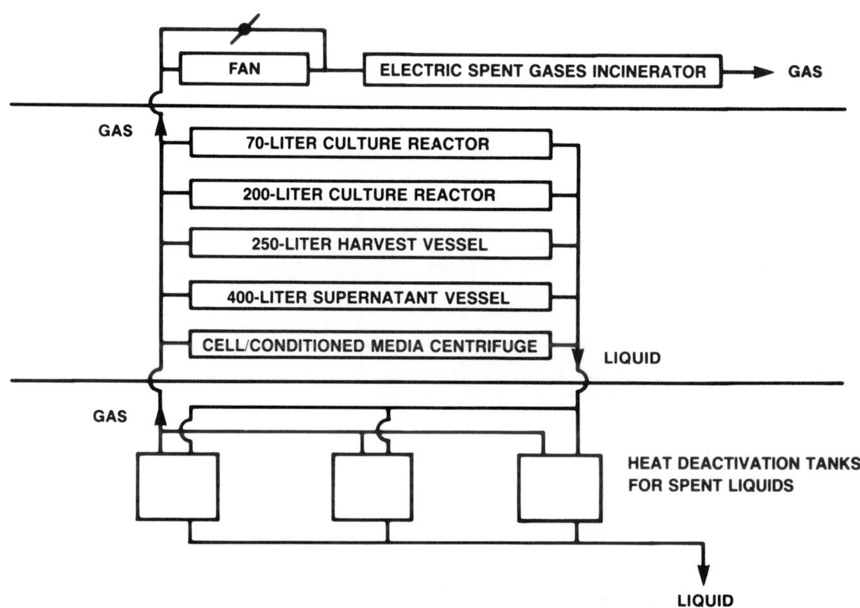

Figure 14. *Waste Liquid and Exhaust Process Gas Biohazard Containment Systems (Ref. 2,3,5)*

aseptic via other methods (2,118). The Amicon ultrafiltration
hollow fiber membranes are rendered aseptic with 100-200 ppm
NaOCl aqueous solution treatment via circulation for about 20
minutes (119). The media mixing tank and pump are rendered
sanitized with hot water and Super-Q^R water rising before
starting media preparation. All vessels were stirred via
marine-type impellers which provide excellent liquid-cell suspension agitation with low shear for the very fragile cells.
The impellers are driven via magnetic coupling, thus eliminating
any vessel shaft penetrations to achieve axenic cultures (3,118,
120,121). New Brunswick Scientific Company (Edison, NJ)
manufactured the system. The total system has been demonstrated
operationally successful in fed-batch mode with no contamination
problems with large culture volume productivity and high cell
density (3-4x10^6 cells/cm^3) of high viability (>93%) and low
material and operational costs (2,3,5). Similar systems have
been installed by the same manufacturer both in U.S. and
Japan (122-125).

The medium preparation section consists of three vessels
and an in-line steam sterilizable Pall cartridge filtration
system as shown in Figure 15. The final sterile filter should
be substituted now with 0.1 µm which was the smallest, available
porosity available in 1976. The working volumes of the medium

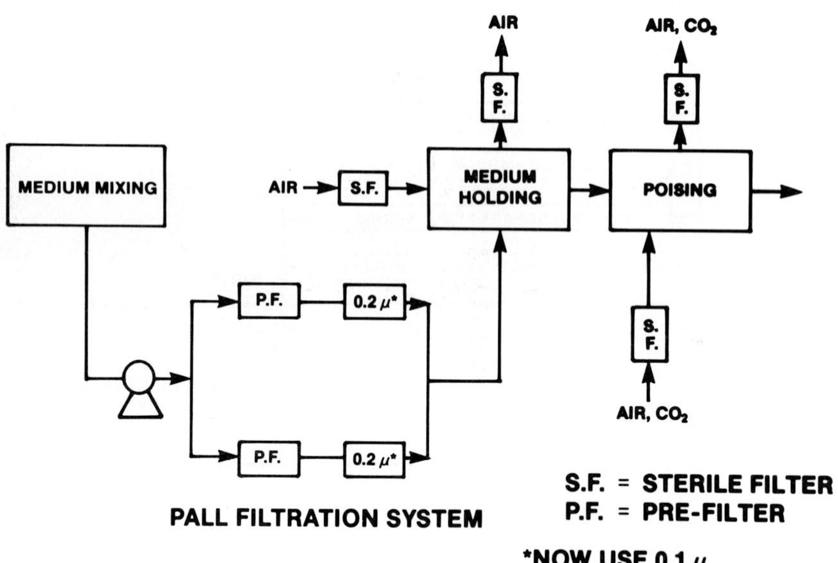

Figure 15. Cell Growth Medium Preparation System (Ref. 3,5)

preparation, medium holding and medium poising vessels are 500, 500 and 250 liters, respectively. About 500 liters of media can be totally prepared in a period of about four hours. The medium holding vessel is kept at 1-2°C when medium is stored in it. Thus, its various components such as serum proteins, amino acids, etc. are prevented from being deactivated. The poising vessel is used to condition the medium to the pH and temperature used in the culturing vessels before it is transferred to them. In this way, when fresh medium is transferred to the cell culture vessels, the pH and temperature of the culture are not disturbed. CO_2 is used to adjust the pH of the medium to the desired level in the poising vessel. The two culture vessels of the system have working volumes of 70 and 200 liters, as shown in Figure 16, with the various control systems. The vessels can be operated either simultaneously or in sequence. The operating conditions for the BW5147 cancer cell line were as follows: 1) temperature 37°C, 2) pH was controlled at 7.0 ± 0.01 by CO_2 and 1 M Na_2CO_3 solution which was fed via a peristaltic pump, 3) sparged air: 70-liter vessel with about 2 liters/min and 200-liter vessel with about 7 liters/min which results in dissolved oxygen, $\underline{O}$, levels between 30-90% saturation, 4) vessel pressure, 4 psig, 5) agitation for either vessel ∿90-100 R.P.M., 6) vessels were harvested every 22-23 hours and the amount of harvested culture is about 50-60% of the working vessel volume. The 70-liter vessel was inoculated from a 12-liter laboratory culture vessel; the]00-liter vessel was inoculated from the 70-liter vessel. All media and culture transfers are made via $1\frac{1}{4}$-inch diameter flexible Teflon lines interconnected to the vessels via appropriate SWAGELOK[R] fittings and via sterile air pressure differential. The operational temperature for heating or cooling of all vessels, except the media mixing tank, is controlled via an ethylene glycol which is circulated in the peripheral jackets of each vessel (2,3,5,118).

For successful cell culture propagation in suspension from a frozen vial to such large 70- and 200-liter vessels, particular attention must be given at each intermediate stage to assure that critical culture parameters are within good operating limits. Suck key parameters are: a) sterility to achieve axenic culture, b) temperature to be 37 ± 0.1°C, for a higher temperature will destroy the cells and temperatures lower than 35°C will put cells in the lag phase (Figure 10); c) pH in range of 6.8 - 7.2 for the BW5147 cell line used as an example and preferably controlled at 7.0 ± 0.01 for optimumun results; d) low agitation just to keep cells in suspension with good mixing within the vessel and thus to minimize shear rate, temperature, pH and concentration gradients of the various growth components including dissolved oxygen, O; e) cell inoculation density ρ_i, to be > 0.25 x 10^6 cells/cm^3 for the 12, 70

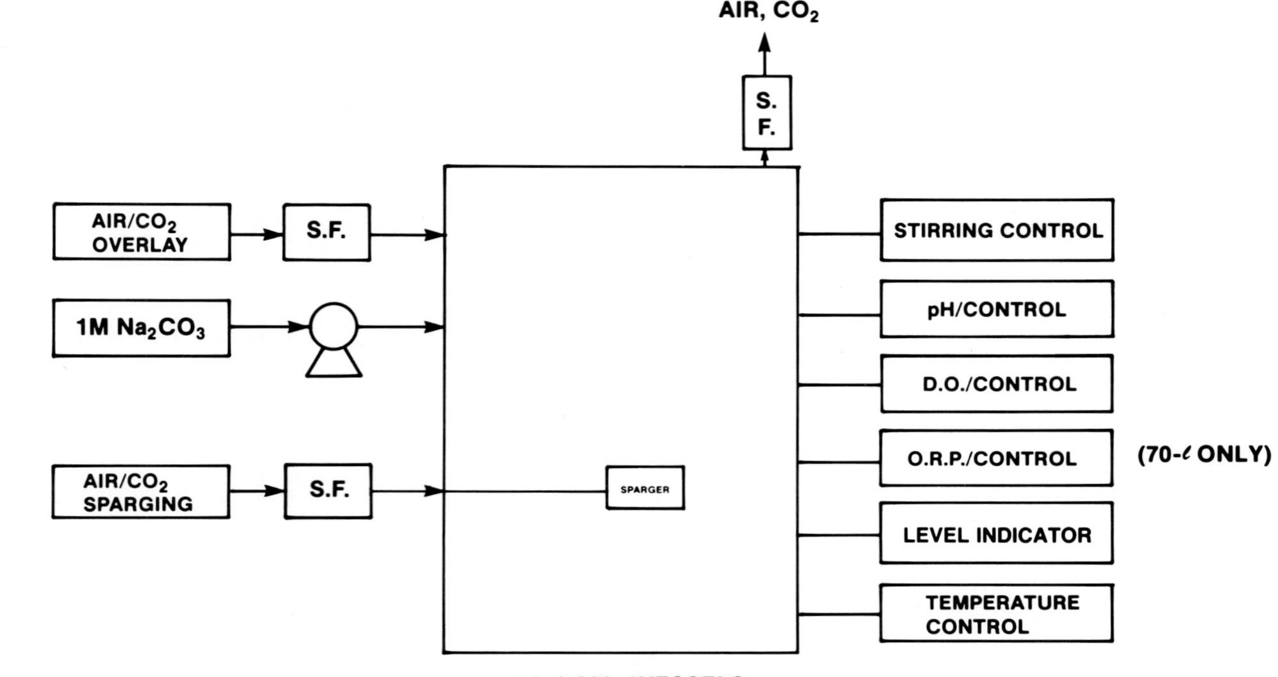

Figure 16. Cell Culture Reactor Instrumentation (Ref. 3,5).

and 200-liter and larger vessels in order to minimize the lag
time and maximize the cell growth time in the exponential phase
(Figure 10) and thus achieve low cell culture cost via high
cell harvest density, ρ_H, of about 3-5 x 10^6 cells/cm^3 and
faster turn-around of the culture for the next harvest in fed-
batch mode of operation, provided the culture is harvested
before reaching the stationary phase (Figure 10)(3,5,33). The
staging of cultures for upscale propagation has been elaborated
previously (3,84).

A summary of a sustained cell culture run with the 70- and
200-liter vessels operated simultaneously over a 42-day period
of the system shown in Figure 12 is given in Table 4. A total
volume of 4,715 liters of culture was collected by harvesting
the 70- and 200-liter capacity vessels 31 and 36 separate times
respectively by removing about 50-60% of the culture volume of
each every 22-23 hours on a daily basis. The total number of
cells produced was 1.11 x 10^{13} starting from approximately
1 x 10^6 cells from a frozen vial with corresponding wet cell
weights of about 10,000 and one gram. The 70-liter vessel was
operated in intermittent runs because the growth parameters
were changed for short studies. The run was terminated because
the air compressor of the cell culture center failed mechanically.
The relative economics demonstrated by this run are very good,
taking into consideration that: a) $1\frac{1}{2}$ man operators operated
the facility; b) 2% volume of either horse or fetal bovine serum
was used for 94% of the culture volume produced; c) total
working vessel volume was 230 liters since the vessels were
operated at 85% working volume capacity in order to allow enough
space above the liquid level for bubbles generated by the process sparged air to collapse and not be entrained with the
exhaust process gas and thus wet the exhaust sterile gas filters;
d) apparent average cell generation time was 18-20 hours instead
of the optimum 10-12 hours; e) vessels were not harvested at
optimum conditions before reaching the stationary phase (Figure
10) as has been elaborated previously for this and other systems
(3,5,33). The operational dynamics of this integrated cell
culture system have been discussed in detail elsewhere (3,5).

Other large-scale suspension bed-batch culture vessel
operation has been reported recently using vibromixing as a
means of stirring (33,126,127). Such culture vessels are
placed in 37°C warm rooms for cell growth and removed every 2,
3 or 4 days in 22°C rooms to be harvested, and then fed fresh
growth media which are usually kept either at room temperature,
4°C, or 37°C and the vessels are not instrumented (33). Some
laboratory size perfusion and continuous suspension systems have

Table 4. *Summary of sustained long-term suspension culture operation of an integrated and instrumented cell culture center (3,5)*

| | Culture vessel | |
	70-liter	200-liter
Total Running time, Days	37	38
Run Duration, Days	11,3,2,16,5	38
Culture/Medium Transfers, No.	44	39
Culture Harvested, No.	31	36
Culture Harvested, Liters	1,158	3,555
Cells Made, No.	2.79×10^{12}	9.38×10^{12}
$\langle p \rangle \times 10^{-6}$, Cells/cm^3	2.37	2.61
Range of $p \times 10^{-6}$, Cells/cm^3	1.41-3.85	1.53-3.56
$\langle$Cell Viability$\rangle$, %	92,44,83,97,96	97
Range of Cell Viability, %	0-99	91-99
Cell Growth $\langle 2X \rangle$ Time[d], Hours	18	20
Operation with 2% Horse Serum, Days	21	23
Operation with 2% Fetal Calf Serum, Days	6	10
Operating without P/S[b], Days	2	5
Cell line	BW5147[c]	BW5147[c]

[a] *Twelve batches of media were made and used during run for a total of 4,900 liters. At end of run, 200 liters of media were left in medium holding tank.*

[b] *P -Penicillin, S= Streptomycin*

[c] *Zwerner/Acton, J. Exp. Med. 142:378, 1975.*

[d] *Generation or doubling time. Optimum for BW5147 is 10-12 hrs*

been reported which show success in achieving cell densities
in the order of 10^7 - 10^8 cells/cm^3. Such high densities are
close to the theoretical maximum for close packing of spheres
of 10-20 μ diameter cells (128). The stability of operation of
such systems has been problematic although good progress has
been made recently (7b,57,129,130).

E.   Anchorage-Dependent Culture

     Cell strains or normal cells propagate for a limited
number of generations and require a surface to adhere and grow
as monolayers as discussed previously. Anchorage-dependent
cells besides being used to produce culture products are also
used extensively to replicate viruses for vaccine production
for both animals and humans (1,4,7b,8,44,57,131). To grow
large quantities of culture of such cells, equipment of
enormous size is required which occupies large space by using
either T-flasks or roller bottles which are shown in Figures 8
and 17 respectively (57). Significant progress has been made
since the late 1960's in developing several systems which have
large surface-to-volume ratio compared to roller bottles, namely:
a) stacked flat plates shown in Figure 18 (57); b) flat bed
hollow fiber reactors shown in Figure 19 (7a); c) microcarriers
as shown in Figure 20 which can be suspended and grown as suspended and grown as suspension culture in reactors stirred with
a paddle as shown in Figure 21 (1,4). Such systems have the
potential for industrial-scale normal cell production as already
reported for microcarriers (1,4,6,132-141).

     Microcarrier cell culture is a major development for the
efficient and economical production of normal cell products and
vaccines (1,4,139). van Wezel is the inventor of microcarrier
culture. In comparison to roller bottles, microcarriers offer
the following major advantages: a) working volume is decreased
by 100-fold; b) anchorage surface area is increased tenfold;
c) labor costs are decreased by about tenfold; d) contamination
rates are decreased by 50-fold (4). In general, about one
liter of microcarrier culture corresponds to about 50 roller
bottles (490 cm^2/bottle) for the same cell strain (4). CYTODEX[R]
microcarriers are made from dextran, and they are functionalized
with a number of different organic surface coating such as
N,N-diethylaminoethyl (DEAE) for specific cell adhesion with the
following physical characteristics: a) density, 1.03 g/ml; b)
average diameter, 180 μm; c) surface area, 6 x 10^3 cm^2/g dry
weight; d) number of beads, 6.8 x 10^6 per g dry weight; e)
swelling, 18 ml/g dry weight; f) microcarrier loading for cell
culture, 3-5 mg dry microcarrier per ml of final culture (4,140).

Figure 17.   Cell Culture Roller Bottle Reactor for Anchorage-
Dependent Cells

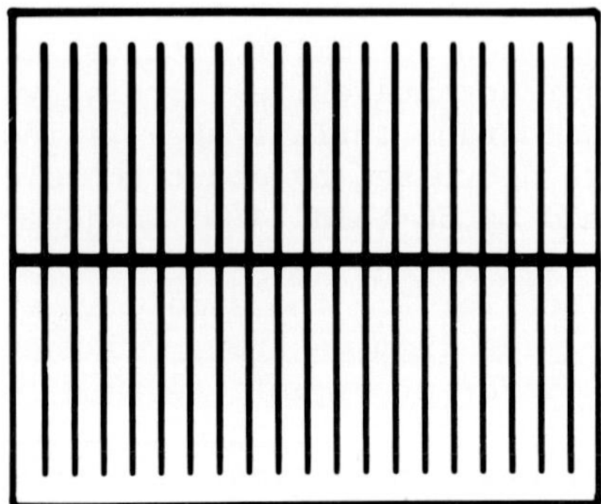

Figure 18.   Stacked Plate Cell Culture Reactor for Anchorage-
Dependent Cells (Ref. 57)

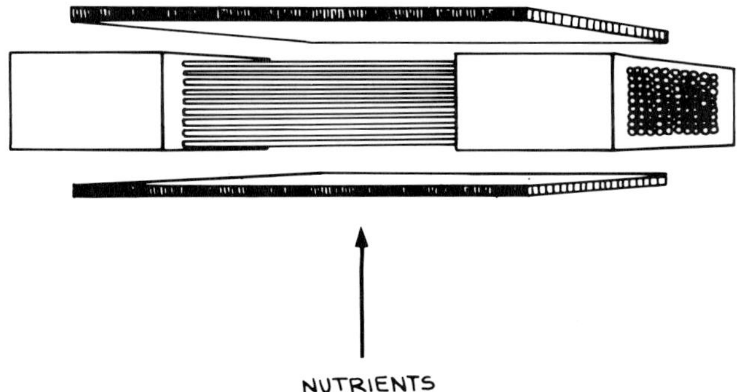

Figure 19. Flat Bed Hollow Fiber Cell Culture Reactor (Ref. 7a)

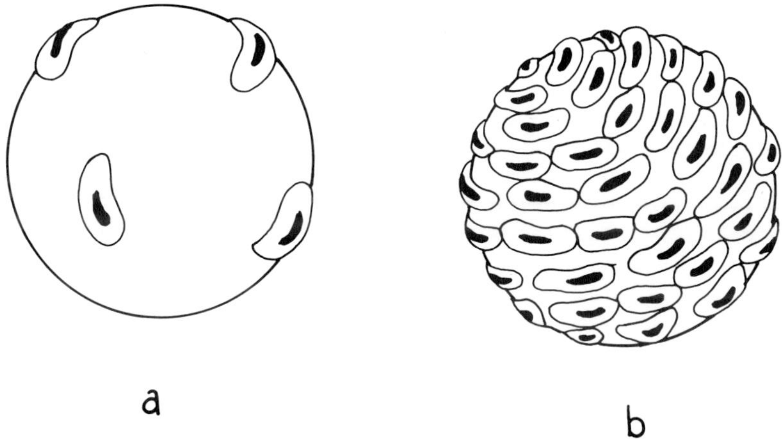

Figure 20. Cell Growth on Microcarriers: a) Microcarrier Inoculated with Few Cells; b) Microcarrier Surface Confluent with Growing Cells (Ref. 4).

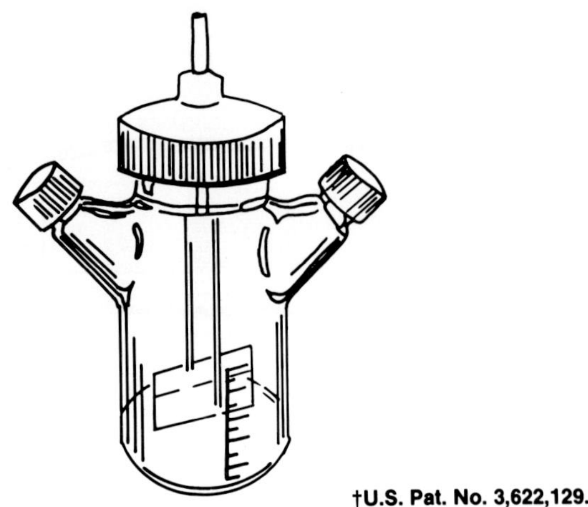

*Figure 21.   Cell Culture µ-carrier Reactor for Growing Cells
             on Microcarriers*

A number of different methods has been used to suspend microcarriers with low shear (4) and one type is shown in Figure 21.
Vibromixing has been used effectively for 100 to 650 liter
reactors (4,140b). Also, continuous systems have been reported
using microcarriers (1,7b,139). The details of microcarrier
culture developments have been described up to the end of 1981
by Pharmacia, manufacturer of CYTODEX^R microcarriers (4).

F.   Cell Growth Parameters

     The quantitation of cell growth parameters has been diffi-
cult because the mammalian cell requires a very complex growth
media including serum and autocrine factors with an environment
that is sterile with closely controlled conditions such as
temperature, pH and low shear field. However, such quantitation
is necessary for upscaling and operation of large culture systems
economically. Significant progress has been made during the last
decade for the definition of such parameters (1-9,33,42-44,57-
59,88,89,111,112,142-145). The parameters to be discussed are
a) Monod cell growth phase diagram; b) transport of energy,
mass and momentum; c) limiting nutrients in the growth medium;
and d) autocrine cell growth factors.

     1. <u>Monod Diagram</u>. The monod cell growth phase diagram
shown in Figure 10 is probably the most powerful tool available
to the cell culturist who uses scientific and engineering methods
to grow homogeneous cultures for both conditioned media and
cells obtained either via small scale or large-scale systems.

The diagram has been partly explained in the first part of this
report. It is important to establish such a growth diagram for
each cell culture reactor run whether it is suspension or
anchorage-dependent culture. There are at least two key reasons
for such quantitation: a) product uniformity, and b) maximum
economic production. Cultures should be harvested before entering the retardation phase (Figure 10: ϕ-4) for among other
reasons in the subsequent culture cycle, the culture enters the
exponential phase with minimum time in the lag and acceleration
phases (ϕ-1, ϕ-2) in fed-batch operations as elaborated extensively elsewhere (3,5,33). The time in ϕ-1 and ϕ-2 for a
culture reactor is also minimized by not changing the reactor
temperature, pH and dissolved oxygen concentration during the
various operational procedures in fed-batch operations (3,5).
Different products can be obtained from cultures in different
phases as discussed elsewhere (33). It is possible that the
variability observed for a transplantation antigen expression
(75) from cells grown in suspension was related to the fact
that they were harvested on a constant harvest period which may
have been in either the exponential, retardation, stationary or
declining phase (3,5).

     2. <u>Transport Phenomena</u>. The transport of energy, mass
and momentum are critical parameters for the efficient and
economic propagation of homogeneous cultures. Puck first
established the temperature sensitivity of mammalian cells for
their growth (42). It was shown recently that productivity can
be increased significantly by appropriate attention to the
temperature of 100-liter capacity cell culture reactors operated
in fed-batch modes in warm rooms and harvested at room temperature (33). The effect of dissolved oxygen transfer for increased
level of harvest cell density has been reported (33,146). The
effect of pH level for improved productivity has been established
(56a,147,148a). The pH level establishes the appropriate
availability of electron transfer via hydrogen ions for the
proper function of the cell enzymes which act as catalysts to
accelerate the reaction rate of the various and many chemical
reactions for the cell to survive and replicate in culture.
The productivity of the suspension culture of Walker 256 rat cell
line was nearly doubled by only adjusting the culture conditions
as dictated by appropriate consideration of the transport of
energy and mass as shown in Figure 22 (33,148b).

     3. <u>Limiting Nutrient(s)</u>. Extensive systematic studies
have not been reported which establish the reason that a cell
culture stops to grow after reaching a relatively high density
as shown, e.g., in Figure 23 (3,5). It is possible that

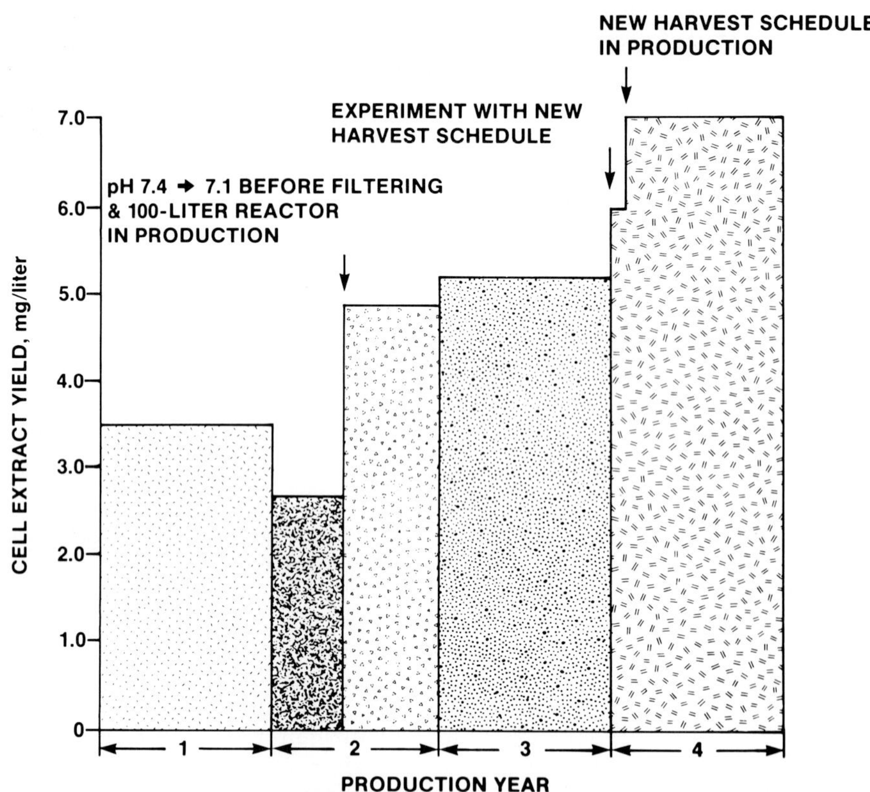

Figure 22.   Effect of Cell Culture Process Parameters on Productivity (Ref. 33)

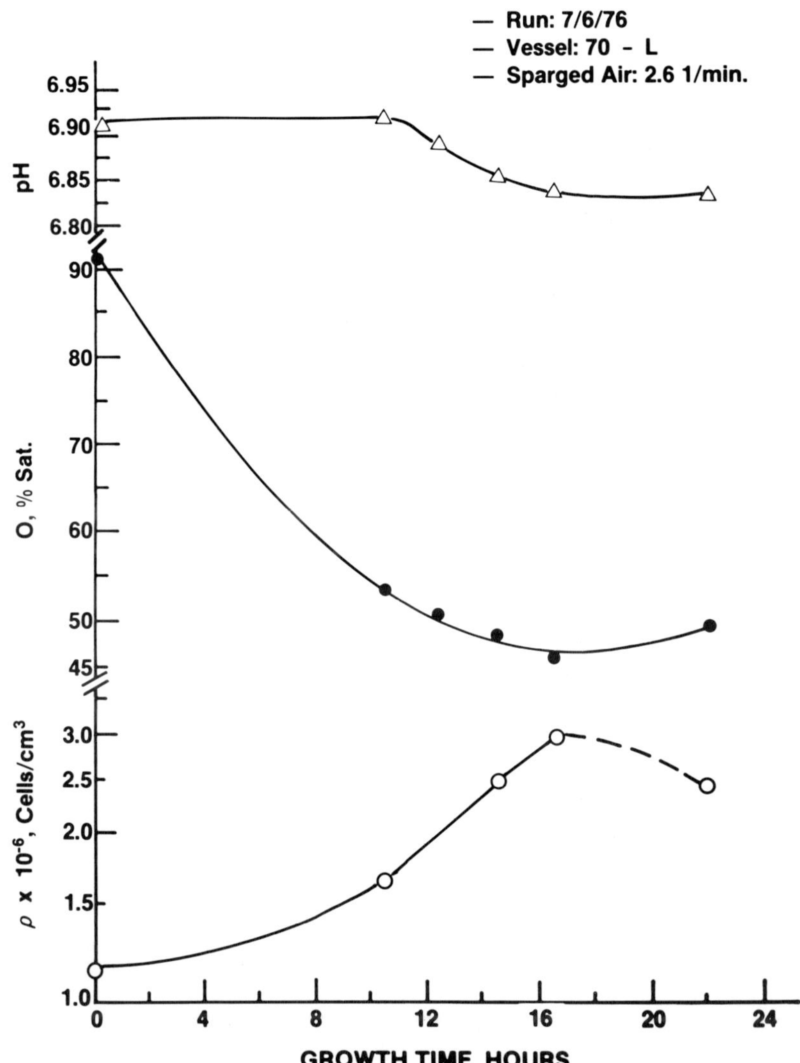

Figure 23. Cell Growth Parameters for a 70-Liter Suspension Fed-Batch Run (Ref. 3,5)

dissolved $\underline{O}$ may be the limiting parameter in most cases for standard growth media formulations demonstrated in limited studies by inference (3,5,33,146). The primary difficulty for establishing the limiting nutrient(s) is related to the complex chemical composition of the growth media which is not completely defined because of the presence of serum; and also because most of the cell culture studies that have been reported were not performed with reactors which were fully instrumental for control or measurement of either temperature, pH, $\underline{O}$, etc.

4. <u>Autocrine Cell Growth Factors</u>. Eagle determined that a certain minimum number of cells ($\sim 10^5$) were required as an inoculum to start up and propagate cultures efficiently (61). This study established that the cells secrete certain factors into the growth media during the start-up of the culture during the lag and acceleration phases (Figure 10) in order to reach exponential growth rate and that a certain concentration of these secreted nutrients is required. A number of cell growth factors has been isolated recently. An important such factor recently isolated is the T-cell growth factor, or interleukin-2 which now allows the growth of T cells and T-cell subclones to be grown in culture (63,68). Previous to the isolation of the interleukin-2, T-cells could not be grown in culture. Other cell growth factors that have been purified to a certain level or isolated and used to stimulate cells in culture for various purposes including more efficient and economical growth are as follows: fibroblast growth factor; platelet derived growth factor; angiogenesis factor; epidermal growth factor; soma-tomedin; fibronectin; various chalones, et. (64-67).

IV. *PRODUCTS*

It is estimated that the human cell has enough genetic information to code for about 1.0 - 1.5 x 10^5 different pro-teins (66b). Only about 1-2 x 10^2 such proteins have been identified and probably only about fifty have been isolated and partially characterized (8,9,97). Hence, many therapeutic and useful products can be derived from human cells. Examples of mammalian cell products are histocompatibility transplanta-tion cell surface antigens and interferons (2,109-112). It is important to select the appropriate cell strain or cell line for a given product for its maximum and economical production for *in vivo* the various cells have been differentiated for only a few or several functions (8,149). The methodology for product acquisition and some products from mammalian cells are examined.

A.   *Acquisition of Products*

The general scheme used to acquire products from cell culture both from cells and conditioned growth media is shown schematically in Figure 24. The first step is to separate the cells from the media via centrifugation (1,000 x g x 10 min). Useful products are found both in the media in which the cells secrete during growth and the cells. Both require special purification methods for product isolation as given, e.g., for one study that two different cell surface antigents (H-2K^k and Thy 1.1) were obtained from the cell plasma membrane and a murine leukemia virus from the media (2,3,78,79).

1.   Conditioned Media. Ultrafiltration (UF) is usually the first partial purification method used to isolate products from conditioned media (CM). UF is a very rapid and economical method, and it does not denature proteins. It can reduce the volume of CM by 100 to 1,000-fold (2,3,150); thus large CM volumes, e.g., 1,000 liters can be reduced to 1-10 liters in three to five hours and operating at 4°C. The resulting small volume can be handled efficiently for further product purification via specialized chromatographic and ultracentrifugation methods (151-153).

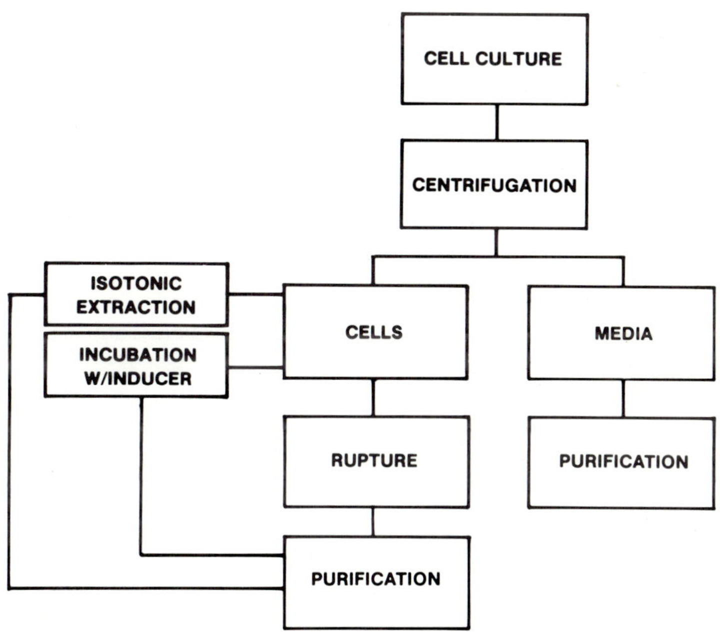

Figure 24 Generalized Scheme for Product Acquisition from Mammalian Cells

2. <u>Cells</u>. Products from cells can be obtained by three methods: a) incubate cells with an inducer, e.g., Sendai virus for interferon production; b) extract cells in an isotonic saline solution; and c) cell rupture. Cells can be ruptured either via sonication, N_2 pressure (1,000 psig) in a bomb and fast pressure release, or by passing them through a small orifice at low pressure ($\sim$150 psig), in the Stansted system. Then the various subcellular components are separated through a series of ultracentrifugation protocols (2,3,154). Subsequently, various chromatographic methods can be used as for CM for product purification and isolation.

B. Mammalian Cell Products

1. <u>Interferons</u>. Three different classes of interferons have been identified. They are leukocyte, fibroblast, and immune interferon induced by various antigens including viruses from leukocyte, fibroblast and T lymphocyte cells respectively (111,112). They have been derived from human cells and other mammals. Within these three classes of interferons, there are probably thirty different types with slightly different amino acid sequence and composition. Interferons are species specific. Interferons are produced by the respective types of cells either during or after cell growth in culture in which the inducer is included or by incubation of the cells separated from the media and the inducer added. All three classes were produced via extensive work by cell culture and tested clinically although the cost was high because of low yields. For fibroblast interferon, yields are 10-20 units per 10^3 cells (111). Large-scale cell culture systems are in operation for the production of lymphoblastoid interferon which belongs to the fibroblast class (123). The respective genes of the three classes of human interferons were isolated, inserted and expressed in *E. coli* and thus are produced via recombinant DNA technology which is much more economical than mammalian cell culture (111,112). Pestka has recently combined succinctly the enormous methodology for the production of interferons via mammalian cell culture and recombinant DNA (111,112).

2. <u>Cell Surface Antigens</u>. The high therapeutic potential of plasma membrane cell surface antigens for organ transplantation feasibility provided a major impetus for their isolation from animal cell lines for initial efficacy testing. From the murine cell line BW5147 cultured in suspension, the Thy 1.1 and H-2K^k cell surfaces antigens were isolated and purified (75,78,79). Later, one of the H-2 moieties was isolated to homogeneity and the amino acid sequence and one-dimensional structure established (155).

C.   Genetic Engineering of Mammalian Cells

The frontier of molecular biology is the area of genetic engineering of mammalian cells with three major goals: a) area of human genetic diseases; b) quantitation of the components and functions of the mammalian cell; c) economical production e.g., of useful glycolated proteins and possibly immunoglobulins which cannot be produced in bacterial or yeast systems. Major initial advances have been made in all three areas and such progress is described elsewhere (156-166).

V.   EPILOGUE

Mammalian cell culture technology has come of age. It is both a key industrial development and a scientific tool with many applications. The concurrent advances in recombinant DNA technology have allowed to insert genes from mammalian cells into bacterial cells and thus produce them much more economically than mammalian cell culture after efficacy demonstration by the latter. However, some proteins, e.g., the glycolated type, may function therapeutically and without antibody formation only in that form. Such proteins will be produced via mammalian cell culture and possible with genetically engineered cells for better economics. The first authoritative textbook on the molecular biology of the mammalian cell has just been published and is authored by James D. Watson and others, and thus establishing a true, quantitative and didactic perspective (8). Mammalian cell culture will continue to support the development of bioscience scientifically, clinically and industrially (167).

REFERENCES

1.   van Hemert, P., Kilburn, D.G. and van Wezel, A.L. (1969).
     Homogeneous cultivation of Animal Cells for the Production
     of Virus and Virus Products. *Biotech. Bioeng. 11*,875-885.

2.   Acton, R.T. and Lynn, J.D. (eds.) (1977). Cell Culture and
     its Application. Academic Press, NY, pp. 1-752.

3.   Acton, R.T. and Lynn, J.D. (1977). Description and Operation
     of a Large-Scale Mammalian Cell, Suspension Culture Facility,
     in *Advances in Biochemical Engineering*, T.K. Ghose, A.
     Fiechter and N. Blakebrough, eds., Springer-Verlag, NY, 7,
     85-110.

4.   Microcarrier Cell Culture: Principles and Methods, Phar-
     macia Fine Chemicals, Uppsala, Sweden, pp. 1-127 (1981).

5.   Harakas, N.K., Lynn, J.D., Acton, R.T., and Bennett, J.C.
     (1979) Industrial Scale Suspension Mammalian Cell Culture
     Center: Operation and Performance Dynamics, 72nd Annual
     Meeting AichE, San Francisco, CA, Reprint No. 24F, pp 1-34.

6.   New Biological Assembly Lines (1978) *Mosaic 9*(6) 26-33.

7a. Ku, K., Kuo, M.J., Delente, J., Wildi, B.S., and Feder, J.
     (1981) Development of a Hollow Fiber System for Large-
     Scale Culture of Mammalian Cells. *Biotech. Bioeng. 23*,
     79-95.

7b. Feder, J. and Tolbert, W.R. (1983) The Large-Scale Culti-
     vation of Mammalian Cells. *Sc. Am. 248*(1) 36-43.

7c. Tolbert, W.R. and Feder, J. (1983) Large-Scale Cell Culture
     Technology, In *Annual Reports of Fermentation Processes,*
     G.T. Tsao et al (Eds) Academic Press, NY 6, 35-74.

8.   Alberts, B., Bray, D., Lewis, J., Raff, M., Robert, K.,
     and Watson, J.D. (1983) Molecular Biology of the Cell.
     Garland Publishing, Inc., New York, pp 1-1146.

9.   Watson, J.D., (1976) Molecular Biology of the Gene, 3rd
     Ed., W.A. Benjamin, Inc., Menlo Park, CA, pp. 1-739.

10a. Schafer, J.A., Personal Communication (October 1976).

10b. Solem, J.C. and Baldwin, G.C. (1982) Microholography of
     Living Organisms *Science 218*, 229-235.

11. Palade, G (1975) The Process of Protein Secretion.
     Science 189, 347-358.

12. Padilla, G.M. and McCarty, Sr., K.S. (Eds.)(1982) Genetic
     Expression in the Cell Cycle. Academic Press, NY.

13. Horrobin, D. (1981) Nucleus or Nucleolus: Which Runs the
     Cell? *New Scientist 89* (1238) 266-269.

14. Mazia, D (1974) The Cell Cycle *Sci. Am. 230*(1) 54-64.

15. Tjio, J.H. and Levan, A. (1956) The Chromosome Number of
     Man. *Separatur Hereditas 42*, 1-6.

16. Wolfe, S.L. (1972) Biology of the Cell. Wadsworth Pub-
     lishing Co., Inc., Belmont, CA.

17. Singer, S.J. and Nicolson, G.L. (1972) The Fluid Mosaic
     Mocel of the Structure of Cell Membranes *Science 175*,720-731.

18. Darnell, J.E., Jr. (1982). Variety in the Level of Gene Control in Eukaryotic Cells. *Nature 297*, 365-371.

19. Marx, J.L. (1981). Gene Control Puzzle Begins to Yield. *Science 212*, 653-655.

20. Hood, L.E., Weissman, I.L., and Wood, W.B. (1978). Immunology. The Benjamin/Cummings Publishing Co., Inc., Menlo Park, CA.

21. Lehninger, A.L. (1982). Principles of Biochemistry. Worth Publishers, Inc., New York, NY.

22. Bloom, W., and Fawcett, D.W. (1975). A Textbook of Histology, 10th Edition, W.B. Saunders Co., Philadelphia, PA.

23. Olander, J.V., Marasa, J.C., Kimes, R.C., Johnston, G.M., and Feder, J. (1982). An assay measuring the stimulation of Several Types of Bovine Endothelial Cells by Growth Factor(s) Derived from Cultured Human Tumor Cells. *In Vitro 18(2)*, 99-107.

24. Folkman, J. and Haudenschild, C. (1980). Angiogenesis *in vitro*. *Nature 288*, 551-556.

25. Stevens, C.F. (1979). The Neuron. *Sci. Am. 241(3)*, 55-65.

26. Marx, J.L. (1981). Coagulation as a Common Threat to Disease. *Science 218*, 145-146.

27. Dustin, P. (1980). Microtubules. *Sci. Am. 243(2)*, 66-76.

28. Lazarides, E. (1980). Intermediate Filaments as Mechanical Integrators of Cellular Space. *Nature 283*, 249-256.

29. Heggeness, M.H., Wang, K., and Singer, S.J. (1977). Intracellular Distributions of Mechanochemical Proteins in Cultured Fibroblasts. *Proc. Nat. Acad. Sci. USA 74(9)*, 3883-3887.

30. Lazarides, E., and Weber, K. (1974). Actin Antibody: The Specific Visualization of Actin Filaments in Non-Muscle Cells. *Proc. Nat. Acad. Sci., USA 71(6)*, 2268-2272.

31a. Ganong, W.F. (1977). Review of Medical Physiology, 8th Edition. Lange Medical Publications, Los Altos, CA.

31b. Guyton, A.C. (1981). Textbook of Medical Physiology, 6th Edition. W.B. Saunders Co., Philadelphia, PA.

32. Castleman, B., and McNeely, V.U. (1974). Blood Plasma, or
     Serum Values. *New Eng. J. Med. 290,* 39-49.

33. Harakas, N.K., Lewis, C.,Jr., Bartram, R.D., Wildi, B.S.,
     and Feder, J., Mammalian Cell Culture: Technology and
     Physiology, in *Cell Culture in Support of Bioscience,* R.T.
     Acton and J.D. Lynn, Eds., Plenum Press, New York, NY
     (in press).

34. Willmer, E.N. (ed). (1965-1966). Cells and Tissues in
     Culture-Methods, Biology and Physiology, Academic Press,
     New York, NY. *1,*1-788, 1965; *2,* 1-809, 1965; *3,* 1-826,
     1966.

35. Harrison, R.G. (1907).Observations on the Living Developing
     Nerve Fiber. *Proc. Soc. Exp. Biol. Med. 4,* 140-143.

36. Carrel, A. (1912). On the Permanent Life of Tissues Outside
     of the Organism. *J. Exp. Med. 15,* 516-528.

37a. Carrel, A. (1923). A Method for the Physiology Study of
     Tissues *in vitro. J. Exp. Med. 38,* 407.

37b. Rous, P., and Jones, F.S. (1916). A Method for Obtaining
     Suspensions of Living Cells from the Fixed Tissues, and
     for the Plating out of Individual Cells. *J. Exp. Med. 23,*
     549-555.

38. Sanford, K.K., Earle, W.R., and Likely, G.D. (1948). The
     Growth *in vitro* of Single Isolated Tissue Cells. *J. Natl.
     Cancer Inst. 9,* 229-246.

39. Gey, G. (1954). Some Aspects of the Constitution and
     Behavior of Normal and Malignant Cells Maintained in
     Continuous Culture. *Harvey Lectures 50,* 154-229.

40. Eagle, H. (1955). Nutrition Needs of Mammalian Cells in
     Tissue Culture. *Science 122,* 501-504.

41. Puck, T.T., and Marcus, P.I. (1955). A Rapid Method for
     Viable Cell Titration and Clone Production of HeLa Cells
     in Tissue Culture: The Use of X-Irradiated Cells to
     Study Conditioning Factors. *Proc. Natl. Acad. Sci., USA
     42,* 432-437.

42. Puck, T.T. (1972). The Mammalian Cell as a Microorganism.
     Holden-Day, San Francisco, CA, 1-219.

43. Pollack, R. (ed.)(1975). Readings in Mammalian Cell
     Culture, 1st edition, Cold Spring Harbor Laboratory, Col
     Spring Harbor, NY, 1-864.

44. *ibid,* (1981). 2nd edition, 1-718.

45. Hayflick, L, and Moorhead, P.S. (1961). The Serial Culti-
     vation of Human Diploid Cell Strains. *Exp. Cell Res. 25,*
     585-621.

46. Reich, E., et al., (eds.)(1975). Proteases and Biological
     Control, CSH Conferences on Cell Proliferation, Vol. *2,*
     Cold Spring Harbor Laboratory, Cold Spring Harbor, NY.

47. Hayflick, L. (1962). Discussion in Analytic Cell Culture,
     National Cancer Institute Monograph No. 7, 63-67.

48. Altman, P.L., and Katz, D.D. (1976). Biological Handbooks-
     Cell Biology, Fed. Amer. Soc. Exp. Biol., Bethesda, MD,
     1, 32-89.

49a. Hay, R., Lavappa, K.S., Macy, M., Shannon, J., and Williams,
     C. (eds.)(1979). Cell Lines, in The American Type Culture
     Collection, Catalogue of Strains II, 2nd Edition, American
     Type Culture Collection, Rockville, MD, 1-114.

49b. Dulbecco, R. (1980). Animal Cell Cultures, in Microbiology,
     B.D. Davis et al.,(eds.), 3rd Edition, Harper and Row
     Publishers, New York, NY 941-966.

50. van Wezel, A.L. (1967). Growth of Cell-Strains and Primary
     Cells on Microcarriers in Homogeneous Culture. *Nature 216,*
     64-65.

51. Cooper, G.M. (1982). Cellular Transforming Genes.
     Science 218, 801-806.

52. Tabin, C.J., et al.,(1982). Mechanism of Activation of a
     Human Oncogene, *Nature 300,* 143-149.

53a. Anatomy of a Human Cancer Gene (1982).*Nature 300,* 103.

53b. Logan, J., and Cairns, J. (1982). The Secrets of Cancer.
     Nature 300, 104-105.

54. Littlefield, J.W. (1982). NIH 3T3 Cell Line. *Science 218,*
     214-216.

55. Jensen, H.M. et al. (1982). Angiogenesis Induced by "normal"
     Human Breast Tissue: A Probably Marker for Precancer.
     Science 218, 293-295.

56a. Eagle, H. (1974). Some Effects of Environmental pH on Cellu-
     lar Metabolism and Function, in Control of Proliferation
     in Animal Cells, B.Clarkson and R. Baserga, eds., CSH
     Conferences on Cell Proliferation, Cold Spring Harbon
     Laboratory, Cold Spring Harbor, NY, *1*,1-11.

56b. McLimans, W.F. (1972). The Gaseous Environment of the
     Mammalian Cell in Culture, in Growth, Nutrition and Metabol-
     ism of Cells in Culture, G.H. Rothblat and V.J. Cristofalo,
     eds., Academic Press, New York, NY *1*, 137-170.

57. Kruse, P.F., Jr. and Patterson, M.K., Jr. (eds.)(1973).
     Tissue Culture-Methods and Applications. Academic Press,
     New York, NY, 1-869.

58. Tooze, J.(ed.),(1973). The Molecular Biology of Tumor
     Viruses. Cold Spring Harbor, NY, 74-268.

59. Jakoby, W.B., and Pastan, I.H. (eds.)(1979). Cell Culture,
     Methods in Enzymology. Academic Press, New York, NY *58*,
     1-642.

60. McLimans, W.F., Davis, E.V., Glover, F.L., and Rake, G.W.
     (1957). The Submerged Culture of Mammalian Cells: The
     Spinner Culture. *J. Immunol. 79*, 428-433.

61. Eagle, H., and Piez, K. (1962). The Population-Dependent
     Requirement by Cultured Mammalian Cells for Metabolites
     Which They Can Synthesize. *J. Exp. Med. 116*, 29-43.

62. Zwerner, R.K., Runyan, C., Cox, R.M., Lynn, J.D., and
     Acton, R.T. (1975). An Evaluation of Suspension Culture
     Systems for the Growth of Murine Lymphoblastoid Lines
     Expressing TL and Thy-1 Alloantigens. *Biotech. Bioeng. 17*,
     629-657.

63. Fathman, C.G., and Fitch, F.W., eds. (1982). Isolation,
     Characterization and Utilization of T Lymphocyte Clones.
     Academic Press, New Yor, NY.

64. Folkman, J., Haudenschild, C.C., and Zetter, B.R.(1979).
     Long-Term Culture of Capillary Endothelial Cells. *Proc.
     Natl. Acad. Sci. USA 76*, 5217-5221.

65. Manning, D.C., Snyder, S.H., Kachur, J.F., Miller, R.J., and Field, M. (1982). Bradykinin Receptor-Mediated Chloride Secretion in Intestinal Function. *Nature 299(5880)*,256-259.

66a. Antoniades, H.N., and Owen, A.J. (1982). Growth Factors and Regulation of Cell Growth. *Ann. Rev. Med. 33*, 445-463.

66b. Rich, A. (1983). Personal Communication, September 19.

67. Baserga, R. (ed.)(1981). Handbook of Experimental Pharmacology: Tissue Growth Factors. Springer-Verlag, New York, NY *57*, 1-630.

68. Morgan, D.A., Ruscetti, F.W., and Gallo, R. (1976). Selective *in vitro* Growth of T Lymphocytes from Normal Human Bone Marrows. *Science 193*, 1007-1008.

69. Kuchler, R.J.(1977). Biochemical Methods in Cell Culture and Virology. Dowden, Hutchinson and Ross, Stroudsburg, PA, 1-113.

70. Waymouth, C.(1956). A Rapid Quantitative Hematocrit Method for Measuring Increase in Cell Population of Strain L (Earle) Cells Cultivated in Serum-Free Nutrient Solution. *J. Natl. Cancer Inst. 17*, 305-313.

71. Tolbert, W.R., Hitt, M.M., and Feder, J. (1980). Rapid Determination of Cell Volume Density in Mammalian Cell Suspensions. *Anal. Biochem. 106*, 109-113.

72. McLimans, W.F. (1979). Mass Culture of Mammalian Cells, in Cell Culture, W.B. Jakoby and I.H. Pastan (eds.), *Methods in Enzymology, 58*, 194-211.

73. Eagle, H.(1971). Buffer Combinations for Mammalian Cell Culture. *Science 174*, 500-503.

74. Monod, J.(1949) The Growth of Bacterial Cultures. *Ann. Rev. Microb. 3*, 371-394.

75. Barstad, P., Henley, S.L., Cox, R.M., Lynn, J.D. and Acton, R.T.(1977). Production of Milligram Quantities of H-2K and Thy-1 Alloantigens by Large-Scale Mammalian Cell Culture. *Proc. Soc. Exp. Biol. Med. 155*, 296-300.

76. Folkman, J., and Klagsbrun, M. (1975). Tumor Angiogenesis: Effect of Tumor Growth and Immunity, in Fundamental Aspects of Neoplasia, A.A. Gottlieb, O.J. Plescia and D.H.L. Bishop, eds. Springer-Verlag, New York, NY, 401-412.

77. Folkman, J.(1975). Tumor Angiogenesis in Cancer, F.F.
     Becker, ed., Plenum Press, New York, NY *3*, 355-388.

78. Zwerner, R.K., Barstad, P.A. and Acton, R.T.(1977).
     Isolation and Characterization of Murine Cell Surface
     Components. I. Purification of Milligram Quantities of
     Thy-1.1. *J. Exp. Med. 146*, 986-1000.

79. Barstad, P.A., Zwerner, R.K., and Acton, R.T. (1978).
     Partial Purification of H-2K Molecules from Large-Scale
     Cell Culture, in Protides of the Biological Fluids,
     H. Peeters (ed.). Pergamon Press, New York, NY *25*, 636-640.

80. Acton, R.T., Barstad, P.A., Cox, R.M., Zwerner, R.K.,
     Wise, K.S., and Lynn, J.D.(1977). Large-Scale Production
     of Murine Lymphoblastoid Cell Lines Expressing Differentia-
     tion Alloantigens, in Cell Culture and Its Application,
     R.T. Acton and J.D. Lynn(eds.) Academic Press, NY, 129-160.

80b. Vaheri, A., Ruoslahti, E., and Mosher, D.F.(1978). Fibro-
     blast Surface Protein. Annals N.Y. Acad. Sci. *312*, 1-456.

31. Glick, J.L (1980). Fundamentals of Human Lymphoid Cell
     Culture. Marcel Dekker, Inc., New York, Ny.

82. McLimans, W.F., Personal Communication, November 1982.

83. McGarrity, G.J. (1979). Detection of Contamination, in
     Cell Culture, W.B. Jakoby and I.H. Pastan, (eds.). *Methods
     of Enzymology 58*, 18-29.

84. Acton, R.T., Barstad, P.A., and Zwerner, R.K. (1979)
     Propagation and Scaling-up of Susepnsion Cultures, in
     Cell Culture, W.B. Jacoby and I.H. Pastan, (eds.). *Methods
     in Enzymology 58*, Academic Press, New York, NY, 211-221.

85. McGarrity, G.J., Murphy, D.G., and Nichols, W.W.(eds.).
     (1978). Mycoplasma Infection of Cell Cultures. Plenum
     Press, New York, NY.

86. Davis, D.B. et al. (1980). Microbiology, 3rd Edition, Harper
     and Row Publishers, New York, NY.

87. Rothblat, G.H., and Cristofalo, V.J. (eds.). (1972 and 1977).
     Growth, Nutrition and Metabolism of Cells in Culture.
     Academic Press, New York, NY *I*,1-471, 1972; *II*, 1-445, 1972;
     III, 1-548, 1977.

88. Sato, G.H. and R. Ross (eds.)(1979). Hormones and Cell
     Culture, Cold Spring Harbor Conferences on Cell Prolifera-
     tion, Cold Spring Harbor Laboratory, Cold Spring Harbor,
     NY. *6.*

89. Sato, G.H., Pardee, A.B., and Sirbasku, D.A. (1982). Growth
     of Cells in Hormonally Defined Media, CSH Conferences on
     Cell Proliferation, Cold Spring Harbor Laboratory, Cold
     Spring Harbor, NY *9,* 1-1208.

90. Katsuka, H.(ed).(1978). Nutritional Requirements of
     Cultured Cells. University Park Press, Baltimore, MD.

91. Davis, D.B. and Dulbecco, R. (1980). Sterilization and
     Disinfection, in Microbiology, B.D. Davis et al.,(eds.)
     3rd Edition, Harper and Row, Publishers, New York, NY
     1263-1274.

92a. Sykes, G (1969). Methods and Equipment for Sterilization
     of Laboratory Apparatus and Media, in Methods in Micro-
     biology, J.R. Norris and D.W. Ribbons(eds.). Academic
     Press, New York, NY *1,* 77-121.

92b. Elsworth, R.(1969). Treatment of Process Air for Deep
     Culture, in Methods in Microbiology, J.R. Norris et al.,
     (eds.). Academic Press, New York, NY *1,* 123-136.

92c. Duberstein, R.(1979). Mechanisms of Bacterial Removal by
     Filtration, *J. Parent. Drug Assoc.,* Sept./October, 250-256.

92d. Howard, G., Jr., and Duberstein, R.(1980). A Case of
     Penetration of 0.2-μm Rated Membrane Filters by Bacteria.
     J. Parent. Drug Assoc., March/April, 95-102.

93. Phillips, G.B., and Runkle, R.S. (1973). Biomedical Appli-
     cation of Laminar Air Flow. CRC Press, Inc., Boca Raton,
     FL, 1-180.

94. McDade, J.J., Phillips, G.B., Sivinski, H.D., and
     Whitfield, W.J. (1969). Principles and Applications of
     Laminar Flow Devices, in Methods in Microbiology, J.R.
     Norris, et al.(eds). Academic Press, New York, NY *1,* 137-168.

95. Tolbert, W.R., and Feder, J. (1982). Sterile Air-Shielded
     Connectors for Aseptic Operations. *Biotech. Bioeng. 24,*
     1885-1887.

96. Cairns, J., et al. (1966). Phage and the Origins of Molecular Biology, Cold Spring Harbor Laboratory, Cold Spring Harbor, NY.

97. Stent, G.S., and Calendar, R.(1978). Molecular Genetics: An Introductory Narrative, 2nd Edition. W.H. Freeman and Co., San Francisco, CA.

98. Edelman, G.M. (1970). The Structure and Function of Anti-bodies. *Sci. Am. 223(2),* 34-42.

99. Good, R.A., and Fisher, D.W. (eds.). Immunobiology. Sinauer Associates, Inc., Sunderland, MA, 1-305.

100. Watson,J.D., and Tooze, J. (1981). The DNA Story: A Documentary History of Gene Cloning. W.H. Freeman and Co.,

101. Kornberg, A. (1980). DNA Replication, 2nd Edition. W.H. Freeman and Co., San Francisco, CA.

102. *ibid,* (1982). Supplement to DNA Replication.

103. Freifelder, D. (1978). Recombinant DnA - Readings from Scientific American. W.H. Freeman and Co., San Francisco, CA.

104a. Köhler, G., and Milstein, C. (1975). Continuous Cultures of Fused Cells Secreting Antibody of Predifined Specificity. *Nature 256,* 495-497.

104b. Milstein, C. (1980). Monoclonal Antibodies, *Sci. Am. 243(4),* 66-74.

105. Industrial Microbiology - A Series of Articles (1981). *Sci. Am. 245(3),* 66-215.

106. Abelson, P.H. (ed.)(1983). Biotechnology. *Science 219(4585),* 609-746.

107. Acton, R.T., Addis, J., Carl, C.F., McClain, L.D., and Bridges, W.F. (1978). Association of Thy-1 Differentiation Alloantigen with Synaptic Complexes Isolated From Mouse Brain. *Proc. Natl. Acad. Sci., USA 75,* 3283-3287.

108. Strominger, J.L. et al. (1981). Biochemical Analysis of Products of the MHC, in the Role of the Major Histocompata-bility Complex in Immunobiology, M.E. Dorf, (ed.), Garland STPM, New York, NY, 115-172.

109. Finter, N.B.(ed.)(1973). Interferons and Interferon
     Inducers. Elsevier Publishing Co., Inc., New York, NY.

110. Stewart, W.E., II (1979). The Interferon System. Springer-
     Verlag, New York, NY.

111. Pestka, S.(ed.)(1981). Interferons, Part A, Methods In
     Enzymology. Academic Press, New York, NY *78,* 1-632.

112. *ibid,* Part B, Methods in Enzymology. Academic Press,
     New York, NY, *79,* 1-677.

113. Osborne, N. (1983). The Brain's Information Technology.
     New Sci. 98(1358), 445-447.

114. Watson, J., Mochizuki, D., and Gillis, S. (1980). T-Cell
     Growth Factors: Interleukin 2. *Immunol. Today 1,* 113-116.

115. Vitetta, E.S.,(ed.)(1982). B and T Cell Tumors. Academic
     Press, New York, NY.

116. Landy, M. (Advisory ed.).(1983). Lymphokines. Academic
     Press, New York, NY, *1-8.*

117. Overview of a Production Scale Animal Cell Suspension
     Culture System, Technical Brochure on University of
     Alabama Medical Center Culture Facility, New Brunswick
     Scientific Co., Edison, NJ, 1-54 (1976).

118. Lynn, J.D., and Acton, R.T. (1975). Design of a Large-
     Scale Mammalian Cell, Suspension Culture Facility.
     Biotech. Bioeng. 17, 659-673.

119. Hollow Fiber Ultrafiltration Membrane Manual, Amicon Corp.,
     Danvers, MA, (1982).

120. Cameron, J., and Godfrey, E.I. (1969). The Design and
     Operation of High-Power, Magnetic Drives, *Biotech. Bioeng.
     11,* 967-985.

121. *ibid,* (1974). Magnetic Drives, *Biotech. Bioeng. Symp. 4(2)*
     821-835.

122. NBS Interferon Production System for Clinical Trials,
     NBS Microwaves, New Brunswick Scientific Co., Edison, NJ,
     14(3), 1-4 (1980).

123. Cell Culture System for the Production of Interferon, New
     Brunswick Scientific Co., Edison, NJ, 1-8 (1980).

124. Weisman, E., Personal Communication, November 1981.

125. General Catalog (No. 1028), New Brunswick Scientific Co., Edison, NJ, C57-C59 (1982).

126. Chemap[R] Vibro-Fermenter, Chemap AG, Mannedort, Switzerland, 1-16, (1982).

127. General Catalog, Contact-Roestvrijstaal Ltd., Ridderkerk, Holland, (1977).

128. Pirt, S.J. (1975). Principles of Microbe and Cell Cultivation. John Wiley & Sons, New York, 1-275.

129. Kruse, P.F., Jr. (1972). Use of Perfusion Systems for Growth of Cell and Tissue Cultures, in Growth, Nutrition and Metabolism of Cells in Culture, G.H. Rothblat and V.J. Cristofalo (eds.) Academic Press, New York, NY *1*, 11-66.

130. Perfusion Cell Culture Reactors in General Catalog, Virtis Co., New York, NY, (1975).

131. Ginsberg, H.S., and Dulbecco, R. (1980). Multiplication and Genetics of Animal Viruses, in Microbiology, Davis, B.D. et al. (eds.), 3rd Edition, Harper and Row, Publishers, Philadelphia, PA, 967-999.

132. Giard, D.J. et al. (1977). Virus Production With a Newly Developed Microcarrier System. *Appl. Environm. Microbiol. 34*, 668-672.

133. Meignier, B. (1979). Cell Culture on Beads Used for the Industrial Production of Foot-and-Mouth Disease Virus. *Develop. Biol. Standard. 42*, 141-145.

134. Crespi, C.L., and Thilly, W.G. (1981). Continuous Cell Propagation Using Low-Charge Microcarriers. *Biotech. Bioeng. 23*, 983-993.

135. Meignier, B., Mougeot, H., and Favre, H. (1980). Foot-and-Mouth Disease Virus Production on Microcarrier-Grown Cells. *Develop. Biol. Standard. 46*, 249-256.

136. van Wezel, A.L, van der Velden-de-Groot, C.A.M., and van Herwaarden, J.A.M. (1980). The Production of Inactivated Polio Vaccine on Serially Cultivated Kidney Cells from Captive-Bred Monkeys. *Develop. Biol. Standard. 46*, 151-158.

137. Sinkey, A.J. et al. (1981). Production of Cell-Derived
     Products: Virus and Interferon. *Annals N.Y. Acad. Sci. 369*,
     47-64.

138. Montagnon, B.J., Fanget, B. and Nicholas, A.J. (1981).
     The Large-Scale Cultivation of Vero Cells in Microcarrier
     Culture for Virus Vaccine Production: Preliminary Results
     For Killed Poliovirus Vaccine. *Develop. Biol. Standard 47*,
     55-64.

139. Tolbert, W.R., Hitt, M.M., Feder, J., and Kimes, R.C. (1982).
     Method of Growing Anchorage-Dependent Cells. U.S. Patent
     No. 4,335,215.

140a.Kuo, M.J., Lewis, C., Jr., Martin, R.A., Miller, R.E.,
     Schuck, J.M., and Wildi, B.S. (1980). Growth of Anchorage-
     Dependent Mammalian Cells on Glycine Derived Polystyrene
     in Suspension Culture. *In Vitro 16(3)*, 218.

140b.van Hemert, P. (1964). The "Bilthoven Unit" for Submerged
     Cultivation of Microorganisms. *Biotech. Bioeng. 6*, 381-401.

141. Levine, D.W., Thilly, W.G., Wang, D.I.C., and Wong, J.S.
     (1980). Cell Culture Microcarriers, U.S. Patent No. 4,189,534.

142. Baserga, R. (1981). Introduction to Cell Growth: Growth
     In Size and DNA Replication, in Tissue Growth Factors, R.
     Baserga, (ed.) Springer-Verlag, New York, NY, 1-12.

143. Ham, R.G. (1981). Survival and Growth Requirements of
     Non-Transformed Cells, in Tissue Growth Factors, R. Baserga,
     (ed.). Springer-Verlag, New York, NY, 13-88.

144. Clarkson, B., and Baserga, R. (eds.). (1974). Control of
     Proliferation in Animal Cells, CSH Conferences on Cell
     Proliferation. Cold Spring Harbor Laboratory, Cold Spring
     Harbor, NY *1*, 1-1029.

145. Paul, J. (1975). Cell and Tissue Culture, 5th Edition.
     Churchill Livingstone, New York, NY, 1-484.

146. Pilot-Scale Production of Lymphoid Interferon Increases
     Yields in Suspension Culture, (1979). *Microwaves 14,(2)*, 1-4.

147. Eagle, H. (1973). The Effect of Environmental pH on the
     Growth of Normal and Malignant Cells. *J. Cell Phys. 28*, 1-8.

148a.Harakas, N.K., unpublished results.

148b. Feder, J., Kimes, R.C., and Tolbert, W.R., unpublished Results for the first 1.5 Years in Figure 22.

149. Zwerner, R.K., and Acton, R.T. (1975). Growth Properties and Alloantigenic Expression of Murine Lymphoblastic Cell Lines. *J. Exp. Med. 142,* 378-390.

150. Harakas, N.K., Bentle, L.A., Mitchell, J.W., and Feder, J. (1982). Plant Scale Concentration/Fractionation of Biological Streams Via Ultrafiltration, in Harnessing Theory for Practical Applications. Proceedings Third World Filtration Congress, Uplands Press Ltd., Croydon, England, *II,* 513-518.

151. Neurath, H., and Hill, R.L.(eds.)(1975). The Proteins, 3rd Edition. Academic Press, New York, NY *1.*

152. Jakoby, W.B., and Wilckek, M. (eds).(1975). Affinity Techniques: Enzyme Purification - Part B, Methods in Enzymology. Academic Press, New York, NY *34.*

153. Scopes, R.K. (1983). Protein Purification: Principles and Practice, Springer-Verlag, New York, NY.

154. Crumpton, M.J., and Snary, D. (1974). Preparation and Properties of Lymphocyte Plasma Membrane, in Contemporary Topics in Molecular Immunology, G.L. Ada,(ed.) Plenum Press, New York, NY *3,* 27-54.

155. Coligan, J.E. et al.(1981). Primary Structure of a Murine Transplantation Antigen. *Nature 291,* 35-39.

156. Anderson, W.F., and Diacumakos, E.G. (1981). Genetic Engineering in Mammalian Cells. *Sci. Am. 245(1),* 106-122.

157. Pavlakis, G.N., Hizuka, N., Gorden, P., Seeburg, P., and Hamer, D.H. (1981). Expression of Two Human Growth Hormone Genes in Monkey Cells Infected with Simian Virus 40 Recombinants. *Proc. Natl. Acad. Sci. 78(12),* 7398-7402.

158. Hamer, D.H. (1980). DNA Cloning in Mammalian Cells with SV40 Vectors, in Genetic Engineering, J.K. Setlow and A. Hollaender (eds.). Plenum Publishing Co., New York, NY, *2,* 83-101.

159. Hamer, D.H., and Walling, M.(1982). Regulation *in vivo* of a Cloned Mammalian Gene: Cadmium Induces the Transcription of a Mouse Metallothionein Gene in SV40 Vectors. *J. Mol. Genet. 1,* 1-16.

160. Gray, P.W. et al. (1982). Expression of Human Immune Interferon cDNA in *E. coli* and Monkey Cells. *Nature 295,* 503-508.

161. Mantei, N., and Weissmann, C. (1982). Controlled Transcription of a human α-Interferon Gene Introduced Into Mouse L Cells. *Nature 297,* 128-132.

162. Hauser, H. et al., (1982). Inducibility of Human β-Interferon Gene in Mouse L-Cell Clones, *Nature 297,* 650-654.

163. Gluzman, Y. (ed.)(1982). Eukaryotic Viral Vectors, Cold Spring Harbor Laboratory, Cold Spring Harbor, NY, 1-221.

164. Schimke, R.T. (ed.)(1982). Gene Amplification, Cold Spring Harbor Laboratory, Cold Spring Harbor, NY, 1-339.

165. Pavlakis, G.N., and Hamer, D.H. (1983). Regulation of a Metallothionein-Growth Hormone Hybrid Gene in Bovine Papilloma Virus. *Proc. Natl. Acad. Sci. USA 80(1),* 397-401.

166. Fox, J.L. (1983). Columbia Awarded Biotechnology Patent, *Science 221, 933.*

167. *Acton, R.T., and Lynn, J.D. (eds.) Cell Culture in Support of Bioscience, Proceedings Second Cell Culture Congress, (in press).*

CHAPTER 8

MICROBIAL BIOMASS FROM RENEWABLES:
A SECOND REVIEW OF ALTERNATIVES

Carlos Rolz

Central American Research Institute for Industry
Guatemala, Central America

I.   INTRODUCTION

Microbes can be grown in renewable and synthetic substrates
and the resulting biomass can be employed as a nutrient source
as produced or after processing through animal feed rations or
in processed food products. The imminent world food crisis
much debated in the technical literature has encouraged research
and development activities related to the production and processing of microbial biomass (MB). However, as yet, its
worldwide commercial production is still low, and the reasons
are mainly economical in nature. The standard process for MB
production as developed for large-scale capital-intensive
operations, consists of a single species, strictly aseptic,
operation carried out in synthetic substrates (gaseous and
liquid hydrocarbons, C_1 and C_2 alcohols), where biomass is
reproduced aerobically, continuously, in a highly diluted
aqueous media (typical cell productivities in the range of
205 kg m^{-3} h^{-1}), under mesophilic temperatures and in large-
scale plants (50-100,000 tons/year). This operation is followed
by cell recovery, washing and drying where a nonviable, easily
stored biomass powder is obtained. In some cases further processing includes protein extraction and concentration, nucleic
acid reduction and development of appropriate functional
properties. In the judgement of some, the product obtained
this way should be called SCP (single cell protein) and any
other product obtained in a different manner should be
identified as MBP (microbial biomass product). This suggestion
is somewhat irrelevant, more important is the acceptance of
the challenge to synthetize alternate process routes, which
due to the incorporation of modifications to the standard process will produce, in a large or small production scale,
competitive products both in enconomic and also nutritional
terms. The use of the standard process with renewable substrates
(by-products, residues, wastes) is not economically attractive
as the many pilot plants built during the last decade have

213

shown. Only a few lead to commercial operations employing cane molasses, wood acid hydrolyzates, sulphite waste liquors and whey as substrates. The first step is to select for each case the most adequate product and correct process strategy and this task is by no means simple due to the many possible process alternatives available as discussed in recent reviews (1-27).

Production costs for microbial biomass produced with standard process are around U.S. $450-550/metric ton, 1977 figures employing synthetic substrates (28). An up-dated figure will probably be around U.S. $700/metric ton. This last value will also apply to MB production from molasses in a tropical country, and when compared to current soybean meal prices still is 2.5 times higher. As a consequence, there is little incentive to produce MB as fodder ingredient or to further process it for human food, unless the prices of conventional fodder ingredients and basic protein foods should rise substantially (29) or the MB is produced cheaper.

Economic analysis of the standard process (22) shows clearly that the most significant production cost is due to raw materials (carbon and nitrogen sources). The use of agro-industrial effluents, by-products, residues and wastes as raw materials looks then promising, as in many cases they can have a negative value (the equivalent cost of treating the effluent in order to minimize pollution). However the scheme is not as simple as other factors come into play. For example, once a waste has demand, as if by a miracle, it suddenly has also a unitary price. Accumulation of the waste in just a few months of the year, the geographical dispersion of the production centers and in some cases their extreme dilution are usual items that could add up to the waste actual cost. It also must be remembered that MB production is only one of the many schemes available to upgrade wastes (1). These substrates can be: a) exported to the world market as high price raw materials (cane molasses), b) used directly or after some treatment as animal feeds (straws, animal manure), c) used as a source of fibers (bagasse, straws, waste paper), and d) employed directly as fuels (sawdust, coffee parchment, bagasse), or by an anaerobic fermentation process (biogas from animal manure, liquid effluents from food processing plants).

After raw materials cost, utilities follow and usually aeration requirements in the biological reactor are the highest, hence suggesting the need for better aeration devices which justify the recent developments in air lift, bubble columns and closed loop reactors (5,30,31). However, there are other alternatives available where aeration is practically eliminated. First the use of algal-bacterial combinations in high rate

oxidation ponds, where MB production from animal wastes or
sewage is usually integrated with the operation of a biogas
digestor and where several process flow sheets are available (1).
Excellent reviews of the state-of-the-art on this subject have
recently been published (32-35). Second, the employment of
anaerobic processes where usually metabolites and biomass are
considered jointly as products, hence defining a multiproduct
variation of the standard process. Examples are: a) the use
of animal wastes as feed materials (coupled to biogas production
or not), b) acid production and bacterial biomass, and c) fuel
alcohol and yeast biomass. In these cases an interesting point
in the economic analysis results from having a multiproduct
scheme (1). Should the price of the produced biomass be fixed
by its price set by the standard process and hence improving
the economics of the other metabolites produced? Or should it
be the other way around, resulting in a cheaper BM product.
For example, what is the adequate price for yeast produced as
a by-product in ethanol-fuel processes? Anaerobic processes
are already providing microbial biomass nutrients to animals
(36-40) (recycling of animal excreta) and also to humans through
the use of microbial cultures in fermented foods: vegetables
(41), dairy (42), legume and cereal products (43,44), and meat
products (45), and in some way or the other, they should be
included when discussing MB production and use.

Another important cost item in the standard process (22)
is the one related with fixed investment, where fermentation,
drying and cell harvesting units are the most expensive. Solid
phase (moist solids) biological reactions might be a way of
reducing reactor costs, although the mechanism for solids
handling, temperature control and the design of the aeration
system might be mechanically complicated. However these
systems still need a lot of study and development work before
optimal geometries are established. The utilization of more
labor is envisaged which favors its application in developing
countries. Recent reviews cover the field quite well (46-50a).
There are various alternatives to eliminate or reduce the drying
requirements (1,5). They include broth evaporation, mixing with
other solids of the feed formula and pelletizing, final drying
or ensiling of the pellets. This is one area that also needs
experimental work.

It is easily seen that the MB products that might result
from these modified processes will differ in characteristics
according to microorganism, substrate and process employed.
Worldwide many alternatives are currently being evaluated and
in this review we will try to discuss recent references related
to the subject in a way that will bring up-to-date our previous
publication (1).

II. *MB PRODUCTION IN SOLUBLE, DISPERSIONS OF INSOLUBLE SUB-STRATES BY SINGLE SPECIES*

Pure culture systems are practically a requirement for its acceptance as food or animal feed. Hence a majority of the research has focused on a single specie for a particular substrate.

1.   *Yeasts on Starch and Sugar Substrates, Processing By-products and Effluents*

Spencer-Martins and van Uden (51) tested the growth of several yeasts in 0.2% (w/v) solutions of soluble starch and found that strains of the genera *Lipomyces* were the best. The yeasts consumed 84 to 100% of the starch supplied and the yields in terms of g biomass/g starch supplied were 0.489, 0.585, 0.384 and 0.442 for *L. starkeyi, L. kononenkoae, L. lipofer,* and *L. tetrasporus. Candida tropicalis* gave a yield of only 0.234, *Hansenula anomala* 0.357, *Schwanniomyces alluvius* 0.376, *Pichia polymorpha* 0.293, and *Trichosporon pullulans* 0.392. *L. kononen-koae* strain IGC 4052 was found to produce an extracellular amy-lolytic enzyme system capable of total starch hydrolysis, as it consisted of an α-amylase, a glucoamylase and a debranching transferase (52), which were subject to catabolite repression (53). Azoulay and co-workers (54,54a) on the contrary isolated by successive subcultures on soluble starch a mutant from *Candida tropicalis* CBS 6947. This strain utilized quite effectivel the soluble starch fraction as a source of carbon. Its gluco-amylase was inducible and intracellular. Continuous cultures at 35°C, pH 4.0 and D = 0.26 h^{-1} in thermally pretreated cassava powder suspensions gave a cell productivity of 3.6 $gl^{-1}h^{-1}$, a yield of 0.55 g biomass/g soluble starch consumed and 62% sub-strate consumption. Thermally pretreated corn powder gave inferior results. Nga et al (55) grew *Endomycopsis fibuliger* and *E. bustonii* in suspensions of cassava flour and ammonium sulphate, obtaining a six and thirteen fold increase of the solids total aminoacids, respectively. Lemmel et al. (56) and Admassu et al. (57) employed *Saccharomyces fibuligera* Y-1062 to treat potato extract simulating processing blancher water. Cell productivities ranged between 0.43 to 0.52 gl^{-1} h^{-1} at 30-32°C pH 4.8. Very high cell yields were reported (0.84-0.95) and the authors explained them by saying that other carbon forms besides carbohydrates were being used by the yeast. An extracellular glucoamylase was produced at a maximum rate at D = 0.22 h^{-1}. Sa-Correia and van Uden (58) reported recently the production of biomass and enzymes in batch and continuous culture of *L. kononenkoae* IGC 4052 parent strain and one derepressed mutant. A μ max of 0.14 h^{-1} was found in batch culture for both yeasts. Biomass productivities were 0.086 and 0.137 gl^{-1} h^{-1}, employing a 0.5% (w/v) soluble starch medium at pH 5.7 and 28°C for the

parent and derepressed mutant, respectively. Enzymes productivities in continuous culture were also higher for the
latter. The derepression of the enzyme permitted the use of
dilution rates near μ max. Starch assimilating yeasts were
tested by Yoshizawa et al. (59) in wastewaters discharged from
miso production with COD values of 10.5-14.0 gl^{-1}. Dremaine
et al. (59a) reported that the addition of wheat bran extract
and ethanol stimulated the growth of *E. fibuligera*. Wilson
et al. (59b) studied the growth of *S. alluvius* IGC 2829 on
soluble starch media at 0.2% concentration. Growth rates of
0.21 to 0.25 h^{-1}, yields of 0.59-0.62 and substrate consumption
of 96.5 to 97.5 were obtained. The same growth rates were
obtained with glucose hence the extracellular amylolytic
enzymatic system was not limiting growth. However it was subjected to catabolite repression and was induced by soluble starch
or its degradative products. Depressed mutants could be
obtained which showed a higher amylase activity. Oten-Gyang
et al. (59c) studied the effect of medium pH on the excretion of
amylolytic enzymes, the yields and growth rates of eleven amylolytic yeasts. When pH was below 3 amylases remained intracellular
and biomass yields were low. On buffered media amylases were
excreted, yields were high but growth rates did not improve.
Touzi et al. (59d) tested 15 yeast strains on five different
starchy media and found *Schwanniomyces castelli* best in terms of
biomass yield, 0.45, and batch growth rate: 0.28 h^{-1}.

Hang (60) and Hang and Woodams (61) grew *Saccharomyces
fragilis* ATCC 12424 on lemonade-processing and sauerkraut-
processing wastewaters. The yeast strain had good settling
characteristics for an effective yeast recycle. A batch productivity of 0.08 gl^{-1} h^{-1} with 87% BOD reduction was reported
in the former; a yield of 0.33-0.37 g biomass per gram of lactic
removed and 95.97% lactic acid removal was reported for the
latter. Hang (61a) used *K. fragilis* in a sequencing batch
(fill and draw) operation with lemonade and orange juice
processing waste effluents, obtaining 91-93% COD reduction.

Oliva and Hang (62) previously reported the continuous
culture of *Candida utilis* in sauerkraut wastewaters. At D =
0.18 h^{-1}, biomass production was 0.30 gl^{-1} h^{-1} and 96% lactic
acid consumption; at D = 0.35 h^{-1} the corresponding values were
0.56 gl^{-1} h^{-1} and 56%. Prior et al. (63) employing *S. fragilis*
were able to reduce 90% the original COD of pressings from spent
grain brewery wastes; on the other hand *C. utilis* grew best in
supplemented pineapple cannery effluent. The biomass yield and
specific growth rate were higher at lower initial substrate
concentration. Maximum biomass was 4.4 g/l when initial substrate was 13.5 g/l. They also studied the growth of *C. utilis*

ATCC 9255 on pineapply cannery effluents as such and mixed with
the water extracts of pineapply peels (63a). Maximum growth
rates of 0.42 h^{-1} and a cell yield of 0.19 were obtained with
the latter and 0.44 h^{-1} and 0.31, respectively with the
former. Taguchi et al. (63) studied the growth of *Torulopsis
candida* batchwise and in continuous culture using soybean whey
as substrate. A 90-92% sugar conversion with a 0.36 gl^{-1} h^{-1}
biomass productivity was obtained in a 22 h batch. In continuous culture at D = 0.3 h^{-1} a 0.44 gl^{-1} h^{-1} biomass productivity with a 0.68 g cell/g sugar consumed were obtained.
The biomass productivity was increased to 3 gl^{-1} h^{-1} with yeast
recycle. Braun et al. (65) reported the growth of *Candida
scottii* batchwise in citric acid production waste effluent,
obtaining a 50% COD reduction and a biomass productivity of
0.29 gl^{-1} h^{-1}. In continuous culture at D = 0.1 h^{-1}, 0.32 gl^{-1}
h^{-1} and 66% COD reduction values were reported for the same
yeast; improving to 2.4 $gl^{-1}h^{-1}$ and 60% COD reduction when
Candida vartiovaardi was employed (a flocculent yeast) (65a).
Instead of drying the biomass obtained they suggested the
concentration of the broth and its utilization in liquid feeds
for pigs.

Kamel (66) used ground dates mixed with four times their
weight of cold tap water as substrate for various yeast strains.
An Ogi yeast gave the best results in shake flasks, obtaining
0.25 gl^{-1} h^{-1} and 56% sugar conversion. Batchwise in a fermentor:
0.65 gl^{-1} h^{-1} and 56% conversion; continuous culture at pH 3.5,
37°C, 0.4 gl^{-1} h^{-1}. Aligedi et al. (66a) obtained the best
growth yields with *Candida* sp. and *Rhodotorula* sp. in date
extracts containing 10-12 g/l of initial sugars and nitrogen
and phosphorus supplemented. Optimum temperature was 28°C and
a range for pH between 4.0 to 6.5 (66b). Guiraud et al. (67)
tested in shake flasks at 28°C, 16 different yeasts on synthetic
inulin. *K. fragilis* CBS 1555 and *K. marxianus* gave the best
productivities (initial rate 12 gl^{-1} h^{-1}, average rate 4.2-5.4
gl^{-1} h^{-1}).

Kajs and Vanderzant (68) tested the growth of *C. utilis* and
S. lipolytica in 1% tallow. At a pH of 4.8 a yield of 0.76 g g^{-1}
and a productivity of 0.32 gl^{-1} h^{-1} were obtained. Zarnescu
et al (69) reported the isolation of *Candida* yeast strains capable
of assimilating animal fats. Martinet et al. (69a) tested the
growth of several yeasts in the solid fraction of palm oil,
giving some strains a dry biomass yield per unit weight of substrate consumed above one. Nakahara et al. (69b) selected
Candida blankii CBS 1898 from eleven yeast strains out of 126
that grew in a refined palm oil plus mineral enriched medium. A
batch yield of 0.78 and a growth rate of 0.37 h^{-1} were observed.

Mudgett et al. (70) studied the growth of *C. utilis* in the juice resulting from alfalfa pressing diluted 1 to 4. A biomass productivity of 0.24 gl^{-1} h^{-1} and a 59% total organic carbon reduction were obtained. Okada et al. (71) tested and screened one hundred and twenty five yeast strains on juice obtained from pressing citrus peels. Two *Candida* sp. gave the best growth results in shake flasks at 30°C, 3 days with nitrogen and phosphorus supplementation. The yields were around 0.52 and the productivity 1.22 gl^{-1} h^{-1}. Mouchet (71a) describes the industrial set up to produce yeast biomass from the effluent from fermented cabbage; *Hansenula anomala* was grown at D = 0.1 h^{-1} to 12-14 b biomass 1^{-1} obtaining 90% BOD reduction. Li Sui Fong (71b) used the heat deproteinized juice from sugarcane tops and leaves extraction.

Alian et al. (71c) tested the growth of *Candida utilis, Candida arborea, Torulopsis magnoliae* and *Saccharomyces fragilis* on mango and orange wastes which were mixed with water, boiled, filtered through cheese cloth and diluted to 8 and 5% total soluble solids respectively. The first three yeasts grew quite well (7.3 to 11.10 g/l after 48 h, 30°C) in a mineral supplemented fruit waste (0.03% potassium phosphate, 0.07% magnesium sulphate, and 0.15% urea) in shake flasks.

Ramirez and Gonzalez (71d) grew *Candida utilis* PPR-291 in cane alcohol slops obtaining 10 gl^{-1} of biomass and BOD reductions of 55-60%, when the slops were supplemented with 0.15% N as ammonium sulphate and 0.10% P as KH_2PO_4. Tauk (71k) grew *Rhodotorula glutinis* and *R. mucilaginosa* in vinasse, vinasse containing molasses and vinasse supplemented with salts and acids. Yeasts were cultivated at pH 4.6 and 30°C and after 30 h growth was better in the enriched media. In another publication (71n) she reported the greatest biomass production with *C. krusei* if phosphate was added and *C. guillermondi* and *C. utilis* if both phosphate and urea were added.

Davy et al. (713) describe the Bioaccelerator process installed in England which grows *C. utilis* on carbohydrate waste streams from a confectionary plant. The plant capacity is 1-1.5 ton of dried biomass per day and removes 60-75% of the influent COD. Plant results are similar to the laboratory scale results reported by Forage (71f).

C. utilis was grown on filtered piggery excrements carbon supplemented with beet molasses obtaining a biomass productivity of 1.58 gl^{-1} h^{-1} and removals of 76% of the original total nitrogen and 84% of ammonium nitrogen (71g, 71h, 71i).

C. tropicalis was grown batchwise and in continuous form in polyethylene pyrolysate at 31°C and pH 5.5. Maximum growth rate was 0.168 h^{-1}, cell yields were between 0.39 and 0.47 and utilization of pyrolysate was between 33 and 49% (71j). A continuous process for the cultivation of *Brettanomyces* sp. in ethanol or acetic acid mixed with carbohydrates has been developed (71l). Moresi and Marchionni (71m) grew *C. utilis* ISS28 on grape must and the yield was optimized by adding nutrients according to a composite design experiment. Ercoli and Ertola (71o) grew *Saccharomycopsis lipolitica* in olive black water obtaining a biomass concentration of 18-26 gl^{-1} and a yield of 0.32.

2.   Yeasts on Acid Hydrolyzed Lignocellulosic Biomass

The growth of yeasts has also been studied in soluble substrates produced from the hydrolysis of lignocellulosic materials. Scientists from Czechoslovakia and Bulgaria continue thier publications regarding the addition of ethanol to wood-acid hydrolyzates as a supplementary source of carbon due to its beneficial effects on the growth of *C. utilis*, *C. tropicalis*, and *C. scottii* (72, 73). Volfova and collaborators (74-75b) selected *C. tropicalis* 2838 as the best yeast strain to grow in milled straw mild acid hydrolyzates and neutralized without requiring any additional vitamins or trace elements. After 30 h the whole product contained 8-10% digestible proteins. The yield was 0.44 g g^{-1} and the growth rate 0.34 h^{-1}. No drop in the growth parameters was found when straw was initially mixed with the solid waste resulting from anaerobic digestion of swine excrements. Hence eliminating the need of inorganic nitrogen salts. The yeast was capable of assimilating from hydrolyzates, first glucose, then xylose and finally arabinose and mannose, but not completely. The yeast cells were cultivated in a thick medium in the presence of straw particles which was added to the fodder mixture either dried or wet. Eliminating in this way the separation, disposal and purification of the wastewater. Semenov and Podgorski (76) used *Candida numicola* and *Candida* sp. K-19-73 with acid hydrolyzates from woody parts of flax and hemp, rice and wheat straw, sunflower stems and oak bark. Malinovskaya et al. (77) found the optimum hydrolysis conditions of wheat straw for yeast growth as 7% HCl, 60 min and 100°C. Yields of 35-36 g biomass/100 g

reducing sugars were obtained with *C. utilis* and *C. tropicalis*
on the pentose fraction of acid hydrolyzates of sunflower seed
husks; the yield increased to 45-48 g g^{-1} for the hexose fraction
(78). *C. utilis* was grown on acid hydrolyzates of banana leaves
(79) and two *Candida* strains isolated from compost on acid
hydrolyzates of brown seaweed (80). Abou-Zeid et al. (81)
pressed the fleshy leaves of onion skins to extrude their juice;
it was aerated to remove volatiles and hydrolyzed with 0.2N HCl
for 4 h at 90°C. The juice with nitrogen and phosphorus
enrichment gave a yeast *(S. cerevisiae)* batch productivity of
0.067 gl^{-1} h^{-1} at 30°C, with 92% of sugar consumption, these
results were superior to those found in a synthetic medium
(81a); Moo-Young et al. (82) have described the Waterloo process,
in which digestor sludge is used as a nutrient source for yeast
growth mixed with acid hydrolyzates of lignocellulosics. The
residual fibers can be sent to the digestor or fed to ruminants
after sodium hydroxide treatment. The yeast cream obtained by
centrifugation and flotation can be directly utilized as
animal feed or can be further processed in a rotary drum drier.
In batch trials at 30°C and pH 4.5 a specific growth rate equal
to 0.28 h^{-1} was found mostly for glucose uptake, and 0.033 h^{-1}
for xylose; also a yield of 0.32 g biomass/g reducing sugars was
reported. More recently, Gonzalez and Moo-Young (83) have
studied the growth of *C. utilis* CM1-23311 in mild acid hydrolyzates of corn stover. The rate of sugars uptake was first
galactose, glucose, xylose, cellobiose, xylose, and finally
arabinose. In a batch reactor 8 g biomass l^{-1} in 13 h were
obtained. In continuous culture maximum productivity (1.8 gl^{-1}
h^{-1}) was obtained at D = 0.22 - 0.26 h^{-1} with close to 100%
substrate utilization. *C. utilis* seems the best yeast to grow
in acid peat hydrolyzates obtained by heating to 160°C for 1 h
and 0.5% H$_2$SO$_4$ (84). Batch growth of *C. utilis* UWO 993 at 30°C,
pH 5.0 gave 10 g biomass l^{-1} in about 20 h (0.5 g biomass
l^{-1} h^{-1}), with a consumption of only 55-60% of the initial
carbohydrates, with a yield of .38-.40 g biomass g^{-1} carbohydrate utilized. In continuous culture a yeast productivity
of 1.24 gl^{-1} h^{-1} at 0.125 h^{-1} and a yield of 0.59 g biomass g^{-1}
carbohydrate utilized were obtained. *C. tropicalis* gave a yield
based on sugars utilized (glucose, cellobiose, and xylose) of
38-62% (86). *C. rugosa* was grown on maize gum, a by-product of
the corn hydrolyzing industry, and a productivity of 0.34 gl^{-1}
h^{-1} was achieved with N, Mg, and K supplementation (86a).
Pozmogova and collaborators (86b) report the growth of
C. scottii and *H. amomala* in continuous culture at D = 0.2 h^{-1}

and 10 g/l of reducing sugars. Yields were between 0.45-0.47
but tended to decrease to 0.37-0.39 at higher temperatures (43c).
Experiments were done also as mixed cultures. Bobleter et al.
(86c) grew *C. utilis* in the hydrothermal degradation products
of lignocellulosic biomass and obtained yields of around 0.5
with consumption of glucose, cellobiose and xylose, and
Khandelwal et al (86d) grew it on prehydrolyzate liquor from
eucalyptus. *C. utilis* QM8240 (86e) was grown on neutralized
hydrolyzates from sulfuric acid treated mesquite wood. The
treatment hydrolyzed 25% of the wood and produced 0.14 to 0.16 g
hexose g^{-1} dry wood. The microorganism utilized 45 to 50% of
the sugars on shake flasks at 37°C for 24 h. The highest biomass
yield and protein contents were obtained by *C. robusta*, *C. humicola*,
and *H. anomala* in undiluted hydrolyzates from sawdust, cotton waste
and rice hulls (86f). Gonzalez and Knackmuss (86g) hydrolyzed
rice hulls with sulfuric acid and different mixtures with nitric
acid and obtained maximum growth rates of *C. tropicalis* QML 7601
between 0.3-0.6 h^{-1}. Davey and Bruce (86h) have found inhibitory
substances in the acid hydrolyzates of piggery slurry. *C. utilis*
was grown on acid hydrolyzed-ammonium neutralized wheat straw for
five days, increasing by 37% the crude protein content and by 14%
the in vitro digestibility of the total mixture (86i).

3.   *Yeast Growth on Whey, Mollasses and Sulfite Waste Liquors*

Research work on the use of whey, molasses and sulfite waste
liquors as substrates continues being published. Teuber (87)
presents a summary of the microorganisms and processes available
for the production of SCP from cheese whey and lists the world
commercial production of yeast biomass as only 15,600 tons
(France, Belgium and United States of America). The following
microorganisms have been tested: *Kluyveromyces fragilis* and
K. lactis; Candida kefyr, *C. utilis*, *C. pseudotropicalis* and
C. intermedia; Torulopsis sphaerica, *Penicillium cyclopium* and
Oospora lactis. de la Gueriviere (88) describes the process
characteristics employed by Bel Industries: deproteinized whey,
K. lactis or *K. fragilis*, continuous culture in air lift-
fermentors, centrifugation for biomass recuperation and then
either: i) concentration and heat shock treatment in a triple
effect evaporator plus spray drying to a powder, or ii) filtration,
heat shock treatment, drum drying and pelletizing. In a series
of articles, Moresi and collaborators studied the optimization
and scale-up from shake flasks to 100 l stirred tank fermentor
of the growth of *Kluyveromyces fragilis* IMAT 1872 (89-94). Best
yields were obtained when deproteinized whey was supplemented with
nitrogen, phosphorus and yeast extract. Optimum pH was 5.1 and
temperature in a range of 31-34°C. Best COD removal was 91% and
biomass yield 0.57 g biomass/ initial lactose. Maximum batch

productivities of 2.16 gl^{-1} h^{-1} were achieved. Initial lactose concentration between 5-24 gl^{-1} and k_La between 1200-2100 h^{-1} gave biomass yields between 10% of maximum value indicating that the surface around the maximum is very flat. Burgess (95) grew *S. fragilis* NRRL 1109 on lactose permeate from whey in a laboratory scale tower fermentor. Growth was oxygen limited and at a maximum oxygen transfer rate of 280 mMO_2 l^{-1} h^{-1} a biomass yield of 0.55 was recorded and 81% and 93% of COD and lactose were removed after 8 h respectively. Castillo and Sanchez (96) studied the growth in shake flasks of *K. fragilis* NCYC 587 in deproteinized whey supplemented with 0.2 $(NH_4)_2SO_4$ and 0.1% yeast extract. A yield of 0.53 g biomass/g lactose utilized and 92% of lactose utilization were reported in 24 h. Nitrogen supplementation of whey and pH effects have been studied recently (97,98) as also the characterization of the β-galactosidase from *K. fragilis* (98a) and *C. pseudotropicalis* (98b,c). Gomez and Castillo (98d) found optimum continuous culture conditons for *C. pseudotropicalis* on whey supplemented with tryptophan and minerals: 4.6 gl^{-1} yeast concentration; 1.4 gl^{-1} h^{-1} yeast productivity and maintenance coefficient 5-10 mg lactose g^{-1} h^{-1}. *S. fragilis* consumed practically all of the lactose in milk as demonstrated by Edelsten et al. (99, 99a) in a process to prepare lactose free milk for lactose intolerant persons. Yaziciouglu et al. (100) tested various yeasts and fungi on whole whey supplemented with 0.5% $(NH_2)_2SO_4$ and 0.1% of KH_2PO_4. Two filamentous fungi *Rhizopus* and *Fusarium* sp. produced more biomass than the yeasts. When *K. fragilis* was grown on whole whey at 35°C for 18-24 h and then the broth steamed to 95°C in order to coagulate the whey proteins, the resulting liquid had a final COD of about 50% of the original material (101) and the resulting biomass was formed of 43% yeast cells and 57% of whey protein (102). This material could be further processed for food or used directly as feed. A similar approach was suggested by Terra (103) who reported a final biomass (yeast plus whey proteins) of 27.7 gl^{-1}. In cottage cheese whey the growth of *S. fragilis* NRRL 1156 at 35-40°C completely depleted lactose in 8-12 h and was able to utilize 20-40% of the whey proteins (104). Osovik et al. reported the growth of *Trichosporum cutaneum* K-1 on heat deproteinized whey (105). A biomass productivity of 3.11 gl^{-1} h^{-1}, a yield of 0.40 based on initial total dry matter and a maximum batch growth rate of 0.45 h^{-1} were reported. Foda (106) found *Debaromyces hansenii* 70-100 the best yeast strain among sixteen different strains tested in sweet or salted deproteinized whey, nitrogen supplemented obtaining 28-30 gl^{-1} of biomass at 24°C for 3-4 days. For optimal growth of *K. fragilis* B15 in swiss cheese whey, nitrogen, phosphorus and vitamins were needed in order to obtain a maximum yield of 0.45 and a total yeast concentration of 22 kg m^{-3}. The yeast did not use whey

protein nitrogen and the most suitable concentration of ammounium
hydroxide was from 1.0 to 1.4 kg m^{-3}. The phosphorus requirement was lower than expected only 0.12 kg m^{-3} and yeast extract
was required at a rate of 1.0 kg m^{-3} (107). An osmophilic
strain of *Turulopsis candida* was grown on deproteinized concentrated at not more than 15% total solids whey. Good lactose
utilization was obtained, the resulting biomass was inactivated
by heating, the whey proteins were added back, the mixture
homogenized and dried (108). The final biomass concentration
in wheys of different calcium content was about the same, 45-55
gl^{-1} (108a) and in whey from different type of cheese a 98-100%
lactose utilization was observed (108b). According to Giec and
Kosikowski (109) *K. fragilis* ATCC 8582 was most productive in
5% lactose permeates; they also tested 10 different lactose
fermenting yeasts in cheese whey and skim milk UF permeates and
found widely different biomass yields. Three *Kluyveromyces*
yeasts were most productive, including *K. fragilis* ATCC 8582.
Bretanomyces anomalus ATCC 10559 and *Candida blankii* ATCC 18735
developed high yield of biomass but required more time to attain
complete lactose utilization, which was enhanced by the addition
of fungal acid lactase (109a). Yields and growth rate were
higher for *K. fragilis* CBS397 than for *K. fragilis* NRRLY1190
when grown on deproteinized whey at pH 4.5 with mineral supple-
ments (110). *Candida pseudotropicalis* gave a yield of 0.29 and
a productivity between 1.12-1.21 gl^{-1} h^{-1} in soft cheese
deproteinized whey, with a lactose consumption of 91% (111).
Andrzejwski et al. (112) obtained productivities of 1.4 to
2.0 gl^{-1} h^{-1} employing *T. cutaneum* when the whey was enriched
with β galactosidase and corn steep liquor; 92.8% of the
lactose was utilized. Vrignaud (113,114) describes the continuous culture technology for *K. fragilis* at 38°C and pH 3.5
on deproteinized (heat plus acid) whey with 25-35 g lactose l^{-1}.
A yield of 0.50-0.53 g biomass g^{-1} lactose initially present
was reported. The biomass was centrifuged, washed, centrifuged
again, filtered, heat treated (83-85°C) and dried. *T. candida*
and *S. fragilis* were isolated from raw whey and showed greater
ability to utilize lactose (115). A more direct yeast screening
by Moulin et al. (116) favored *B. anomalus* CBS77 or *Candida
blankii* CBS1898 as both yeasts gave high yields of biomass,
0.44 and 0.52, respectively, and had few poly-unsaturated acids.
However, from their results *Pichia polymorpha* CBS186 and
P. pseudopolymorpha CBS2008 seem also to satisfy both requirements and follow close behind the two strains mentioned before.
Mohandas and Srinivasan (117) showed that *S. fragilis* accumulated
maximum amounts of fat when grown in static cultures in
unsupplemented cheese whey medium. Data reported also by Moon
and collaborators (118-120) and is discussed in (2). Dluzewski
et al. (121) showed a yield increase by the addition of inorganic
nitrogen and phosphorus, but a decrease in yield was observed

if a heating pretreatment of whey was done (5 min at 134°C).
A dialysis fermentation technique employing UF pretreated whey
has been published by Lane (122). The operation was cyclic,
each cycle lasting 24 h. Starting with an inoculum of 1 g
biomass l^{-1} after four cycles it was 90 g biomass l^{-1}. Reducing
sugars consumption was about 84%. Abreu et al. (123) obtained
16.6 gl^{-1} of *K. fragilis* biomass on buttermilk containing 2%
lactose and 1.3% protein. Vananuvat and Kinsella (124, 125)
employed spray dried crude lactose from UV and RO cottage cheese
permeates as substrate in batch and continuous cultivation of
S. fragilis at 30°C and pH 5. A 90% lactose consumption in 8 h,
a yield between 0.55-0.62 and a batch productivity of 0.14 gl^{-1}
h^{-1} were reported. In continuous culture at a dilution rate of
0.23 h^{-1}, a 2.5 gl^{-1} h^{-1} productivity, a 0.70 yield (g biomass/g
lactose initially present), 10.45 gl^{-1} biomass concentration in
the effluent were also reported. General descriptions of whey
utilization alternatives or yeast production plants have been
published (126-130) as also an economic analysis of a commercial
yeast plant (131). Several patents have been granted for this
technology (132-140). Kappeli et al. (140a) report the growth
of *T. cutaneum* GCM70698 on ultrafiltrate from milk in laboratory
and pilot plant bioreactors at pH 4.5 and 30°C. Maximum growth
rate from laboratory batch data was between 0.35-0.40 h^{-1} and
yield on lactose ranged between 65 and 70%. The rather high
values in yield are due to uptake of urea also as a carbon
source. Growth became oxygen limited after approximately 8 h,
time which oxygen demand exceeded the maximum transfer rate of
the bioreactor of 133 m mol O_2 l^{-1} h^{-1}. Fed batch culture
was successful at the pilot plant level although growth rates
were lower due most probably to oxygen limitations. Whey from
cheddar cheese manufacture adjusted to pH 6.0 was used as
substrate for the growth of *K. fragilis, Torulopsis sphaerica*
and *C. utilis;* more than 90% lactose consumption was obtained
with all yeasts but a higher total solids reduction with the
former (140b). Poncet and Jacob (140c) compared the growth of
Candida curvata CBS570, *Candida shehatae* CBS5813 and *K. fragilis*
in acid and sweet whey supplemented with ammonium sulphate.
Better growth rates were obtained with sweet whey and *C. curvata*.
Gonzalez and Berry (140d) report on the induction of β-galacto
sidase in *K. lactis.* Sandhu and Waraich (1403) tested thirteen
yeast species belonging to nine genera in cheese whey supplemented
with minerals and yeast extract; *Wingea robertsii* proved to be
the best strain in terms of high yield and shorter growth period.

    Fodder yeast production from cane molasses is a commercial
process in many tropical countries. In Taiwan improvements on
continuous fermentation have been described (141,142). Strains
of *S. cerevisiae* have been studied in detail in Egypt (143)
where isolation and screening (144 and levels of nitrogen and

phosphorus required (145) have been studied, recommending as a
conclusion an incremental feeding or fed batch system in order
to optimize between cell yield and substrate utilization (146).
Estevez and Almazan from Cuba report results that show the
excellent biomass yields obtained in continuous culture of high
test molasses (as such or in slops-molasses mixtures) and crude
sugarcane juice (147, 148). Molasses quality in terms of its
suitability for fermentation has been reported (149). An
interesting alternative from the standard process has been
developed in Cuba, where the yeast microbial biomass is
produced as a suspension in sugar syrups. Sugar juice is
divided into two processing parallel schemes. In one a syrup
is obtained by evaporation; the other is used to grow aerobically yeast. Then the two are mixed and further concentrated.
The ratio carbohydrate:protein is variable according to the
animal feeding requirements. Typical values are: 72% total
solids and 10-12% crude protein (150, 151). Kontramin 210 was
found effective in controlling foam in an evaluation of
different antifoaming agents (151a). Moreira et al. (152)
grew *R. gracilis* as well as *C. utilis* in molasses enriched with
urea and inorganic phosphorus. Mian et al. (153) have shown
that for *Candida utilis* there is a threshold sugar level in
continuous continuous culture beyond which there is a
fermentative pathway that may be similar to a Crabtree effect.
The sugar level is around 20 g/l. Below this range, aerobic
metabolism predominates irrespective of the dilution rate up
to 0.4 h^{-1}. Productivity of biomass above this sugar value
declines due to ethanol accumulation which causes earlier washout. Later Rickard and Hewetson (153a) have shown that the
four regulatory responses to glucose found in *S. cerevisiae* are
also found in *C. utilis* and the differences are found in the
degree that these regulatory responses are operative. Peschard
and Viniegra (154) presented a cost analysis for yeast biomass production from cane molasses in small plants. These
could be economically feasible if yeast was further processed
in order to obtain as the main product yeast nucleotides.
Konstantinova and Lippert (155) studied the possibility of
using mixtures of nitrogen-containing industrial waste products
and molasses for the cultivation of *C. utilis*. Heisel describes
a new process (Garrido-Schick technology) to produce yeast from
alcohols or molasses (156, 156a). El Sheikh Idris and Berry
(157) report a study in Sudan in which various thermotolerant
yeast strains were isolated from sugar mills. Their biomass
yield was determined at 40°C in a medium that had 0.5% sugars
as molasses and 0.17% urea. The highest yields (0.38-0.42) were
obtained by *C. albidus, H. polymorpha, S. chevalieri, Rhodospor-
idium toruloides.* Higher biomass yields, protein and nucleic
acid contents were exhibited by the yeast strains at 35 and 30°C
(157a). A two-stage continuous system was operated with

R. glutinis at 32°C, pH 4.5 with a 20 gl^{-1} as molasses medium.
By operating at D$\leq$ 0.1 h^{-1} a biomass yield of 0.30-0.33 was
obtained with the following composition: 30% of fat and protein
and 4% of RNA (158). The growth of *Endomycopsis* sp. in molasses
diluted with seawater has been patented (159). Chahal et al.
grew in a medium with 50 gl^{-1} sucrose and potassium nitrate as
nitrogen source *R. glutinis, L. starkeyi, C. utilis* and *S. cere-*
visiae and found a lipid content in the microbial biomass between
6.3 and 53.6% (159a). Under nitrogen-limiting conditions and
continuous culture the specific lipid production rate,
employing *R. glutinis* NRRL Y-1091 in a 30 gl^{-1} glucose medium,
was 0.012 g lipid g biomass $^{-1}$ h^{-1}(159b). Hsiao et al. (159c)
have studied the sequential uptake of mixtures of glucose,
xylose, xylulose and xylitol by *S. cerevisiae, S. pombe, C. utilis,*
and *R. toruloides.* Moebus and collaborators at the Institut für
Mikrobiologie in Kiel have developed a technique for growing yeast
cells in form of solid particles in a gaseous fluidized bed to
which is sprayed the growth medium (159d). Ethanol can also
be produced (159e). However growth rates are almost an order
of magnitude slower than rates of yeast in submerged culture
(159f).

Candida utilis is the yeast being employed by industry in
the microbial biomass production from sulphite waste liquors
(160-163). Barta (164) reported much better biomass yields with
Cryptococcus diffluens and Chaudry et al. (165) have given
results with *Candida rugosa, C. tropicalis* and *C. papapsilosis.*
Nitrogen supplementation in the form of urea or ammonium sulphate
and phosphorus as phosphoric acid optimized biomass yield and
substrate consumption (166). Yeast extract addition resulted
in a 20-40% yield increase (167). The addition of ethanol as a
second substrate has been experimented by Rychtera et al. (167a)
and Setzermann and Zimmer (167b). Aldonic acids in their
lactone form as they occur in WSL are not assimilated by yeasts
but can be converted to aldonic acids by bubbling air into the
WSL adjusted to pH 8 (168). Sugar uptake has been studied in
a solution simulating the hexose content of WSL (169) and also
from a solution of monosaccharides, acetic acid and ethanol
(170). Maekawa et al. (171) reported a continuous culture
biomass productivity of 0.9 gl^{-1} h^{-1} with around 60% of substrate utilization. Camhi and Rogers (172) in a two-stage
continuous culture system improved substrate utilization to 81%
(as measured by BOD) with a 0.95 gl^{-1} h^{-1} biomass productivity.
Recently Gold et al. (1981) have reported biomass productivities
in the range of 5-6 gl^{-1} h^{-1} with a cell recycle one stage
continuous system (having a cell concentration in the reactor
of around 25 gl^{-1}) and an 83% substrate utilization measured
as reducing sugars.

4. *Yeasts on Enzymatically Hydrolyzed Lignocellulosic Biomass*

Enzymatic hydrolyzates have been also employed as substrates for yeast growth. Revah-Moiseev and Carroad (171) have shown that the yeast *Pichia kudriavzevii* assimilated chitin hydrolyzates from shellfish waste. The hydrolyzates contained mainly glucosamine and N-acetylglucosamine and were obtained by the chitinase of *Serratia marcescens* QMB1466 at 50°C and pH 6. The yeast productivity at 37°C and pH 4.6 was 0.13 gl^{-1} h^{-1} at 0.33 h^{-1}, substrate consumption was 88% and Y= 0.45. A process scheme including a cost analysis was also published (171a,b). A patent was issued to grow *C. utilis* on an enzymatically hydrolyzed cassava suspension (172). Solomon et al. (173) grew batchwise *C. utilis* Y-1084 on previously hydrolyzed corn dust. The hydrolysis was carried out by adding a heat stable α-amylase from *Bacillus licheniformis* and amyloglucosidase from *Aspergillus niger*. A detailed analysis of growth parameters is given both for filtered and unfiltered corn hydrolyzate medium. Araujo and D'Souza (174) hydrolyzed previously hammer milled and alkali treated rice straw with cellulases from the fungus *Aspergillus terreus* Thom; five yeasts were then gorwn in the hydrolyzates at shake flask level. *C. utilis* and *S. cerevisiae* produced the highest cell concentrations and the best substrate consumption, around 90%. Nishio and Nagai (175) hydrolyzed mandarin orange peel with macerating enzymes from *A. niger;* the batch cultivation of four yeasts in a jar fermentor at 30°C was tested with the following results: *S. cerevisiae* (Y = 0.51), *C. utilis* (Y = 0.48), *Debaromyces hansenii* (Y = 0.69), and *Rhodotorula* (Y = 0.70). The higher yields of the last two yeasts might be due to the assimilation of other sugars that were not accounted for by chemical analysis as pointed out by the authors. Acid hydrolysis (15 min at 120°C with 0.8N H_2SO_4) was also tested but the yield of reducing sugars was lower than the enzymatic hydrolysis and the yeast yield was also lower in the acid hydrolyzates. In another patent, a cellulase solution was added to *C. utilis* as also a cellulosic substrate and a simultaneous cellulose hydrolysis and yeast growth occurred (176). Kolarova and Farkas (177) have recently shown that practically all of twenty six yeast strains were found to be sensitive (cell wall weakening and lysis) during prolonged (20 h) incubation with the crude extracellular cellulolytic enzyme complex of *Trichoderma reesei* QM 9414 and its mutants M6 and MHC22. However under yeast growth conditions these effects were less pronounced. The end product enzyme inhibition that the coupled saccharification/yeast growth offers might be offset by the instability of the yeast cell. On the other hand, it might be seen as an *in situ* cell wall degradation scheme that might improve the nutritional characteristics of the product. Beja da Costa (178) grew four yeasts

on the sugar syrup obtained by enzymatic hydrolysis of pretreated *Eucalyptus globulus*. *C. utilis* did not use part of the
xylose in the medium; *C. maltosa* did not use cellobiose;
Hansenula holstii did not use part of the cellobiose and xylose
present; *C. tropicalis* used all the sugars in the syrup and
gave the best yield (Y = 0.62). Weckstrom and Leisola (179)
grew *C. tropicalis* and *C. utilis* in shake flasks in bisulphite
spent liquor which had been previously enzymatically hydrolyzed.
The hydrolysis was carried out in a column reactor with
immobilized *A. niger* hemicellulase and as a result the reducing
sugars were increased about two times to 27 gl^{-1}. The yeasts
consumed about 58% of the reducing sugars and yields of 0.43 and
0.33 were reported respectively. Nojiri (179a) has presented
data in the continuous culture of *C. utilis* on previously
dextrinized cassava starch. However, saccharification was done
simultaneously in the three fermentor continuous system by
adding in the feed a saccharogenic amylase. The dextrinization
was also done enzymatically with enzymes from *B. subtilis*. A
biomass productivity of 3.17 gl^{-1} h^{-1} and a 90% substrate consumption were obtained when the operation was done at 35°C,
4-4.5 controlled pH, concentrations in the feed of 4.0-8.6-4.5
and 1.5 gl^{-1} of dextrin, ammonium nitrate, potassium phosphate
and magnesium sulphate respectively, and a dilution rate of
0.17 h^{-1}. Kamikubo et al. (179b) grew *C. utilis* AKU 4570 and
S. cerevisiae AKU 4100 in shake flasks on an enriched medium
containing the enzymatic hydrolyzate of microcrystalline cellulose, filter paper, alkaline treated rice straw and newspaper.
The hydrolytic enzymes were obtained from *T. reesei* QM9414 and
P. filamentosa FERM-P1797. *C. utilis* had a higher yield and
consumed more substrate than *S. cerevisiae,* which presumably
could not assimilate cellobiose, cellotriose and xylose and
arabinose present when rice straw was the substrate for hydrolysis.
Musenge et al. (179c) reported that *C. utilis* NRRLY-100 gave a
higher yield on cassava hydrolyzate (0.54) than on molasses
(0.46) after 40 h at 30°C and pH 4.5. The hydrolysis was carried
out by adding commercial enzymes and after 48 h a 98% dextrose
equivalent solution was obtained. Moo-Young et al. (179d)
obtained good growth of *C. utilis* from mixtures of anaerobically
fermented cattle manure liquor and barley straw acid hydrolysate.
A protein productivity of 77 mg l^{-1} h^{-1} in a 32 h batch was
reported. However 30-48% of the reducing sugars were not
utilized by the yeast mainly galactose and arabinose. Yoon et al.
(179e) used an enzymatic hydrolyzate of pretreated rice straw
as substrate for the growth and fat production of *Rhodotorula*
gracilis, with an initial C/N ratio in the culture medium of
43.6, 9 gl^{-1} of biomass with 54% of lipids was obtained and a
lipid yield of 10.3 g per 100 g carbon source.

Taniguchi et al. (179e) used the enzymes of *Pellicularia filamentosa* to hydrolyze sodium chlorite pretreated straw and obtained the following growth results on the sugar solutions: *C. tropicalis* 70% sugar consumption and 6.8 g biomass l^{-1}; *Torulopsis xylinus* 76%, 10.1 gl^{-1}; *Trichosporon cutaneum* 76.1%, 10.1 gl^{-1} and *C. guillermondi* 74%, 7.6 gl^{-1}.

C. utilis ATCC 9226 and *C. tropicalis* ATCC 1369 utilized 93 and 84% of the sugars obtained from the enzymatic saccharification of sulphite pulp (179f). Morrison et al. (179g) grew *C. utilis* on sugars released from feedlot waste fibers alkali-pretreated and hydrolyzed with cellulases from *T. viride*.

5.   Fuel Ethanol and Yeast Biomass

Saccharomyces cerevisiae produced under anaerobiosis as a by-product from ethanol from di- and monosaccharides is probably until today one of the most underutilized microbial biomass sources. There is an average yield of 0.1 kg of dry yeast biomass per liter of pure ethanol produced in the yeast fermentation of dilute sugar solution either batchwise or in continuous (1). In actual practice part of this yeast, after acid pretreatment, is recycled to the process. Depending on the operational practice, yeast cell mortality values are around 30-40%, hence a biomass bleed always is necessary in the recycle stream. Hence, ethanol-fuel production in a large scale should also be seen as a microbial biomass production scheme. The aerobic batch growth of *S. cerevisiae* with a sugar as the limiting carbon and energy source is diauxic. The first phase of sugar utilization is characterized by a period of respiration repression or Crabtree effect. Recently various authors have modelled the process under different approaches (180-184) and also have postulated different control strategies for fed-batch processing in order to optimize biomass formation (185-192c). There is practically no recent analysis published in the process optimization for the simultaneous production of both products, although Haraldson and Bjorling (193) have studied the behavior of 20 yeast strains at high osmotic pressures under aerobic conditions. We believe that in such a process the carbon substrate will be distributed in three ways: a) for ethanol production, b) for biomass formation, and c) leftover in the media. After alcohol stripping, which might form an external loop with the fermentor-BIOSTIL and ATPAL processes (194), the biomass and residual carbohydrate could be used as a concentrated suspension in animal feeding alternatives (1). dos AnjosMagalhaes et al. (195) have reported that the addition of urea or ammonium sulphate improves the amino acid content of *S. cerevisiae*. Maximum content of total amino acids was obtained when 0.6 gl^{-1} of urea were added. The uptake of amino

acids from the mash by the yeasts is influenced by the degree of
oxygenation (196). Tanner et al. (197) have shown that
cultivating *S. cerevisiae* with 0.6M sodium chloride increased
four times the free lysine concentration in the cell although
glucose uptake and ethanol in the medium were lowered approximately by half.

6.   *Filamentous Fungi in Submerged Culture*

As explained before (1) filamentous fungi lend themselves to
a low pH protected, batchwise production scheme which has been
tested with various fungal strains and different agro-industrial
wastes and effluents. Research along these lines contunues
worldwide obtaining biomass yields, substrate consumption and
growth data.

Balagopal and Maini (198,199) found *Aspergillus niger* NRRL
330 and a *Rhizopus* sp. superior in terms of mycelial weight and
protein production in a 25 gl^{-1} cassava starch waste liquid media.
Muindi and Hanssen (199a) grew *Trichoderma harzianum* in a slurry
of cassava root meal. An input of 100 g of this material with
2.4% of crude protein gave about 30 g of an enriched product with
about 37.6% crude protein on dry basis. Fermentation time was
60 h, 23°C, pH 4.0-4.2 on a 4% slurry. Eklund et al. (78)
reported a yield of 0.63 for *Paecilomyces varioti* in the pentosan
fraction of sunflower seed husk sulphuric acid hydrolyzate and
a yield of 0.94 from the hexose fraction. This figure suggests
that the fungi was also employing as substrates other compounds,
like organic acids present in the medium. Worgan (200) summarized
data of the final biomass concentration obtained by *Fusarium
semitectum,* 28.7, 22.8, 22.9 and 14.6 gl^{-1} in palm waste, citrus
molasses, lucerne leaf liquor and maize leaf liquor, respectively,
and by *Aspergillus oryzae,* 19.5 and 16.6 gl^{-1} in olive waste and
lucerne leaf liquor respectively. Imrie and Righelato (201)
have shown that although the biomass yield from sucrose for
Fusarium sp. MR was about 0.45 and rather constant as pH was
varied from 3 to 6 and growth temperature from 25 to 35°C, the
growth varied drastically. It was optimum at around 35°C and
pH above 5. The growth rate decreased at acid pH values
(around 3) and growth almost stopped at 40°C. The oxygen demand
for growing cultures at 10-20 gl^{-1} biomass concentration was in
the range of 0.1-0.2 mol l^{-1}h^{-1} and if this was not met the
batch time to reach maximum biomass increased greatly but the
biomass yield was not affected. Fuska and Kollarova (202)
obtained from 12 to 14 gl^{-1} after 72 h incubation in a sawdust
hydrolyzate medium employing *Penicillium resticulosum, Mycelium
sterilium, Gibberella fujikuroi,* and *Coprinus* sp.; hexoses were
faster utilized than pentoses. Deshpande and Joshi (203) used
deproteinized leaf extract as substrate and Ghewande and

Bansode (204) found *Phytophthora rubra* the highest protein
producer among five plant pathogenic fungi. Gewaily (205)
found *Aspergillus sydowi* 308 the best fat producer obtaining
a concentration of 24.08%. Nga et al.(55) grew mutants of
A. niger in suspensions of cassava flour and ammonium sulphate, obtaining a 10 to 20 fold increase of the solids total
amino acids. *Penicillium notatum* and *Penicillium digitatum*
grew well on media from potato processing wastes, a biomass
yield of 9-24 gl^{-1} was obtained (206). Four cellulolytic
fungi: *Myrothecium verrucaria*, *Paecilomyces* sp, *Gliocadium*
sp, and *Aspergillus terreus* were grown in a 50 ml basal
medium, pH 5.5, plus one gram of sugar beet pulp for seven
days at 28C (207); the first one showed the highest amount
of protein in the biomass. In the course of batch cultivation
of *Aspergillus awamori* at 30C the increase in the concentration of the extract of sugar beet cossettes from 2 to 4% resulted in more than three-fold increase in mycelial yield,
and a further increase to 8% increased the yield by 13%. Protein content was the highest at a 4% carbon source concentration and the growth was diauxic, obtaining a maximum specific
growth rate of 0.214h^{-1} for the first part (208). A diauxic
growth was obtained previously with *Rhizopus cohnii* cultured
in bread medium supplemented with corn steep liquor (209);
pH control in this system resulted in a higher mycelial
yield. The highest mycelia and protein yields were always
obtained at an aereation rate of 1.0 1 l^{-1} h^{-1} for various
fungi but some of them were susceptible to shear stress,
optimum 460 rpm *R. cohnii*, *Mucor* sp, and others were less,
Rhizopus microsporus 700 rpm, *Actinomucor repens* 940 rpm (210).
Maximum protein production was by *Mucor muceido,* 16.75 gl^{-1}
(210a). The most economic protein production was by *Rhizopus*
at low oxygen transfer rates (210b). Rosenberg et al. (211)
used *Trichoderma viride* QM6a to enrich a 10% slurry of previously hydrolyzed oat hulls with fungal biomass with the
objective of producing a balanced animal feed. The incuba-
tions were done in shake flasks at 25°C and pH 5 and the culture dry weight reached a maximum value of a 17.5% increase
in 14 days, with a concomitant 75.8% enrichment of crude protein. *Aureobasidium pullulans* grew well in acid treated wheat
straw increasing two fold the amount of protein (211a). Jones
and Bu'Lock (212) tested several filamentous fungi in the
supernatant remaining after the expressed juice from sugar
beet tops and meadow grass had been heat deproteinized and
supplemented as required by glucose or ammonium sulphate.
A. niger, A. oryzae and *Paecilomyces* sp performed very well
and although reducing sugars were preferentially utilized
some non carbohydrate carbon was also used along with both
ammonia and non ammonia nitrogen. For the fast growing

strains apparent yields were 0.66 - 0.76 on total carbohydrates and 0.21 - 0.24 on the total solids. Cahal et al.
(159a) grew in a medium with 50 gl^{-1} sucrose and potassium
nitrate as nitrogen source *Fusarium oxysporum, Aspergillus
luchuensis, A. niger* and *Oospora* sp and found a lipid content in the microbial biomass between 8.9 - 27.5%. Similar
results were obtained earlier by Bhatia and Arneja (213) who
studied the metabolism of the different lipid fractions of
F. oxysporum under batch cultivation. Ghai et al. (214) grew
Chaetomium cellulolyticum ATCC 32319 in a slurry dispersion of
delignified canning sag (leaves of mustard and spinach) and
obtained a final protein content of 20-25%. Abraham and
Srinivasan (215) studied the growth of *Penicillium fre-
quentans, Aspergillus nidulans* and *Fusarium* sp in deproteinized whey in shake flasks at 28C for 4 days. The first two
fungi produced more protein, the third one more fat. The
fungal biomass increased when whey was supplemented with
another carbon source like molasses, an ammonium salt or
yeast and malt extracts. The rotating disc fermentor was
suitable for growing filamentous fungi including representatives of the genera *Aspergillus, Rhizopus, Mucor* and *Penicillium,* The fermentor exploits their surface adherent properties and data have been presented which indicates the
continued viability of the fungal cells in the innermost
layers of the disc (216). With *A. niger* grown in a medium
containing 2 gl^{-1} glucose, a glucose uptake rate of 1 gl^{-1}
h^{-1} was observed with a liquid residence time of 4.7 h and
complete glucose removal. After three days of continuous
feed the interdisc spaces eventually blocked and the reactor
filled up with mycelium. Biomass stripping was required with
regrowth occurring from residue entrapped on the disc surfaces. The stripped mycelium was separated from the fluid
(217). de Gonzalez and de Murphy (218-220) in Puerto Rico
studied the growth of *Aspergillus phoenicis* (locally isolated from spent wash), *A. oryzae* NRRL 2220 and *Aspergillus
flavus* NRRL 3518 in rhum spent wash. An average biomass
concentration of 14.2 gl^{-1} was obtained after 48 h incubation
and a BOD reduction of 56%. Quinn and Marchant (221) grew
Geotrichum candidum in malt whiskey distillery waste. Ek
and Eriksson (222) employed *Sporotrichum pulverulentum* ATCC
32629 in wastewaters, from fiber board mill. The wastewater had a 2-3% dry matter and contained a mixture of monomeric and oligomeric water soluble sugars, low molecular-
weight acids and lignocellulosic particles. Continuous culture at 10 l scale and maintained at pH 4.2, 39C and DO > 10%
saturation gave biomass productivities of 0.19 to 0.35 gl^{-1}
h^{-1}, COD reductions between 48 and 52% and reduction of suspended particles between 38-88%. Malfait et al. (223)

employed a pilot scale column loop reactor to grow *Monascus purpureus* ATCC 16365 in a rice starch medium which had been previously enzymatically hydrolyzed. With controlled pH at 6.0, a biomass concentration of 16 gl^{-1} and a yield of 0.4 were obtained; values that were higher than those obtained in a stirred tank configuration. Barker and Worgan (224) tested various filamentous fungi on effluents from the process of palm oil extraction and selected *A. oryzae* IMI 44242. Inorganic nitrogen supplementation was required in order to optimize biomass yield and COD reduction. Values obtained in shake flasks at 30C for 72 h were 0.5 to 0.6 and 73 to 91% respectively. The growth profile in a fermenter showed a sequential utilization of protein, lipid and carbohydrates and most of the nonbiodegradable material was found to be water soluble carbohydrate and nitrogenous materials, possibly Maillard reaction products, and polyphenols. Gibriel et al. (225) grew *Aspergillus terreus* and *A. niger* in a medium containing 15 to 20 gl^{-1} of milled wheat bran and inorganic nitrogen supplementation in shake flasks for up to 15 days obtaining a residual biomass with about 40% crude protein. Garg and Neelakantan (226) isolated *A. terreus* GN 1 and grew it on 1% alkali-treated bagasse slurry in shake flasks for seven days. The highest crude protein content obtained of the biomass (mycelium plus unfermented bagasse) was around 20% at 30C and pH 4. This microorganism was selected from an initial screening of 12 fungi (226a). More recently data was reported on its growth on 1% alkali-treated bagasse in a 10 l reactor (226b). After 4 days a product with 10 times more protein was produced utilizing 73% of the initial cellulose present. Untreated bagasse was less degraded, although a product with 7 times more protein was obtained after three days (226c). Hot water extraction or low pressure steaming of lignocellulosic material produces solutions containing phenolics, free carbohydrates and lignocarbohydrate complexes (227). Recently Miller and Srinivasan (227a) grew *A terreus* ATCC 20514 in alkali treated solka floc and sugarcane bagasse obtaining 80-88% cellulose batch consumption in 30-36 h and 78-84% in continuous culture at 35-45C. Schmidt et al. (228) grew the fungi *P. varioti* on steamed extracts from birchwood oat husks and wheat straw. Milstein et al. (229) studied the growth of *Aspergillus japonicus* OM-4 both, on the liquors obtained from hot water and steam treated wheat straw and in 5 gl^{-1} slurries of treated straw. Free reducing sugars in the liquor were depleted within a day, however phenolics and phenolic-carbohydrate complexes were more resistant. The fungus rapidly degraded lignin and cellulose from the slurries. *Penicillium cyclopium* was grown on deproteinized whey batchwise and in continuous culture (230,231). Biomass

productivities of 0.25 and 2 gl^{-1} h^{-1}, yields of 0.68, a cell
maintenance coefficient of 0.005 g g^{-1} h^{-1} and substrate con-
sumptions of 84 and 85% were obtained respectively. Foda
(232) reported 23, 30.1, 23.5 and 20.6 g biomass l^{-1} of *A.*
flavus, A. niger, A. ochraceus and *A. terreus* in salted (50
gl^{-1}) whey medium after 7-9 days at 28C in shake flasks.
Martin (233) grew *Morchella esculenta* NRRL 2603 in a peat
acid hydrolyzate in a bench fermentor at 24C, 1 vvm, 200 rpm
and an initial pH of 7.0. A yield of 0.37 and a substrate
consumption of 45 to 52% were obtained. *Cephalosporium*
eichhorniae ATCC 3825 isolated by Gregory et al. (234) was
grown batchwise in 4-50 l fermentors at 300-500 rpm, pH 3.8
and 45C in a 40 gl^{-1} acid-heat gelatinized cassava starch
slurry coming from dried cassava meal or chips, obtaining
biomass productivities in the range of 0.58 to 0.98 $gl^{-1}h^{-1}$,
a carbohydrate consumption of around 95% and yield in terms
of cassava meal used of around 42% (234a). Ground fresh
cassava was inhibitory (235). The Pekilo process in which
P. varioti is grown in sulfite waste liquors has been described (236,237). Betucci and Perez (237a) tested 104
isolates on the residual sulfite liquors from sugarcane
pulping and found *Cladosporium herbarum* the one that not
only produced more biomass but was able to use the soluble
lignin sulfonated compounds present. Eaton et al. (237a)
used *S. pulverulentum (Phanerochaete chrysosporium)* to reduce the COD of kraft bleach effluents. Saquido et al.
(237c) grew *A.niger* and *A. foetidus* in ripe and unripe
banana (whole crushed fruit) slurries in jar fermentors
at 30C obtaining products with 20-30% crude protein.

The growth in submerged culture of higher fungi continues to be studied worldwide (238-243). Labaneiah et al.
(244,245) studied the optimum growth temperature and the
effect of the carbon source on the growth of *Morchella*
crassipes NRRL 2686, *Agaricus bisporus* IRJP, *Lentinus edodes*
IFRI 257/c, *Pleurotus ostreatus* IFRI 958 and *Polyporus*
sulphureus IFRI 954. They also have reported data on the
growth of *A. bisporus* and *M. crassipes* on citrus peel extracts supplemented with glucose and inorganic nitrogen.
Biomass yields of 0.40 and 0.45 and productivities of 1.66
and 1.31 g biomass l^{-1} h^{-1} are reported (245a). The presence of organic nitrogen was optimal for the growth of
Monascus sp (246). Carroad and Wilke (247) have shown that
pellet forming lignin degrading *Polyporus versicolor* ATCC
12679 and *P. ostreatus* ATCC 9415 grew at a rate proportional
to the two-thirds power of the cell mass, however it was not
the best or simplest model as one based on a growth rate
directly proportional to the cell mass. Duvnjak et al. (248)

studied the growth of *Agaricus campestris* and *Morchella hortensis* in diluted whey plus potato infusion medium for 3-6 days at 24C, obtaining around 20 gl^{-1} biomass concentration. They also supplemented whey with corn steep liquor, yeast extract, casein or asparigine and checked the biomass yields at various pH values for 5 days at 24C (248a). Kurtzman (249) was able to produce fruiting bodies in submerged culture of *Pleurotus sapidus* in alfalfa brown juice medium supplemented with 0.1% safflower oil at 20C. Jauri and collaborators at IARI in New Delhi studied the growth of various filamentous and higher fungi among which were *Macrophomina phaseoli, Polystictus xanthopus, Rhizopus arrhizus, A. flavus, T. viride, M. mucedo, Rhizoctonia melongina, Polystictus affinis, Coprinus aratus* and *P. ostreatus* in three agricultural wastes sugarcane bagasse, wheat straw and cow dung added as suspensions at 15 gl^{-1} in Chahal's medium. *R. melongina* and *P. ostreatus* produced more biomass in sugarcane bagasse at 35C, pH 5, 10 days and 27C, pH 5, 14 days respectively and *C. aratus* with wheat straw at 35C, pH 7, 14 days (250). A media optimization was done selecting urea phosphate as source of nitrogen and phosphorus and supplementation by magnesium, zinc and iron and acetic and ascorbic acid as growth promoters (251-254). Vecher et al. (255) obtained 8 gl^{-1} of biomass of *Pleurotus cornucopia, P. ostreatus* and *P. pulmonarius* in protein free potato juice plus glucose medium. Ghosh and Sengupta (256) studied the effect of vitamins, hormones and fatty acids on the growth of *Termitomyces clypeatus, Panafolus papillionacens, Gymnopilus chrysimyces, Coprinus lagopus, Lentinus squarrosulus, Volvariella volvacea* and *Agaricus bisporus*.

The use of slurries of lignocellulosic biomass as substrates for fungal growth in submerged culture is an alternative that many research groups around the world are developing, for example see reviews of Gupta et al. (257) and Sadana et al. (258). Griffin et al. (259) grew *T. viride* in feed lot waste fiber obtaining 38% of total solids reduction. Brown and Fitzpatrick (260) on shredded, heated and milled waste paper with *T. viride* obtained a solid material with about 11% protein on dry weight basis; also a fungal biomass yield of 0.5 and a biomass productivity of 0.066 gl^{-1} h^{-1} were reported. De Menezes (261) found that *M. verrucaria* gave the maximum biomass yield but *T. viride* the best productivity. Janus et al. (262) found very disappointing results with alkali pretreated, ball milled rice hulls and *S. pulverulentum*. Lerpido-Barraquio et al. (263) reported that glucose was a better substrate than cellobiose for *T. viride* QM6a. Babitskaya et al. (264) selected

Trichoderma lignorum, Trichoderma koningii, P. notatum,
Pencillium verruculosum, and *A. awamori* among 200 strains as
the best protein producers from alkali treated and pulverized
wheat straw. Peitersen (265) studied the continuous cultivation of *T. viride* on 4 - 11 gl^{-1} slurries of pure ball
milled wood cellulose at pH 5 and 30C reporting a biomass
yield of 0.83 and a maintenance coefficient of 0.034 g
cellulose g cell biomass$^{-1}h^{-1}$. Peitersen and Andersen (266)
tested barley straw alkali pretreated by the Kolding process
as substrate for *T. viride* QM9123. A product containing
18-24% protein and 30% lignin was obtained in 4-7 days fermentation and employing slurries between 20-85 gl^{-1}.
Geethadevi et al. (267) found *Aspergillus carneus* as the best
protein producer in alkali treated straw. Chahal et al.
(267a) found the fungus *Cochliobolus specifer* the most
efficient fungus for protein synthesis from wheat straw.
Sidhu and Sandhu (267b) grew *Trichoderma longibrachiatum*
Rifai (IMI 228288) in pretreated sugarcane bagasse and found
that it grew faster, produced a higher biomass yield and a
better substrate conversion on delignified (sodium chlorite)
bagasse.

At the Department of Microbiology of the Punjab Agricultural University, Ludhiana, India the search for adequate
fungal microorganisms for the bioconversion of various agricultural wastes has been reported in a series of publications.
Wheat and rice straws, cane bagasse, ground nut shells, banana
peel, citrus fruit pulp and spent grain waste have been tested
with *T. viride, M. verrucaria, P. funiculosum, C. specifer,*
Chaetomium globosum, A. niger, Penicillium chrysogenum,
Pantalotia sp, and several others (268-274). Several fungi
were isolated which grew rapidly on alkali treated, rice
straw, bagasse or cellulose powder, obtaining a 12-20%
cellulose conversion to crude protein in five days at 30C
(274a). *Penicillium funiculosum* was grown on acid hydrolyzed-
ammonium neutralized rice straw for seven days, increasing
the protein content by 180% and the *in vitro* digestibility by
31% (86i). Rao et al. (274b) employed *Penicillium*
janthinellum to growth in alkali treated or in alkali treated
plus acid neutralized wheat straw. In the former a higher
cellulose consumption (90%) and final protein in the product
(30%) were obtained, however in the latter a better protein
yield from the original straw was attained (20%).

The white-rot fungi are the most efficient microorganisms to degrade lignin and all other wood components through
an enzymatic machinery capable of oxidizing phenolic compounds and hydrolyzing the hollocellulose matrix.

S. pulverulentum ATCC 32629 was able to degrade completely
a 20 gl^{-1} suspension of waste fibers at 38C, pH 4.6 and a
dilution rate of 0.04 h^{-1} (275). However, it has been shown
recently (276) that for this fungi as also for *Phlebia
radiata, Phlebia gigantea* and their cellulaseless mutants
need cosubstrates for energy and growth. *S. pulberulentum*
grew well on an acid hydrolyzate of piggery slurry obtaining
about 20 g of dry biomass per liter of slurry hydrolyzed at
90C for 3 h with 5% sulfuric acid (276a). Daugulis and Bone
(277) grew *Pleurotus sapidus* A-241, *S. pulverulentum* A-387,
and *Polyporus anceps* DAOM 21401 in suspensions of alkali
treated and ball milled bark from white cedar and silver
maple, obtaining after four days at 30C, 136 and 116 mg crude
protein per g of bark for the last two fungi respectively.
In continuous culture with an optimum substrate concentration
of 10 gl^{-1} at 30C, protein productivities were greater with
pine and cedar barks than with maple. No lignin was utilized
and the bark was limiting (278).

Thermophilic microorganisms have also been studied.
Bellamy (279) grew *Thermoactinomyces celluloseae* in delignified feedlot waste fiber. Armiger et al. (280) and
Humphrey (281) presented experimental data and a growth
model for *Thermoactinomyces* sp on pure cellulose, reporting
the following parameters: maximum growth rates h^{-1} 0.4 - 0.5,
0.2 and 0.48; maintenance coefficients g g^{-1} h^{-1} 0.02 - 0.08,
0.027, 0.038 and yields 0.4 - 0.5, 0.27 and 0.44 for cellulose, cellobiose and glucose as substrates respectively.

C. cellulolyticum is a thermotolerant fungus (282) which
showed excellent growth rates and biomass-protein productivities when cultivated in acid and alkali pretreated sawdust
of mixed hardwoods (283), purified cellulose or partially
delignified (alkali plus peracetic acid treatments) sawdust
of mixed hardwoods (284) and purified spruce wood pulp (285).
The effect of substrate pretreatment is an important parameter
in the fungus growth rate, substrate consumption and biomass-
protein productivities. This has been shown for wheat straw
(286), hardwood sawdust (287) and a review is given by Chahal
and Moo-Young (288). Most of the experiments have been done
employing slurries at 10 gl^{-1}. Alkali pretreated wood
residues gave a crude protein productivity of around 50 mg
l^{-1} h^{-1}. They increased 75-86 if the substrate had been
steam pretreated (Stake and Iotech processes) and were maximized to 100 plus if the substrate had been chemically delignified. Kraft pulp mill rejects appeared very promising
substrates (289). Alkali pretreated wheat straw plus
anaerobic liquor gave crude protein productivities in the

range of 178-200 mg l^{-1} h^{-1}. Cellulose utilization was poorer
in steamed wood at atmospheric pressure, 12.4% utilization
and best in delignified wood, 90% utilization (290). Cattle
manure (291) and mixtures of swine manure and straw (292)
have also been used as substrates for *C. cellulolyticum* growth
in submerged culture. From these works the concept of the
Waterloo process emerged as an integrated scheme for the bioconversion of agricultural, forestry and animal manure into
an animal feed supplement (293,294). Several process
scenarios were simulated by Moo-Young et al. (295) which show
that a 5 ton per day Waterloo plant processing US$33 t^{-1} corn
stover and US$10 t^{-1} manure will have a 13-32% discounted
cash flow return and a net product value in US$ t^{-1} of 147 -
159. Fahnrich and Irrgang (296) have also cultured *C. cellu-*
lolyticum in pure cellulose and waste newspaper. Recently
Hetch et al. (296a) grew it on glucose and pure cellulose substrates avicell and sigmacell. Gacesa et al. (296b) from the
Institute of Microbiological Processes and Applied Chemistry
at Novi Sad have reported work on the Waterloo process regarding batch growth on alkali-pretreated wheat straw,
kinetics of *Chaetomium* characterizing its diauxic growth on
straw in submerged culture, the use of anaerobically digested
manure as a source of nitrogen and its physiological and
morphological growth characteristics on solid substrates. A
biomass containing 32% protein was obtained after four days
growth at pH 5.0 in a nitrogen supplemented milled orange
waste medium at 15 gl^{-1} concentration (296c).

7.   *Solid Substrate Fungal Growth*

In a vegetal ecosystem filamentous and higher fungi grow
in close association with higher plant matter usually in a
solid substrate matrix, either as parasites and mycorrhizal or
truly saprophytic which feed on plant wastes (297). In comparison with submerged growth, the growth on the surface of a
moist solid matrix is subject to the influence of unique physico-chemical properties like substrate solubility, particle
size, shape and solid surface structure which make the process
more difficult to control (48,50,50a). In actual practice
solid substrate fungal growth is used widely in producing fermented foods and in the production of edible mushrooms from
ligncellulosic biomass (43-47,298-309). However, research
data have been published recently with the objective of
developing a practical and economical process to produce enriched fungal biomass products from starchy or lignocellulosic
biomass as animal feeds. In the latter case a higher fungi
might be preferable and the following scheme is possible:

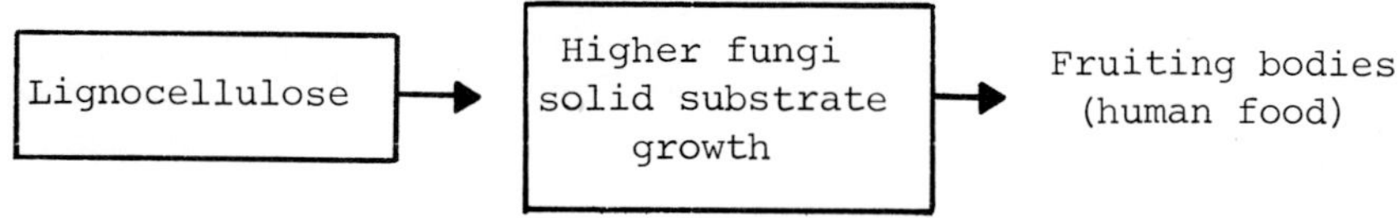

Lignocellulose plus mycelium
(animal feed)

Zadrazil (310) studied the growth of *Pleurotus ostreatus*,
Pleurotus florida, *Pleurotus cornucopiae* and *Pleurotus
eryngii* on wheat straw and found that a carbon dioxide concentration between 22 to 28% and the addition of organic
nitrogen sources stimulated mycelium growth. About 10% of the
initial weight of straw was dry fruiting body biomass, 70%
was turned into carbon dioxide and water and 20% was residual
compost. The three polymeric components of the lignocellulosic matrix, including lignin, were biodegraded. Zadrazil
(311) found that the growth of *Pleurotus sajor caju* on wheat
straw was greatly enhanced by the addition of up to 30% of
soybean or alfalfa meals. The dry fruiting bodies yield as
% of the initial substrate dry weight was increased 3 times
and the loss of organic matter about 50%. Later (312) he
showed that the growth of *Pleurotus* species and *Stropharia
rugosoannulata* increased the *in vitro* digestibility of wheat
straw, due to the enzymatic liberation of soluble substances
as well as physical changes in the straw. Lignolitic activity
and *in vitro* digestibility were strongly dependent on fungal
species and the substrate (313). *Pleurotus* sp *Florida* and
S. rugosoannulata showed good lignin decomposition and increased the *in vitro* digestibility of beech sawdust, reed,
rape and sun flower straw after growth at 25C for periods
of up to 60 days. All the fungi tested decreased the digestibility of rice husks caused by the high SiO_2 contents of the
same. Ammonium nitrate supplementation up to 0.25% of wheat
straw lead to increased digestibility after the growth of
Kuehneromyces mutabilis, *Lentinus edodes*, *P. eryngii*, *P.* sp
Florida, *P. salmoneostramineus* and *Stropharia rugosoannulata*
for up to 90 days (314). Higher concentrations induced lower
digestibilities which might be explained by the different

lignin and cellulose metabolic rates. Recently Zadrazil and
Brunnert (315) have shown that high temperatures, prolonged
processing time and extremes in substrate water content had
an adverse effect on the progress of the solid state fungal
growth. Under optimum conditions some fungal species almost
doubled the *in vitro* digestibility of straw. Reviews of this
work have been presented by the author (316,317a).
Rajarathnam et al. (318) obtained from 30 to 40 g of dry
fruit bodies of *P. flabellatus* per kg of dry straw in about
15 days at 22-28C and relative humidity of 55-75%. In this
period there was a 36% loss of organic matter of which 4% was
converted into fruiting bodies. In the spent straw holocellulose was reduced by 70% and lignin only by 24%. Total
nitrogen was reduced by 63% and nitrogen uptake was higher
during fruit body formation. Platt et al. (319) were able
to reduce by 50% the dry weight of cotton straw by growing
Pleurotus sp *Florida* for 21 days. Lignin was reduced by
65% and the water soluble substances were increased almost
three times, hence it was assumed that the lignocellulose
polymer had been degraded to low molecular weight phenolics.
A flavonoid type compound was present in cotton straw and
was at least partially responsible for the increased growth
of *Pleurotus* on this substrate (319a). Kaneshiro (320) employed a feedlot waste fiber-straw mixture and wood sawdust
for the growth of *P. ostreatus*. Lignin decreased up to 40%.
Cellulose decreased also to the same levels, but only for
the feedlot waste fiber-straw mixture, with the sawdust it
was practically not biodegraded. Virk et al. (321) grew
P. ostreatus on poultry droppings. Danilyak (322) reported
the optimum pH, from 3.8 to 4.8, and temperature, from 35
to 60C, for the cellulolytic enzymatic complex of *P.
ostreatus*. Ginterova et al. (323) studied the relationship
between *P. ostreatus* and *A. flavus* in aflatoxin production.
Many of the *Pleurotus* species classified still are contro-
versial and may be considered tentative (324) as macroscopic
and microscopic characteristics are unreliable for classification (325). With respect to fruiting temperatures two
groups have been defined: a low temperature which fruits only
below 15C and one that tolerates up to 30C and which is more
adapted to the tropics (326). Singh and Rajarathnam (327)
describe briefly the cultivation of *Pleurotus eous* and its
physical composition. Ginterova and Gallon (328) have discussed the possibility of *Pleurotus ostreatus* fixing
atmospheric nitrogen when grown on straw. According to the
data presented, only 0.02% of the total nitrogen in the fungal
biomass would have been fixed from the atmosphere, which re-
inforces the concept that this fungi is an excellent nitrogen
scavenger from lignocellulosic biomass. Chang et al. (328b)

tested the growth of *P. sajor caju* on rice straw with 1% lime
and their mixtures at 25C in cylindrical sacks. Three weeks
were needed for mycelium growth under PVC sheet covering.
At this time the sack was opened and 5 days later the first
flush of mushrooms were harvested. Flour flushes were harvested with 5 to 7 days intervals. Lindenfelser et al. (328b)
grew *P ostreatus* NRRL2366 in wheat straw alone or with addition of inorganic nitrogen or oat meal. The fungi degraded
lignin and cellulose from 10-45 and 15-55% respectively after
36 days. The greater values obtained in the mixtures. Free
reducing sugars in the substrate and its *vitro* digestibility
increased two-fold to three-fold after a 90 day period.
Khanna and Garcha (328c), Singh (328d), Bano and Rajarathnam
(328e), and Chibenjaphol (328f) report of *Pleurotus* grow in
paddy straw in India and Thailand respectively. Recently
Streeter et al. (328g) found that the growth of *P. ostreatus*
for 28 days on wheat straw improved the *in vitro* digestibility
by only six percentage units, however when incubated with
E. carotovora alongside at 50% solid moisture for 56 days
the *in vitro* digestibility increased by fifteen percentage
units (328h).

Matteau and Bone (329) grew *Polyporus anceps* in alkine-
steam pretreated wood shavings in small jars and in a liquid-
recirculating pocked solid tower bioreactor. Salts, ammonium
chloride and yeast extract were added. The use of the latter
reactor was more effective in maximizing substrate utilization
which was 70% and protein production in the residue, up to
17%. Holocellulose degradation accounted for the weight loss
as lignin was not attacked, due probably to the high concentration of ammonium salts. Adhikary et al. (330) grew *P. hir-
sutus* on rice straw at 30C for three weeks. The fungi growth
doubled the susceptibility of the straw residue to the *in
vitro* attack of cellulase and xylanase compared with the
control. Substrate weight loss was about 23%, which could be
reduced to less than 10% by the addition of up to 0.5 g of
ammonium sulphate per g of straw. There was a concomitant increase in protein and the susceptibility to enzymes was not
altered, however no analytical data were provided for the biodegradability of the lignocellulosic matrix.

The presence of ferulic acid in the spawned compost enhances laccase production by *Agaricus bisporus*, protein biosynthesis and carpophore production. However, chlorogenic
acid inhibits them (330a).

Gao et al. (330b) studied the growth of 73 strains of
basidiomycetes on wheat straw and selected two strains, identified as *Calvatia* sp, for their superior growth on non-auto-
claved wheat straw and their performance on repeated batches
with inoculum from the previous one. A 16% weight loss, a
5.5% crude protein content and 13% and 24% lignin and cellulose decrease respectively. Latham (330c) found increases in
digestibility of chopped barley straw of 5-11% after incubation with *Schizophyllum commune* and *Polyporus anceps*.

Volvariella volvacea or paddy straw mushroom is a fungus
of the tropics and subtropics cultivated in China and in other
Asian countries. Yields range from about 1 to 45 kg fresh
mushrooms per 100 kg of dry lignocellulosic material (331) and
productivities from 24 to 198 kg fresh muchrooms m^{-2} $year^{-1}$
(331,332). Wang et al. (333) and Chang and Steinkrus (334)
have studied the cellulolytic enzymes. β 1,4 glucosidase,
β 1,4 glucan glucanhydrolase and β 1,4 glucancellobiohydro-
lase activities were found, partly cell free and partly cell
bound, they were strongly induced and some weakly constitutive and the highest activity was found after 5 to 14 days
of inoculation. Their optimum pH was from 5 to 7, very little
found below 4, and their optimum temperatures from 40 to 65C.
Change and Yau (334a) studied the growth of *V. volvacea* and
P. sajor-caju on cotton waste, paddy straw and used tea-leaves
compost. Yields on one flush were for the former 5.6 kg m^{-2}
and an average biological efficiency of 36. The values for
the latter on three flushes were 1.34 kg per bag and 88 respectively. Santiago and Peberdy (334b) have described a
laboratory technique for fruit body development. Hayes and
Lim (334c) describe the technology for growing *V. volvariella*
and *A. bisporus* in rice and wheat straws. Typical yields are
10 kg and 50 kg of fresh mushroom biomass per 100 kg of fresh
straw respectively.

Ibrahim and Pearce (334c) studied the growth of eleven
white rot fungi on barley and pea straw, sugarcane bagasse
and sun flower hulls, for 21 days at 14-25C. The fungi that
reduced lignin and increased the *in vitro* dry matter digestibility were: *Peniophora gigantea* in barley straw and sugarcane bagasse; *Ganoderma lucidom* in pea straw and sun flower
hulls.

Gold and Cheng (335) found that fruit body formation in
P. chrysosporium (S. pulverolentum) was subjected to strong
catabolite repression by glucose in the presence of physiological levels of nitrogen. It was relieved under nitrogen
limitation. This fungi biodegrades lignin effectively.

Zezula et al. (336) grew the fungi *Coprinus* sp in steamed and cut straw plus 30 g urea and 1.6 g acid potassium phosphate per kg at 32C for 10-12 days. A total weight loss of 20-24% and a consumption of all urea was observed. The solids had 160% more nitrogen than the urea-straw input so the authors suggested nitrogen fixation. The product was tested on swine with favorable nutritional response. Lignin and cellulose are biodegraded by *Coprinus fimetarius* (337), most of the members of the genus *Coprinus* which have been studied are apparently light dependent in order to complete their life cycle, a phenomenon that is being studied with *Coprinus congregatus* (338) and *Coprinus macorrhizus* (339-343). Burrows et al. (343a) grew *Corpinus cinereus* in barley straw at different pH's for 15 days. Cellulose and hemicellulose were utilized. At the end of the treatment digestibility values had decreased below the orginal level. However, at 10 days they were maximal and had increased 27% at the optimum pH. Reviews have appeared for the solid state cultivation in lignocellulosic substrates of *Tremella fuciformis* (344), *Auricularia* sp (345), *Stropharia rugoso-annulata* (346), *Pholiota nameko* (347), *Flammulina velutipes* (348,348a), *Kuehneromyces mutabilis* (349), and *Lentinus edodes* (350,351). Also reports for *P. ostreatus* (351a) and *Gyromitra esculenta* (351b,351c). Reviews of a truly solid state growth, as it occurs in soil in close association with plant roots, have been published for *Tricholama matsutake* (352) and *Tuber* spp (353). In both the fructification of mycorrhizal forms produce a highly delicate human food. Recently, Reade and McQueen (353a) have shown that the growth, after sterilization of poplar shavings enriched with minerals, for 8 weeks of *Polyporos anceps* increased the *in vitro* digestibility from 30 to 72%, 64% with *Ganoderma applanatum* and 62% with *Phanaerochaete chrysosporium* after 4 weeks; 61% with *Polyporus versicolor* after 3 weeks and 42% with *Formitopsis ulmarius* after 4 weeks.

The growth of *T. viride* (*T. reesei*) in solid culture and its cellulase production has been studied by several workers. Toyama and Ogawa (354) found that rice straw was the best substrate and maximum enzyme activities (filter paper) were obtained at 25C for 4 days. When an excessive organic nitrogen source was available (rice bran, vegetable oil cakes) the enzyme activity decreased remarkably as the incubation period was prolonged. Vilela et al. (355) found that in a 30% rice straw-rice bran mixture the maximum endo β glucanase (CMC ase) activity was obtained after 15 days at 29C. On the other hand, Qadeer et al. (356) and Jabbar and Ilahi (357) found for the same enzyme, wheat bran a superior substrate

with maize bran, rice husks or bagasse after incubation at 30C
for 72 h. The inclusion of readily metabolizable sugars in
the medium reduced enzyme formation, which may explain the low
yield of cellulase on cane bagasse due to the residual sugars
in it. *T. viride* was grown in solid culture in previously
acid hydrolyzed water hyacinth (358). Nishio et al. (359)
found that *Taralomyces* sp produced maximum amounts of endo
β glucanasa (CMC ase) and exo β glucanasa (Avicelase) in wheat
bran under solid culture at 60% substrate moisture content and
45C and between 60 and 70 and 50C respectively. Sinha et al.
(360) have reported a cellulase inhibitor present in the solid
culture of *A. terreus* in wheat bran. The authors explained
the presence of the inhibitor in the culture with two possibilities: a) specific compounds present in the wheat bran
may induce the fungus to produce the inhibitor, and b) the
inhibitor may be a breakdown product of wheat bran itself.
Solid state culture has the potential to produce special substances which rarely appear in liquid sultures (359). Theja
et al. (359a) have found that several strains of *Aspergillus,
Penicillium* and *Sporotrichum* gave in solid culture on wheat
bran more filter paper activity and reducing sugars than in
submerged culture.

Chahal et al. (361) grew *C. cellulolyticum* ATCC 32319 in
solid state at 37C and pH 6.0 in alkali pretreated, wheat
straw and corn stover obtaining a final product with 19% (dw)
and 24% (dw) of protein and a cellulose utilization of 63 and
48% respectively. They calculated a protein productivity of
0.13 gl^{-1} h^{-1} in a 25% (w/v) solid state batch culture. Ghai
et al. (214) grew the same fungi in solid state fermentation
of delignified canning sag obtaining a final total biomass
with a protein content of 9-12%. Ulmer et al. (362) present
results of the solid culture of the same fungus on steam
treated feedlot waste fibers. Steam treatment increased the
amount of soluble reducing sugar and the cellulase reactivity
however, it did not increase the protein content of the product beyond that obtained with fermentation of untreated
fibers, which was around 12% (w/v). The final fermented product showed a lower *in vitro* rumen digestibility compared to
unfermented controls. However, it has to be kept in mind
that this only measures fiber digestibility. The loss in this
parameters was compensated by a gain in protein (363). Sekita
et al. (364) have suggested that *C. cellulolyticum* to be
identical to *C. thielavioideum* and they proved the production
of mycotoxins by the fungi in rice cultures by cytotoxicity
tests using Hela cells and TLC.

Sethi and Grainger (365) grew *A. niger* in a solid medium containing gelatinized mangostone powder supplemented with inorganic nitrogen and phosphorus at 35°C for 3 days. The final product contained 17.3% and 12.1% of crude and true protein. Bugdan y Sivers (366) grew *G. candidum* in untreated and acid hydrolyzed corn cobs for 5 days at 22-25°C; in the product there was an increase of sugars, organic acids and protein. Wicklow et al. (367) studied the substrate utilization by *Cyathus stercoreus* in wheat straw and found that approximately 20% of the cellulose and 45% of the lignin were degraded through the activities of this basidiomycete. Recently, Zadrazil et al. (367a) found that the white rot *Ganoderma applanatum* is the key fungi that biodegrades wood, producing "palo podrido" in Chile. Lignin contents decreased to values as low as 1% and the *in vitro* rumen digestibility increased twenty fold. A late appearing, unidentified basidiomycete was most effective in degrading lignin and rendering the cellulosics components of wheat straw more susceptible to hydrolysis by cellulases (367b). Zadrazil and Brunnert (367c) compared the performance of two white rot fungi on the degradation of wheat straw and other residues. *S. pulveruluntum* was quicker than *Dichotimus squalens*, however, the latter, after long incubation, preferentially degraded lignin and thus improved the *in vitro* digestibility of the substrate. Amounts over 0.25% of ammonium nitrate had an unfavourable effect on degradation.

Rao et al. (1983) grew *Pestalotiopsis versicolor* on alkali pretreated rice, wheat and rye straws, cane bagasse and wheat bran obtaining three times as much cellulolytic enzymes in solid substrate than in liquid culture.

Eriksson and co-workers have patented a biopulping process where through solid state culture of *Pheniophora cremea* in 66 days they remove up to 75% lignin and less than 25% of the cellulose (368). This development suggests a multipurpose scheme where a good quality fiber and a possible feed product could be produced simultaneously. Pilon et al. (358a,b) showed that the growth of nine higher fungi in mechanical pulps for seven weeks increased their water retention values, property related to the quality of the paper. Bar-lev et al. (368c) employing *R. chrysosporium* ATCC 34540 also showed the fungal growth reduced energetic requirements for secondary refining of thermomechanical pulps.

Raimbault and co-workers at ORSTOM and IRCHA in Paris have developed a technique where filamentous fungi grow in solid state culture in starch biomass. *A. niger* was found to

grow in cassava at optimal rate under the following conditions: substrate moisture 50-55%, 35°C, a nitrogen source
comprising 60% ammonium and 40% urea and 2 x 10^7 spores/g
substrate (369). Five strains of filamentous fungi grew
equally well in cassava giving a product with 15 to 18% protein and 25 to 35% residual sugars after 24 h. Seven strains
isolated from banana grew well and enriched banana flour up
to 20 to 24% protein (370). The amylolytic enzymes produced
by *A. niger* on cassava starch were found to be mainly gluco-
amylase which was more thermostable than the similar one produced in liquid cultures (371). Vezinher et al. (372) have
grown *Rhizopus arrhizus* CBS 285-55 in a wheat substrate with
the addition of inorganic nitrogen; of 100 g of fresh material
with 12.5% protein content a product of 75 g with 23% of protein was obtained. The solid substrate moisture varied from
67 to 75% and oxygen mass transfer into the pastry product was
a problem. Kokke (373) employed *R. arrhizus* CBS 324-35 and
shopped carob pods as substrate and after 3 days at 30°C a
product containing 7% protein was produced. The same had considerably lower tannin content and still contained 73-85% of
the original carob sugars. Plating and Cherry (374) fermented
cottonseed flour for four and seven days with six selected
filamentous fungi: *Aspergillus sojae, R. oligosporus, M.
purpureus,* and *Actinomucor elegans.* Due to fermentation the
proteins in cottonseed flour were converted to polypeptides
of varying sizes with different solubility characteristics and
free amino acids. Improved extractability was noted in non-
storage and storage protein fractions. Kamalakannan and Mut-
lag (375) have shown that the growth of *R. oligosporus* NRRL
2710 releases and inactivates the trypsin inhibitors of black
gram. The chemical changes of fermented foods have been reported for Ogir (376), Gari and Lafon (377), Kogi and Miso
(378,379), Yokotsuka and Sasaki (380) have shown that myco-
toxin contamination in these types of foods is minimum in
pure-cultured processes employing either *A. oryzae* and *A.
sojae.* Lequerica et al. (381) studied the growth of *A. niger*
on ground orange peels and found a total dry weight loss of
42% and a total protein increase of 73%; the product had 20.0%
crude protein on dry basis. Bajracharya and Mudgett (382)
grew *Aspergillus* sp QM9994; *A. niger* QM877; and *R. nigricans*
QM387 in pressed alfalfa residues and found that up to 50%
more protein could be recovered from alfalfa relatively to the
unfermented material. Moreover, the fermentation process increased cakes that could be suitable for direct ensilage for
animal feeding. In subsequent publications the authors showed
that control of the atmospheric oxygen and carbon dioxide
contents during solid substrate fermentation of rice by *A.
oryzae* stimulated amylase productivity (383,383a). Gibb and

Walsh (384) have shown that the growth of eight fungi is reduced as oxygen concentration decreases. Carbon dioxide, up
to 4% (v/v) was generally stimulatory and never inhibitory.
In solid substrate fermentations by fungi the estimation of
the fungal biomass is difficult and can be approximated by
carbon dioxide evolution (385) or by chemical analysis of
chitin (386); comparisons have been given (387) and a mathematical model for growth has been discussed (388). Navahara
et al. (388a) have reported that the respiratory quotient
during an 80 h rice koji culture of *Aspergillus oryzae* was
rather constant and close to unity.

Smith et al. (389) have shown that *Fusarium solani* can
biodegrade the antifungal phytoalexin phaseollidin into a
hydrate which has a lower antifungal activity.

Inoculation with mesophilic or thermophilic fungi of a
solid lignocellulosic waste, like wheat stalk and straw, results in a slight increase of carbon decomposition and organic
nitrogen over the control, after natural composting where
material was turned over every two weeks (389a,b).

Kindu et al. (389c) have defined an optimal moisture content region for solid state culture; below this level there
was enzyme inhibition due to products formed and above this
level there was greater enzyme diffusion away from the substrate.

8.   Photosynthetic Bacteria

As mentioned in the previous review (1) photosynthetic
bacteria are capable of carrying out two of the basic reactions on earth, namely nitrogen fixation and photosynthesis at
the expense of solar energy. A great deal of interest exists
on these microorganisms as new and detailed knowledge of the
photochemical mechanisms has emerged by studying photochemical
reaction centers of these bacteria (390). The phototropic
green and purple bacteria are included in this group as also
the cyanobacteria or blue green algae with whom the former
have major physiological and biochemical differences (391).

Aiba et al. (392) studied the growth of *Rhodopseudomonas
spheroides* S under continuous culture as a photo-organotroph
and as a chemoorganotroph using acetate as a hydrogen donor
and carbon source. The growth quantum yields were obtained.
Gobel (393) also presents similar data for *Rhodopseudomonas
capsulata* and *Rhodopseudomonas acidophila* and autotrophic and
heterotrophic specific growth rates for the same two bacteria

plus *Rhodopseudomonas rubrum*. Erickson (394) has calculated
true growth energetic yields for *R. spheroides* and found them
higher in oxygen limited cultures than in glucose limited
cultures. Aiking and Sojka (395) found for *R. capsulata*
growth under anaerobic, photosynthetic conditions in continuous culture, that photopigments and RNA levels were
affected by growth rate regardless of light intensity and
that these were not affected by the latter factor at a constant dilution rate. True protein levels were around 55%.
Aretz et al. (396) report a specific growth rate of 0.19 h^{-1}
for *R. capsulata* in anaerobic, photosynthetic conditions employing ammonium salts as nitrogen source, whereas with uric
acid or other purines it was only 0.13 h^{-1}. Sasaki et al.
(396a) grew *R. gelatinosa* and *R. gelatinosa Al* under aerobic-
dark conditions in wastewaters from a miso factory. Optimum
temperature, maximum specific growth rate, maximum biomass
produced and COD removal after 40 h were for *R. gelatinosa*:
35°C, 0.19 h^{-1}, 9.3 gl^{-1} and 81% respectively; for *R. gelatinosa Al*: 25°C, 0.17 h^{-1}, 10.6 gl^{-1} and 79%.

Under certain circumstances these bacteria can produce
as a growth by-product molecular hydrogen and growth conditions have been investigated to enhance its production as
also to extend the period of its production (397-401). Vrati
and Verma (401a) used the supernatant of the anaerobic digestion of cow dung to growth *R. capsulata* to a value of 4
gl^{-1} with a protein content of 61%.

The cyanobacteria *Spirulina* still commands a great deal
of attention due to its high protein content and to the fact
that it grows under alkaline conditions, where the high pH
of the culture ensures adequate supply of carbon dioxide for
photosynthesis and the alkalinity prevents invasion and development of competing microorganisms, creating in a way a
"proteced" pure culture (402). Moreover these blue-green
nitrogen fixing algae also are potential fertilizers (403-
406). Although it is known that some freshwater cyanobacteria
produce phycotoxins (alkaloids, polypeptides or pteridines)
which have caused unpredictable poisoning of livestock and
wild life. The implicated blue-greens were: *Microcystis
aeruginosa, Anabaena flow-aquae,* and *Aphanizomenon flosaquae,*
although the poisonous threshold value of the algae floc will
differ depending on the species and strains of algae and
species of animal involved, and is a function of the environmental parameters that affect growth (407-409). de la Fuente
et al. (410) presented data where a blue-green alga

Mycrosystis sp grown in a Guatemalan volcanic lake was fed
to rats as the only protein source or when the material supplied 25% of the total protein in a corn-algae diet. A high
mortality was observed in the former case whereas the protein
quality of the cereal was significantly improved in the
latter.

Santillana (411) and Durand-Chastel (411a) have described
the process and product characteristics of *Spirulina* grown on
natural alkaline brines of the Sosa Texcoco company in Mexico
City. They used race way ponds and the water was supplemented
with fertilizer to increase biomass production. Depth in the
ponds was less than 0.5 m and they had baffle systems, water
recirculation and mixing equipment. The algae were recovered
by screens, as *Spirulina* has its filaments arranged in an
elongated helix. Clement et al. (412) from Institut Francais
du Pétrole, who pioneered the *Spirulina* development, have
published a very detailed review article in which the principal factors involved in the design of a photosynthetic
reactor (of the open pond type) were discussed. Specifically
the relationship between growth rate and operating conditions,
absorption of carbon dioxide in gas air lifts and the required
stirring. Faucher et al. (413) employed pretreated sea water
to grow *Spirulina* maxima in laboratory bottles and in an open
pond type bioreactor with agitation provided by an airlift
(mixtures of CO_2 and air). Nitrate at 3.0 gl^{-1} was added as
a source of nitrogen. Urea could be substituted for nitrate
only up to 0.2 gl^{-1}, as above it inhibited growth, mainly due
to decomposition to ammonia in the alkali medium. Ayala and
Bravo (413a) described a medium containing seawater and salt-
peter.

A system for the utilization of sewage for algal biomass
production in order to accomplish the double purpose of harvesting microbial biomass and reclaim water has been reviewed
by Richmond and Preiss (414a,b). According to the authors,
three major technical aspects are involved in developing
commercial systems for the cultivation of blue-green algae:
the bioreactor or open pond characteristics, the algae harvesting and its preservation. Note that the authors do not
mention the fact that by far, the system is not pure culture,
but a complex bacterial population develops. However, the
purpose is to harvest pure cyanobacteria, an operation that
could be the bottleneck of the system. Presently only the
filamentous *Spirulina* is readily filtered on a 300 mesh screen
but care must be taken in order not to damage the cells so as
to prevent a leakage of cellular matter (414a). Growth response of *S. platensis* to solar irradiance and air temperature

could be controlled by manipulation of pond depth, population
density and degree of mixing (414c). Raising the temperature
and maintaining 16 gl^{-1} of bicarbonate seemed to favor
Spirulina over *Chlorella* (414d). In *Oscillatoria* self-
aggregation might happen spontaneously and rapidly (415).
S. maxima was grown in batch culture in a 200 l minipond with
mechanical agitation at 30°C in an alkaline medium, pH 9.2,
and raw manure solids to a concentration of 0.4%. Total suspended solids reached 3 gl^{-1} (416). The use of the algae
Spirulina is a key part in an integral process for municipal
wastewater treatment and animal wastes reclamation systems
(416a). Work continues at the CFTRS in Mysore, India to
develop a rural oriented methodology for the production of
S. platensis. A production rate of 8-12 g m^{-2} day^{-1} has been
reported (417,418a). Efforts have been made to utilize local
raw materials like bone meal as minerals source, urine, blood
or biogas effluents as nitrogen sources and carbon dioxide
enriched gas from composting of cow dung as a source of carbon (418). Soong (418b) has briefly described the cultivation
of *Spirulina* in digested hog manure in Taiwan.

Cultivation of *Spirulina* and other blue-green algae under
sterile conditions in pilot photobioreactors has been discussed by Juttner (419,420) and Leduy and Therien (421,422).
The latter scientists built an annular-baffled reactor with
liquid recirculation and agitation by airlift action of the
gases. At a dilution rate of 0.24 day^{-1}, 0.91 gl^{-1} day^{-1} of
S. maxima biomass was produced, which was equivalent to 14.5
g m^{-2} day^{-1}. The former references describe special tube
pilot designs of 30 and 110 l. Ogawa and Aiba (423) report
data of specific rate of carbon dioxide assimilation by *S.
platensis* and from these data they have calculated a conversion efficiency, defined as energy retained as cell material/
light energy absorbed, in the order of 10%. van Eykelenburg
(424) has found an abrupt metabolic change at 17-20°C of *S.
platensis*. Above this temperature cyanophycin was replaced
by polyglucan as the abundant storage material and a more
tightly coiled trichome existed; the organism was able to
grow with nitrate as nitrogen source up to 15.5 gl^{-1}.

Washout studies of *Anacystis nidulans* from steady state
chemostats under light limitation (0.03 - 0.15 h^{-1}) or carbon
dioxide limited cultures (0.03 - 0.17 h^{-1}) have been reported
by Karagouni and Slater (425). Frishknecht and Schneider
(426) have shown, for the same cyanobacteria, that the photosynthesis rates measured as oxygen production in the temperature range of 25-45°C are generally higher than the carbon
dioxide uptake and hence the photosynthetic quotient exceeds

one, specially with nitrate as nitrogen source. Under anaerobic conditions a much higher carbon dioxide evolution is
found, an indication of decarboxylations relating to anaerobic
catabolism. n-Alkanes in blue-green algae were formed by
decarboxylation of saturated fatty acids (427). The transport through the cell membrane was the rate limiting in phosphate incorporation (428). Parrot and Slater (428a) found
that protein content was greater in carbon dioxide limited
than light limited continuous culture, but in both declined
as dilution rate increased above 0.10^{-1}. RNA increased and
DNA increased with dilution rate. Amino acid uptake involves
cyclic photophosphorylation as shown for leucine in *A.
nidulans* (428b).

Contreras et al. (429) have reported growth of *Oscillatoria* sp in sea water under high light intensities (over
50 J m^{-2} s^{-1}) without algal bleaching and subsequent lysis for
the cells. The facultative anoxygenic photosynthesis in *O.
limnetica* with sulfide as electron donor has been studied by
Oven and Shilo (430) and the effect of oxygen by Friedberg
et al. (431). Padan (432) has published a review article discussing the importance of this reaction in the ecology of
alkaline and neutral waters and hot sulfur springs. Yamaoka
et al. (433) have studied the growth of a thermophilic *Synechococcus* sp in the temperature range between 50 and 60°C.
Under light saturation conditions (>9.4 W m^{-2}) the growth
kinetics of *O. agardhii* under nitrate limitations was equally
well described by the Monod and Droop equations (434,435), the
cyanobacteria had a two enzyme system for nitrate uptake
(436). van Liere and Mur (437) gave data of the growth
efficiency or yield and the specific maintenance rate constant
for continuous cultures in light limiting conditions of *O.
agardhii*. Light intensity affected the former and pH of both
parameters. Carbohydrates were synthesized and stored at
very high rates during photoperiods and were metabolized in
the dark (438). Ahlgren (439,440) modelled the growth of *O.
agardhii* in nitrogen or phosphorus limited chemostats and gave
values of the parameters for the Monod and Mitscherlich-Baule
equations. During multiple nutrient limitation five equations
were tested (441) to support the concept of a limited multi-
plicative effect, as it was proved that as soon as one nutrient
is in shorter supply, then it alone seems to control growth.
An *Oscillatoria* sp isolated from a *Spirulina* sample from lake
Chan grew better in indoor shaken and aereated cultures. Its
biomass could be collected on a sieve six times longer than
the one used for *Spirulina* and although crude protein, fat and
ash were about the same, it contained more pigments (441a).

Matterassi et al. (442) showed that *Anabaena cylindrica* had similar growth rates and protein content employing different nitrogen sources. Nitrate inhibits ammonium uptake and vice versa (443). Sahu et al. (444) showed that many hexoses and pentoses promote growth of *Anabaena* sp in light and in the dark. Gotta et al. (445) by isolating mutants of *Anabaena* strain CA impaired in nitrogenase activity and growth on nitrogen and have postulated that there may be several sites in the heterocysts sensitive to O_2 and that the protective mechanism involves several different phenomena. The toxic metal ions (copper, cadmium and lead) inhibited growth of the algae *Anabaena* strain 7120 (446). Selenium significantly reduced phosphate uptake by *A. cylindrica* thereby inhibiting growth (446a). Verstraete et al. (446b) studied the stimulation of growth by iron over *A. flos-aquae.*

Kruger and Juttner (447) studied carbon dioxide fixation mechanisms in *Mycrosystis* under different light intensities. Reserve material usually is formed under sufficient light and synthesis of essential compounds necessary for cell maintenance takes place under limited light conditions. Okada et al. (448) present an equation to estimate the phosphorus specific uptake rate by *M. aeruginosa* with two constants and the intra- and extracellular phosphorus concentration. Lang and Brown (449) discuss the same effect in batch and continuous culture of *Synechococcus*. Sicko-Goad et al. (450) studied the effect on phosphorus uptake in *Plectonema boryanum* and its subsequent incorporation into polyphosphate. Rogerson reported on the nitrogen fixing growth of nonheterocystous *P. boryanum* (451).

Photorespiration was undetectable in the blue-green algae *Coccochloris peniocystis* (452) and it photosynthesized optimally over the pH range of seven to ten. An acidic environment restricted its photosynthetic capacity by a direct pH effect on the carbon dioxide fixing enzymes (453). Evidence for bicarbonate transport were also published (454).

Reports of *Anabaena* sp nitrogen fixing properties and hydrogen evolution have recently proliferated mainly due to its interesting metabolic activities (nitrogen fixation under aerobic conditions) as also due to its potential as a biofertilizer in natural ecosystems and its energy production through biophotolysis. A group of selected references is given in Table 1.

Table 1. Some references on Anabaena sp

Nitrogen Fixation Mechanisms	Hydrogen Evolution/Utilization	Nitrogen Fixation In Natural Ecosystems
Jacobs and Lind (455)	Bothe et al. (460)	Pearl (473)
Abreu et al. (456)	Daday et al. (461)	Reynaud and Roger (474)
Murry and Benemann (457)	Strandberg (462)	Schanz et al. (475)
Stratton and Corke (458)	Belkinand Padan (463)	Padhy (476)
Apte and Thomas (459)	Peterson and Wolk (464)	Rice et al. (477)
Husgrave et al. (459a)	Jeffries et al. (465)	Hable and Alexander (478)
	Hallenbeck et al. (466)	Keller and Pearl (479)
	Lambert et al. (467)	Ito et al. (480)
	Miyamoto et al. (468)	Majid and Khatun (481)
	Asada et al. (469)	Kapoor and Sharma (482)
	Antarikanonda et al. (470)	Barraquio et al. (483)
	Strandberg (471)	Bors et al. (484)
	Berchtold and Bachofen (472)	Collins and Boylen (485)
	Miura et al. (472a)	Al-Maadhidi and Henriksson (485a)

9.   Aerobic Bacteria on Lignocellulosic Wastes

Thayer et al. (486) evaluated the growth of *Cellulomonas* sp ATCC 1399; *Pseudomonas* sp JM127; three *Bacillus* sp; four *Brevibacterium* sp; one *Serratia* sp and one unidentified bacteria, on mesquite wood ground to sawdust and enriched with ammonium sulphate, yeast extract and salts. All of the bacterial isolates compared favorably with *Cellulomonas* in their abilities to hydrolyze mesquite wood and produce protein of comparable quality as judged from amino acid analysis. Short term rat feeding showed the absence of toxicity problems in some of the species. Most of the cultures required about 72 h to reach maximum nitrogen content. The growth of *Brevibacterium* sp JM98A was optimum at a pH range of 6.6 to 7.2 at 37°C (487). Cellulose consumption was 0.067 gl^{-1} h^{-1} in a 72 h batch culture and 0.031 gl^{-1} h^{-1} in a 12 h - 12 cycles semicontinuous culture. Cellulase (β 1,4 glucanase and β 1,4 glucan cellobiohydrolase) activity was higher in the batch culture than the semicontinuous culture. Reducing sugar consumption rate was higher in each of the 12 h cycles than in the 72 batch culture. The ratio of the amount of protein produced by the semicontinuous to the batch processes was around 3, however, in batch culture much more substrate was biodegraded. Bench and pilot fermentor studies were carried out with *Pseudomonas* sp JM127 (488) varying the range of substrate concentration from 10 to 75 gl^{-1}. At 40 gl^{-1}, protein produced (in gl^{-1} h^{-1}) and substrate hydrolyzed (in %) were 0.03 and 41.8% and 0.049 and 35.5% in the 14 l and 3750 l fermentors. Volumetric power consumption was 10 times lower in the bigger fermentor, around 3 watts l^{-1}. The bacteria produced both cell bound and cell free CMCase and filter paper cellulase when growth in glucose, CMC and filter paper (489). It hydrolyzed shredded mesquite wood, cotton-ginning trash, bovine rumen contents and to a minor degree mewquite lignin. Tests on rats showed that the net protein utilization of intact cells was inferior to casein, however, when mechanically homogenized, the NFU and nitrogen digestibility were improved (489a).

Wani and Shinde (490) did screening studies for bacteria that could degrade wheat straw and found that *Cytophaga rubra, Cellulomonas ude, Bacillus macerans* and *Serratia hiliensis* decomposed straw within 60 days. Bomar in Karlsruhe has published his studies on *Cellulomonas* sp growth on straw obtaining up to 200 g biomass kg^{-1} straw in three days (491). Sodium hydroxide pretreatment increased yield to 310 g kg^{-} ,

but extra processing cost increased ten fold and the resulting wastewater was a problem (492,493). Refuse of agriculture and food industry can be used as raw materials for the
process (494,495a).

Srinivasan et al. (496) grew a strain designated LC-10
of *Cellulomonas* sp ATCC 21399 on glucose and obtained a 14
fold increase in biomass to a final of 29.4 gl^{-1} in 10 h.
The biomass yield was a function of dilution rate in continuous culture experiments employing a glucose, ammonium sulphate, salts and vitamins enriched media at 30°C (497).
Blancas et al. (497a) increased the protein content of
henequen pulp from 7% to greater than 20% with a bacterial
strain TE-B1025. Growth was for 72 h in a 2% suspension with
0.3% ammonium sulfate addition.

Ramasamy et al. (498) tested the capacity of an aerobic
mesophilic *Pseudomonas* sp isolated from activated slude to
biodegrade various natural lignocellulosic residues. Biodegradability, as expected, was mostly influenced by the lignin content and the crystalline nature of the substrate.
Alkali pretreatment of wheat straw, wood scrappings and
cotton fibres, surprisingly, did not improve the biodegradation. The best incresae in protein content was with wheat
straw, 8 fold in 15 days. Hagget et al. (499) have shown
that mutant derivatives of a strain of *Cellulomonas* (CSI-1)
were able to degrade crystalline cellulose of cottonwood
which had not been subjected to any specific pretreatment.
Shake cultures with pH control and 33°C accumulated significant amounts (0.57 gl^{-1}) of glucose and cellobiose after 5
days, being their enzymes, then, partially resistant to the
repressive effects of glucose and cellobiose.

Lee and Humphrey (500) determined from continuous culture on glucose the growth yields of *Thermoactinomyces* sp as
0.42 g g^{-1} glucose and a specific maintenance of 0.024 g
glucose g^{-1} cell h^{-1}. A method for the rapid screening of
cellulolytic streptomycetes, their catabolic repression of
the cellulase production by glucose and the amounts of enzymes produced, was recently published (501). Lee et al.
(502) obtained 4.77 gl^{-1} of *Cellulomonas* sp 34 biomass in
shake flasks for 48 h employing a 3% slurry of bamboo shoot
husks as substrate.

Hitchner and Leatherwood (503) used a cellulase derepressed mutant of a *Cellulomonas* species, isolated from the
soil by Steward and Leatherwood (504), to biodegrade pure
cellulose powders as 0.5% slurries in a 250 l fermentor at

30°C and using urea as a nitrogen source. The maximum batch
growth rate was at pH 7 and it was approximately 0.2 h^{-1}. The
bacterial biomass was airflotated and its net protein utilization value when the protein was supplemented with methionine
was 50.4%.

Choudhury et al. (505) explored various experimental conditions to optimize sugar production from alkali pretreated
sugarcane bagasse employing the *Cellulomonas* mutant CSI-17.
By increasing the culture temperature to 37°C under aerobic
conditions, resulted in the release of 23.7 g reducing sugar
l^{-1} from 75 g pretreated bagasse l^{-1} after 48 h incubation.
Addition of the metabolic inhibitor sodium azide increased
this figure to 25.6 gl^{-1}. If the incubation at 37°C was done
under anaerobic conditions, 22.8 gl^{-1} were also produced.
Over 80% of the sugars were released within 18'h of incubation
and 75% of all reducing sugars were glucose, cellobiose and
xylose. Rickard and Peiris (506) showed also that the mutant
CSI-17 was improved over CSI-1 in its ability to hydrolyze
xylan containing substrates, including sugarcane bagasse holocellulose and hemicellulose.

Enriquez (507) employed alkali pretreated sugarcane
bagasse and a *Cellulomonas* strain II bc and found that temperature should be around 32-35°C and pH below 6.5 to obtain a
70% utilization of bagasse and a yield of 0.5 g biomass per g
bagasse utilized. A diauxic growth was observed, achieving
growth rates of 0.18 h^{-1} from the start to 32 h, 0.16 h^{-1} up
to 72 h. In the early stage of the fermentation the crystallinity index of bagasse increased, suggesting that the
amorphous regions of the same were being attacked and the
hemicellulose with preference. Later in the process the
crystallinity achieved a steady state and then decreased,
which indicated that the most complex structure of bagasse
was being attacked (508). A maximum biomass productivity of
2.1 $gl^{-1} h^{-1}$ was obtained when the bagasse concentration was
40 gl^{-1}. The biological value in rats of the product was 62
and nitrogen digestibility 90% (508a).

Kim and Winpenny (509) employed *Cellulomonas flavigena*
KIST321 and a waste paper fraciton separated from domestic refuse as the substrate. The maximum batch growth rate was 0.1
h^{-1}. In continuous culture at a dilution rate of 0.06 h^{-1},
the cell concentration was 3.7 gl^{-1}. The maximum volumetric
productivity was 0.26 g biomass $l^{-1} h^{-1}$ using a medium containing 14 gl^{-1} of waste paper. Cell flocculation by heat
treatment and protein additions were explored. The organism
was found to require only thiamine for growth in mineral salts

medium containing simple sugars or cellulose (510). Growth
rate on glucose was 0.32 h^{-1}, while the values on cellobiose,
xylan and amorphous cellulose were 0.23 h^{-1}. On crystalline
cellulose it was considerably slower 0.11 h^{-1}. Hence, it
seemed probable that the rate-limiting step in cellulose
utilization was the decrystallization process.

Berg (511) found that all CMCase in *Cellvibrio fulvus*
was cellbound when grown on mono or disaccharides, but grown
on cellulose gave cell-free cellulase activity. Breuil and
Kushner (512) reported that *C. gilvus* produced cellulase only
when grown in the presence of the polymer, hence it was induced. Cellulases of a *Cytophaga* sp (ATCC 29474) were found
intracellularly both in the soluble portion of the cell and
on the membrane (513). Cellobiose induced the production of
a two component cellulase system in a *Bacillus* sp (514) and
with cellulose it also induced the production of cellulases
in *Pseudomonas* sp 1, which was best at 30°C and pH 6.5 - 7.0
(515). Ramasamy and Verachtert (516) studied the localization
of cellulase components in a *Pseudomonas* sp isolated from
activated sludge. Endoglucanases were produced extracellu-
larly only when grown on cellulose. When grown on other substrates intracellular endoglucanases and aryl- β - glucosidases were formed. The location of the endoglucanase was
different depending on whether the growth substrate was cellulose or cellobiose. Beguin et al. (517) studied the extracellular cellulases of *Cellulomonas* strain II bc and *Cellulomonas flavigena* (ATCC484); they found one class of enzymes
tightly bound to cellulose and the other was found free in the
culture supernatant fluid. Kolanskaya (517a) found an inverse
relationship between growth rate and enzyme production; parameter that was also influenced by the DO and yeast-extract
concentration of the medium. Starvation conditions stimulated
enzymatic synthesis. Stoppok et al. (518) reported that when
C. uda was grown on polymers, mono or disaccharides, endo
β 1,4 glucanases were extracellular inducible enzymes but β
glucosidases were formed constitutively and found to be cell
bound. Formation of the former was induced by cellobiose
and repressed by glucose. Its activity was inhibited by
cellobiose but not by glucose. On the latter, glucose was
a non-competitive inhibitor. Choi et al. (519) studied the
production of cell bound and extracellular enzymes in *Cellulomonas* CSI-1 and Haggett et al. (520) studied mutants of the
same with high levels of β glucosidase. Rickard and Laughlin
(521) showed that the CSI-1 microorganism produced an extracellular xylanase and had a β xylosidase activity inside the
cell. The enzyme preparation from the mutant derivative CSI-
17 was very resistant to end product inhibition by xylose,

glucose, cellobiose and ethanol (522), and it had been improved over the original parent strain, with respect to xylanase and β xylosidase activities (523). Recent studies by
Rickard (524,525) have pointed out that glycoside activity
towards removal of the side chains is the limiting factor in
the effective hydrolysis of the basic xylan polymer, hence
a mutation program should be addressed towards enhancing the
production of these enzymes by *Cellulomonas*. An endoxylanase
from *Cellulomonas* was purified and its optimum pH was 5.5 -
6.5 and optimum temperature 50-55°C (526). A *Cellulomonas*
sp isolate showed strong hydrolyzing activity on crystalline
cellulose (526a). Stegmann et al. (526b) have shown that
exo 1,4 β glucanases in *C. uda* are inducible but repressed
by glucose and cellobiose, whereas their activity was only
inhibited by glucose. Endoglucanases and glucosidase are
constitutive.

   Hagerdal et al. (527, 528) found that the CMCase and
Avicelase activities in *Thermoactinomyces* were extracellular
and produced associated with growth; the β glucosidase activity, however, was an intracellular soluble enzyme. Stability
of the enzymes, their production, the hydrolysis of the substrates by its action and the growth of *Thermomonospora* sp
on the sugars were also published (528a,b,c,d). Continuous
cultivation in a deep jet bioreactor was successful,
however, care was needed to adjust the circulation pump
speed in order to limit the shear stress exerted on the
organism (528f). Fennington et al. (528g) were able to
separate three distinct components of the extracellular endo-
β-glucanase produced by *T. curvata*.

   Skowronska (529) found exo and endo cellulolytic activity
in 40 soil strains of *Streptomyces*. Sreenath et al. (530)
found extracellular endo and exo xylan hydrolases in *Streptomyces*. Klueppel et al. (531) studied the simultaneous production of cellulase and glucose isomerase by *S. flavogriseus*
and the cellulase complex of a mesophilic *Streptomyces* strain
(532). Loginova et al. (533) studied the cellulolytic and
xylanase activity of *S. afghaniensis*. Yasui and collaborators
(534-536) studied the induction and inhibition of *Streptomyces*
sp. Ishaque and Kleupfel (537) reported results of the production of xylanolytic enzymes by *S. flavogriseus*. Uchino
and Nakane (538) studied a thermostable xylanase from a thermophilic acidophilic *Bacillus* sp. Vance et al. (539,540)
studied the extracellular enzymatic complex of *Sporocytophaga
myxococcoides* NCIB 8639 and its growth and cellulose biodegradation in an air-lift fermentor.

Esteban et al. (540a) have identified two xylanases and one xylosidase when *Bacillus circulans* WL-12 is grown on xylan. Ohba and Ueda (540b) studied the production of pullulanase from *Aerobacter aerogenes*.

At this point it is appropriate to make reference to a review article by Bisaria and Ghose related to the various aspects of the biodegradation of cellulosic materials: the substrate, microorganisms, enzymes, and products (541) and also point out that not only microbial biomass is the final product but in recent years extracellular metabolites are looked upon. Hence, multi-product processes might be developed based on lignocellulosic substrates.

However, a great deal more work has been done by studying the production, the kinetics and characteristics of cellulolytic and xylanolytic enzymes from fungi, and a large majority on *Trichoderma reesei*, as can be seen in Table 2.

10. *Hydrogen Bacteria*

Hydrogen, carbon dioxide and nitrogen are the most chemically simple sources for microbial biomass production. Aerobic hydrogen bacteria are facultative chemolithoautotrophs (634) which can be isolated from any type of soil, where they might play a role (635) in the oxidation of hydrogen coming from anaerobic decomposition or nitrogen fixing biochemical processes (636). By enrichment and isolation studies new nitrogen fixing hydrogen bacteria (637) and thermophiles (638-640) have been identified. A strict thermophile and obligate autotroph has been recently described (641).

The growth of *Alcaligenes hydrogenophilus* under autotrophic conditions has been studied under batch (642) and continuous culture (643). A kinetic model has been developed and its constants determined. The optimum cell productivity was around 0.3 gl^{-1} h^{-1} with a gaseous consumption of close to 30% (644). About the same cell productivity was reported by Morinaga et al. (645) for *A. eutrophus*. Kester et al. (646) showed that *Hydrogenomonas eutropha* Z11 grew well under batch and continuous culture under 5-7% carbon monoxide. Under autotrophic conditions the concentration of metabolites excreted by these bacteria was low (647), in *A. eutrophus* they are maximized when growth is limited by ammonia and oxygen supply (648).

Table 2. References on cellulolytic and xylanolytic enzymes from fungi

Fungi		Fungi		Yeasts	
1. Trichoderma reesei	542-567i	6. Humicola insolens	600-601	Cryptococcus albidus	626-632
Trichoderma koningi	568-570	7. Sclerotium rolfsii	602	Trichosporon cutaneum	633
Trichoderma lignorum	571	8. Schizophyllum commune	603-605		
Trichoderma sp	572-573d	9. Chaetomium cellulolyticum	606		
2. Aspergillus niger	574-579	10. Pyricularia oryzae	607		
Aspergillus terreus	580-583	11. alaromyces byssochlamydoides	608-609		
Aspergillus phoenicis	584	12. Sporotrichum thermophile	610-611a		
Aspergillus fumigatus	585-585a	13. Pellicularia filamentosa	612		
Aspergillus versicolor	586	14. Eupenicillium javanicum	613-614		
Aspergillus oryzae	587	15. Fusarium sp	615		
Aspergillus wentii	588	16. Scytalidium lignicola	616		
Aspergillus foetidus	589	17. Geotrichum candidum	617-620		
Aspergillus sp	590	18. Trametes hirsuta	621-625		
Aspergillus nidulans	590a	19. Fusarium gramineum	625a		
3. Penicillium funiculosum	591	20. Myrothecium sp	625b		
Penicillium janthinellum	592-593	21. Phymatotrichum omnivorum	625c		
Penicillium verruculosum	594	22. Trametes hirsuta	625d-g		
Penicillium wortmanni	595	23. Fusarium solani	625h		
4. Volvariella volvacea	596	24. Thermoascus aurianticus	625i		
5. Irpex lacteus	597-598	25. Thrichoderma auereoviride	625j		
Other brown rot	599	26. Talaromyces emersonii	625k		

11.   Thermophillic Non-Photosynthetic Bacteria

As pointed out by Zeikus (649) microbial biomass production from thermophilic bacteria appears at first of limited
value because of their lower cell yields relatively to their
mesophilic counterparts. However, other reasons may be as or
more important: faster growth rates, minimal contamination
from non-thermophilic microflora and less expensive cooling
and product recovery costs. Nevertheless, very few studies
have been conducted specially with the aerobic (or facultative aerobic), chemoheterotrophic, non-spore forming bacteria
like *Thermus aquaticus* (650), *Thermus thermophilus* (651),
Thermo microbium sp (652), *Flavobacterium thermophilum* (653)
and *Thermus flavus* (654). These bacteria not only are widely
distributed in natural thermal ecosystems but also proliferate
rapidly in man-made thermal environments (655). Recently,
Cometta et al. (656) have studied the behavior of *T. aquaticus*
in continuous culture under different vessels, techniques and
mediums. The maximum specific growth rate was dependent on
the medium, pH, type of vessel and growth technique (batch or
continuous), nevertheless, a very high value of 3.5 h^{-1} was
reported at 75°C and pH 7.3 in a complex medium (peptone and
meat extract). Lower values of 1.62 h^{-1} were found when glucose was the carbon source. A cell yield of 0.4 g g^{-1} was
reported for glucose and a maintenance coefficient for oxygen uptake of 16 mMol g^{-1} h^{-1}. The cell yield on the complex
substrate was very low, 0.02 g g^{-1}, due to incomplete substrate utilization. Cell concentrations were in the range of
0.1 - 0.2 gl^{-1}. Sonnleitner et al. (657) grew *T. thermophilus*
in continuous cultures at dilution rates up to 2.65 h^{-1} at
75°C and pH 6.9 on a complex medium. Again very low yields
(0.12 g g^{-1}organic carbon utilized) and incomplete substrate
utilization were found. Cell densities ranged from 0.05 to
0.16 gl^{-1} and a maintenance requirement for oxygen in the
range of 10-15 mMol g^{-1} h^{-1}. Growth was found to be substrate
inhibited. Although growth was very rapid, its low growth
efficiency, low cell yields and incomplete substrate utilization might diminish the industrial potential of this bacterium. Both *T. thermophilus* and *T. aquaticus* were rather
sensitive to plastic materials found usually in accessories
of bench fermenters (657a). Yoshizaki and Imahori (658)
have shown that glycolysis-gluconeogenesis in *T. thermophilus*
is regulated by a unique mechanism that they called "coupled
regulation," moreover, they also showed that although all the
enzymes of the glucolytic pathway were present one of the
pentose phosphage cycle was absent.

Kuhn et al. (659) found a maximum productivity of 1.3 gl^{-1} h^{-1} at 65°C, pH 6.5 and D=1.9 h^{-1}, in a 2 gl^{-1} glucose medium previously optimized by the chemostat pulse and shift technique for a *Bacillus* sp. Complete substrate utilization was observed and a cell yield of about 0.3 g g^{-1} was found. The maximum specific growth rate for *Bacillus caldotenax* was clearly temperature dependent (660). However, maintenance coefficients were very high relatively to the corresponding values for mesophilic microbes, up to 10 times higher. Although this difference could be the result of an increased turnover of cell material or a bigger requirement of energy for osmotic work, more data were needed to explain it completely.

12. *Anaerobic Bacteria: Acidified Food Products*

There are numerous food products that are preserved and owe their distinct organoleptic characteristics to the activity of lactic acid bacteria (*Lactobacillus, Leuconostoc, Pediococcus* and *Streptococcus*) (661-661a). Moreover, their essential amino acid contents compare well with aerobically grown bacteria (1). Hence, they are, along with mushrooms, fungal fermented foods and certain algae, the microbial biomass products directly eaten as food by human beings. This fact has stimulated research activities for developing non-traditional acid food products from many raw materials:

A. *Products from oilseeds.* Soybeans and peanuts, prepared as milk type suspensions, have been studied (661b). Mital and Steinkraus (662) and Patel et al. (663) have made recent assessments of the state of the art for soybeans. Researchers agree that lactic cultures improve the acceptability of soybean products, although lactic acid production does not eliminate the original and objectionable beany flavor and the bacteria do not consume totally the oligosaccharides responsible for the flatulence associated with soy products. Hence, appropriate heat treatments, sugar fortifications and additives are necessary for flavor and aroma improvements. Pinthong et al. (664) found that *Lactobacillus bulgaricus*, supplemented with glucose and yeast extract, partially removed n-hexanal, an aldehyde that contributes to the beany flavor; also, it produced an acceptable acid product with 3.8 pH in 24 h at 43°C. However, mixed fermentations were needed with *Lactobacillus fermenti* in order to remove raffinose totally but only partially stachyose. However, the authors concluded that a greater reduction could be achieved by altering the levels of supplementation and starter preparation. Patel and Gupta (665) grew a mixture of *Streptococcus thermophilus, Streptococcus diacetilactis* and *Lactobacillus bulgaricus*, and found that acid production as well as flavor and texture characteristics of the final product were substantially improved by lactose fortification. The addition of whey, instead of lactose, produced a slightly weaker consistency but better flavor than lactose-enriched soy milk. Beuchat and Nail (666) and Bucher et al. (667) have fermented peanut milk. The former employed two strains of *L. bulgaricus* and *L. acidophilus* with the addition of 2% lactose and in 3 days at 37°C produced a custard-like product having 0.38-0.53% acidity at pH 4.76 - 4.43. However, the ferments due to phase separation appeared undesirable as yogurt simulates, hence sucrose and flavoring agents were added to resemble the consistency of butter milk. The latter tested 19 different lactic strains on peanut milk. *L. acidophilus, L. brevis, L. xylosus, L. plantarum, Pediococcus acidilactici,* and *S. thermophilus* produced the highest acidity levels after 48 h. The addition of 1% glucose, sucrose, whey, tryphose or yeast extract increased acid production by most cultures. Stern et al. (667a) tested eight different NRRL *L. acidophilus* strains on nine carbohydrate substrates obtaining different fermentation patterns. Raffinose and stachyose were the two substrates hardly utilized. Kraidej (667b) reported problems in texture stability and a residual beany flavor.

   B. Cereals and legumes. In African, Middle-East and Mid-Asian cultures it is usual to ferment cereals and legumes alone or in combination. A lactic fermentation develops quickly (668) which not only increases the relative nutritive value (669) but gives it its special organoleptic properties (670,670a,b,c,671). *S. thermophilus* and *L. bulgaricus* imparted mild agreeable flavor to yogurt-cereal mixtures (672) which could substitute rice or maize in "atole" formulations with very good acceptance of the product by adults and children in rural communities in Mexico (673). Lactic acid beverages from cereals have been patented (674, 675). Ulloa and co-workers (675a,b) have described the fermented corn products commonly found in Mexico. Nout (675c) does the same for Kenya.

   C. Tubers. Most of the work has been done with cassava, specifically on production of "Gari" in Africa (676a,b) and "Polvilho Azedo" in Brazil (677). In both products the fermentation process is acidic in nature (lactic and acetic acids are produced from the soluble carbohydrates sucrose, glucose and fructose) and provides the characteristic flavor and texture of the same. Also, Ikediobi and Onyike (678(have demonstrated recently that linamarase activity present

in cassava is enhanced by the initial drop in pH, hence
accelerating the detoxidification process or hydrolysis of
the constituent cyanogenic glucosides. *Lactobacillus* sp,
Leuconostoc sp and *Streptococcus* sp have been isolated and
tested for their growth in cassave (679,680,681).

Note: In the products mentioned in sections 2.12.B and
2.12.C there is usually a mixed culture present: yeast/
lactobacillus (681a). It is presented here only to have a
more complete picture of the acid food processes.

D. *Vegetables*. Green tomatoes in slices in brines were
fermented by *Lactobacillus plantarum* (682,683). Kozup and
Sistrunck (684) used the same microorganism but with green
beans. *L. cellobiosus* fermented totally the sugars found in
green bean juice and produced lactic, acetic acids, ethanol
and mannitol. It could ferment 37.4 gl^{-1} of sugars producing
a pH of 3.5 - 3.9 which was adequate to preserve the product
for 12 months at 24°C in sealed jars (684a).

E. *Fruits*. Maitra et al. (685) have reported studies
of lactic fermentation of green mango slices and in order to
obtain adequate acid production it was necessary to maintain
about 15% of salt. Aegerter and Dunlap (686) tested five
lactic bacteria in green and ripe banana purees; there was a
pH reduction in 7 days at 37°C from 4.8 to levels around 3.5.
There was phase separation as soft solids floated in a clear
liquor. The color was medium to light creamy brown and the
odor was rather pleasant. Mature tropical fruits with its
rich aroma are indeed excellent substrates for developing
yogurt-type products as they would also provide sugars easily
metabolized by lactic bacteria (sucrose, fructose and glucose).
Specific cultures need to be identified for different fruits
in order to develop the proper organoleptic characteristics.

F. *Animal products*. Katamine et al. (687) have employed 1% glucose to increase five times more the acid production of whole egg. Moon et al. (688) tested forty-four
strains of lactic acid bacteria for its inhibitory action
against pathogens on fresh de-headed shrimp. Filtrates were
generally more inhibitory than cell suspensions, however,
overall behavior indicated that acid fermentation as preservation may not be practical. Raa (688a) has proven experi-
mentally that by the addition of lactic acid bacteria and a
fermentable sugar to fish, an acidified product can be obtained with a good nutritional value which resists spoilage.

The use of lactic bacteria in producing acid dairy products is a well established industry comprising more than 20% of all fermented food products in the industrialized world (689). Changes are brought upon by the actions of microorganisms upon flavor, texture and nutritive value of the resulting products (690). A recent review has been published by Tamine and Deeth (691) on yogurt, its production technology and biochemistry and nutritive and therapeutic aspects. Again, although it is a mixed culture process, it is briefly discussed in this section to complete the picture of acid foods. de Haast et al. (692) have discussed the topic of yogurt quality. Continuous culture experiments have been reported by MacBean et al. (692a) and Reichart (693). Driessen presents a detailed study of the protocooperation of yogurt bacteria (*S. thermophilus* and *L. bulgaricus*) in continuous cultures (693a). Oberman and Libudzisz (694) studied strains of *Streptococcus diacetilactis* and *L. casei*. Zbikowski (695) has informed on the products obtained by employing *Bifidobacterium bifidum* and *Lactobacillus acidophilus* instead of *L. bulgaricus* and *S. thermophilus* currently employed in all yogurt manufacture. The product obtained had different physiochemical properties and showed a higher protein efficiency ratio than milk or natural yogurt. *L. acidophilus* is of important interest due to its healthful aspects (696) and dietary use (697,698). Sozzi and Smiley (699) have shown that about 35% of *L. bulgaricus* and *S. thermophilus* strains have an uncharacteristic resistance pattern to antibiotics and they recommend not to use them commercially in order to reduce an unnecessary distribution of antibiotic-resistant strains. Within a small concentration range, as found in food systems, oleo resins and spices usually stimulate fermentation activity of *L. plantarum* (700,701) and *Pediococcus cerevisiae* (702).

The homolactic streptococci are true homolactics only when carbon substrate is in excess. Carbon limitations, either in continuous culture (703,704) or in agar gels (705) make the cells produce, beside lactate, formate, acetate and ethanol. Golz et al. (706) have reported the oxygen utilization mechanisms in the aerobic growth of *L. plantarum* in glucose, lactate and pyruvate. This growth resulted on a higher cell yield and was coupled with the generation of ATP. Acetate was the main product of the aerobic oxidation of pyruvate by *L. bulgaricus* (707). Zielke et al. (708) reported the positive effect on growth by an aqueous extract of *Senedesmus acutus* and Mikhlin and Radina (709) on the same effects but of acid and alkaline hydrolysates of thermophilic methane fermentation effluents.

13. Anaerobic Bacteria: Acidified Feed

Several papers have been published relevant to the
Michigan State University process in which whole or deproteinized whey is fermented with *Lactobacillus bulgaricus* at
pH 5.5 and 44°C under non aseptic conditions, the acid neutralized with ammonia and the broth used as protein supplement
for finishing cattle. Crickenberger et al. (710) have shown
that the acidified whey product is comparable biologically
to soybean meal as a protein source for finishing corn diets
of steers. Stieber and Gerhardt (711-714) have modelled the
continuous process including a dialysis system. Forney and
Reddy (715) successfully fermented acid hydrolyzed starch
obtained from the wash water of potato chip manufacturing.
Hietala et al. (716) have shown that during the fermentation
of animal wastes by *Lactobacillus plantarum* alpha hydroxyacids
(propionic, lactic, isovaleric and isocaproic) were formed,
which exerted an antimicrobial effect more prominent at lower
pH values when the acids were undissociated. Selected strains
of *Lactobacillus acidophilous* and *L. bulgaricus* grew well on
moist jojoba meal, on rotating bottle fermentors on solid substrate media at 26°C for 21 days. The bacilli reduced the
levels of cyanotoxicants present in the meal rendering it non
toxic to mice, poultry, sheep and cattle (717). Green et al.
(718) showed that *Lactobacillus casei, L. plantarum* and *Pediococcus cerevisiae* were able to grow on fish hydrolyzates.
Moon (719) used a *L. plantarum* inoculum in solid substrate
(silage) fermentations of potato processing wastes and de-
watered green bean wastes mixed with peanut hulls. The inoculum decreased pH, increased lactic acid and decreased
acetic and other volatile acids, and decreased competing
microglora. The relative nutritive value of sorghum was increased significantly by an acid fermentation (720,721).
Fermented and flavored pet food has been described in two
patents (722,723).

In a previous publication (1) emphasis was given to the
advantages found in the anaerobic route to biodegrade lignocellulose. Fascinating research has been published recently
on the mode of action of these bacteria *in vivo* systems like
the rumen, in which *Ruminococcus albus, R. flavefaciens* and
Bacteroides succinogenes are the main cellulolytic bacteria
acting together with acid formers, protozoa (724) and anaerobic phycomycetous fungi (725), or *in vitro* systems, mainly
with *Clostridium thermocellum* or *Acetivibrio cellulolyticus*
(726,727). The former are truly mixed continuous cultures and
their interactions as such will be dealt with later on; the
latter are mainly pure culture studies. Forsberg et al. (728),

Forsberg and Groleau (729) and Groleau and Forsberg (730)
have studied the cellulase complex of *B. succinogenes*. An
extracellular exo β 1,4 cellobiohydrolase and endo β 1,4 glucanase and β glucosidase extra and intra-cellular and an
extra cellular xylanase were present. The enzymes were
associated with the cell envelope or membrane fragments, they
seemed to be inducible with the exception of β glucosidase
which was constitutive. Yu and Hungate (731) found four different extracellular exo-β 1,4 glucanase in R. albus. Petti-
pher and Latham (731a,b) found in batch and continuous culture experiments with *R. flavefaciens* mainly cell associated
exo and endo β glucanase and xylanase, pectinlyase, pectin-
methylesterease and β xylosidase activities. The enzymes
being constitutive and catabolite repressed. It seems that
there is a very strong interaction between these bacteria and
the lignocellulosic polymer. From microscopical observations
it seems that the hydrolytic enzymes are either, a) adsorbed
to the fibers when the bacteria is also attached strongly to
the surface, b) they are contained in outer membrane packages
or subcellular vesicles, or c) are within an exopolymer gel
between bacteria and fiber, or they are in all three forms
plus truly extracellular enzymes in the bulk.

Khan and co-workers at the NRC in Ottowa (732,733) and
Mackenzie and Bilous (734) have shown five major enzymes in
A. cellulolyticus: β glucosidase, and exo β 1,4 glucan cellobiohydrolase, two endo β 1,4 glucanase and a xylanase. The
first was mainly cell associated, the others were detected in
the supernatant. Glucose and cellobiose did not inhibit
enzymatic action. Allcock and Woods (735) found two inducible, not repressed by glucose, extracellular endo β 1,4
glucanase and β glucosidase in *Clostridium acetobutylicum*.
Many researchers have studied the enzymatic complex of *C.
thermocellum* in which the exo and endo β glucanases and xylanase were in the supernatant as well as cell associated.
No β xylosidase was detected. There was also no enzymatic
inhibition by glucose or xylose up to 30% concentration and
cellobiose up to 10% (735-742).

A two stage microbial biomass production system, on
which in the first, anaerobic bacteria could degrade to soluble compounds the lignocellulosic matrix of agro-wastes,
followed by an aerobic bacterial, yeast or fungal growth
phase, has not been tried.

14. Green Algae

    Becker (743a) distinguishes three different algae pro-
duction systems: a) a selected algae strain grown in fresh
water, mineral nutrients and additional carbon sources, b)
several species of algae and bacteria grown on sewage or industrial wastewater, c) a selected algae strain in an enclosed system grown in a completely autotrophic medium. The
first two cases comprise the so-called high rate oxidation
pond systems (HROP) as developed by W. J. Oswald and collaborators in California and field tested in many developing
countries (1). In the second case the main objective is
water treatment and if conditions are met the use of the
mixed flora biomass as animal feed. Enrichment of a pure
culture during solids separation might be used and hence,
for the purposes of a concise presentation, all three cases
will be considered below. Yield optimization and species
control are two of the most important determinants of success
in algae mass culture technology (743b).

    Oswald (744) has summarized the operational data of the
Napa, California two 0.1 Ha (1000 m^2) HROPs and the Manila
two 100 m^2 HROPs fed with domestic sewage. Taiganides et al.
(745) describe the system installed in Singapore for the
treatment of the wastewaters from a standing pig population
and comprise four 125 m^2 pilot ponds and two 500 m^2 demonstration HROPs. At the University of Stuttgart a 70 m^2
HROP has been tested to treat the effluent from an activated
sludge plant (746). The algal population characteristics
in maturation ponds in Southern Africa receiving secondary
treated effluents from conventional sewage works have been
assessed by Shillinglow and Pieterse (747a). The algae present were dominantly members from the Clorophyceae, and they
were probably obtaining most of their carbon dioxide from the
biocarbonate-carbonate system owing to the high pH conditions
existing in the pond. Carbon dioxide limitations as some
Zooplankton grazing were the main causes of population declines. Shilo (747b) also mentions parasitism and predation
by lytic bacteria, phycovirus and cyanophages. Hill and
Lincoln (748) have modelled the algal growth and behavior
in a HROP and compared the simulated data with experimental
data from a 200 m^2 HROP operated with a pig waste effluent.
Recent work from the University of California Engineering
Field Station in Richmond, California (749,750) has described
a controlled bioflucculation methodology for harvesting
micro-algae from municipal wastewater HROPs. Biological
flocculation can be achieved either by: a) isolation of

nonflocculant cultures in secondary ponds or, b) by provision
of continuous paddlewheel mixing in HROPs operated under
moderate to high dilution rate regimes. Edwards (751a) describes the operation of a 200 m^2 pilot plant operating in
the Asian Institute of Technology, Bangkok.

The description of patented apparatus, processes and
feeds for algae production have been described by Watson
(751b) and the state of the art has been presented by Dilov
(752) and Soeder (753).

At the Sherman Environmental Engineering Research Center,
Technion, Haifa in Israel a 120 m^2 pilot plant HROP has produced on the average 34 g m^{-2} day^{-1} of biomass in which algal
population varied as *Micractinium, Scenedesmus, Oocystis,
Franceia,* and *Euglena* were predominant in certain periods
(754a). External changes were observed in certain algae
which were related to the pond operational conditions (745b).
Usually the raw sewage BOD was reduced by 94% and the algae/
bacteria ratio was 1 : 100 (755). Separation of the biomass
from the pond effluent, a key critical operation (756,757)
was studied by froth flotation employing ozone-enriched oxygen
as carrier gas, producing a froth with up to 8.5% solids
(758). Azov et al. (759a) have suggested an optimal operational strategy to insure year-around photosynthetic operation
and high algal biomass productivity by varying retention time
with changes of pond depth at constant pond area. A model
to optimize the operation has been described by Oron and
Shelef (759b). Hard detergents in domestic wastewater did not
affect algal growth, complete lysis of algal cells was observed within a few days at non-ionic detergent concentrations above 100 mg l^{-1} (760). Carbon usually limits biomass
production in HROPs. Free carbon dioxide concentration in
the liquid is the major determinant parameter. High concentrations of free CO_2 are provided through bacterial respiration (761), although Dor and Svi (762) have found that
bacteria usually interfere with algal growth. Free ammonia
toxicity becomes important only at high pH (762a).

Pond liquid mixing improves algal growth and pond productivity. One way of providing mixing is by using air lift
pumps as described by Persoone et al. (763). The mixing behavior of the pond can be modelled by either a dispersed-plug
flow or by a tank in series models, as shown by Miller and
Buhr (764). Engineering-economic evaluations and other economic considerations have been published (765,766).

Through the German Agency for Technical Cooperation (GTZ), the Federal Republic of Germany supported research and development activities in various developing countries on the growth of mainly *Scenedesmus* sp on inorganic fertilizers solutions and CO_2 in HROPs (767). At the CFTRI in Mysore the algae *S. acutus* 273-3a was pilot tested in ponds of an outdoor cultivation area of 223 m^2 (768), their cost projections however, were in the order of US $1500 per ton (769). Sinchumpasak (770) informs of their experiments at AIT, Bangkok with the same algae in 3 x 25 m^2 and 6 x 87 m^2 HROPs; daily productivity was reduced to less than 15 g m^2 day due to parasitic infections. At Trujillo in Perú, the algae was *S. acutus* var. *alterans* which gave yields with yearly variations between 17 and 25 g m^2 day (771); water evaporated at an average rate of 6 l m^2 day (772). Chemical, nutritional and biological evaluations of the dry biomass produced at the different sites were described by Becker (773) and Payer et al. (774,775). In order to increase the algae growth rate, the addition of carbohydrates like sugarcane molasses in combination with carbon dioxide was investigated by Shamala et al. (776), maximizing with this technique the biomass yield both in indoor as in outdoor mixotrophic cultures. With glucose as the sole carbon source, the carbohydrate content was higher in the dark than in the light period (777) and its metabolism was also different (778).

Gons and Mur (779) have shown that for *S. protuberans* light limited growth, substantial amounts of energy were diverted from growth as a result of maintenance requirements, especially at lower growth rates. The specific maintenance was temperature dependent, increasing with it. Under phosphorus limitation the biomass yield was found to change markedly with growth rate (780). Gavis et al. (781) have shown for *S. quadricauda* that the relative number of cells growing as coenobia of different cell number where functions of growth rate when grown axenically in nitrate limited cultures. In *S. obliquus* at alkaline pH, bicarbonate assimilation was the major carbon uptake (782). *S. acutus* could adapt up to a maximum of 50 micromolar copper ions and depending on the concentration in the medium, an enrichment of up to 1000 fold within the cells was possible (783). The growth rate of *S. quadricauda* was influenced by the concentration of copper and zinc and both were toxic in excess predicted by an inhibition model (783a). Steady state growth rates of *S. olbiquus* and *Chlorella vulgaris* under bicarbonate limitation were well described by the Monod equation, however, the internal nutrient concentration was not correlated with the Droop equation as growth rate did not influence the biomass chemical

composition but only cell size; higher steady state growth
rates lead to bigger cells (784). Urea was by far the most
suitable nitrogen source for maximizing algal yield when it
was supplied in combination with carbon dioxide in a proportion that lead to an equilibrium pH near its optimum (785).

Livansky et al. (786) from the Institute of Microbiology,
Biotechnological Laboratory at Tebron have published culture
data of *Scenedesmus acutus,* with contamination by *Chlorella
vulgaris,* in wastewater from sewage treatment plant processing
city sewage and a hoggery effluent. The removal of nitrogen
and phosphorus was optimal at a dilution rate of 0.3 day^{-1},
whereas optimum biomass production at about 0.1 day^{-1}. Dihoru
et al. (787) established a polyculture of *Chlorella* and *Scenedesmus* in diluted wastewaters from a bullrearing complex.
Concentrations of 15 g l^{-1} of *S. acutus* were obtained in
spring waters enriched with commercial fertilizers and bicarbonate (788,789). Barna et al. (790) found the wastewaters from pharmaceutical, ceramic and beer industries good
substrates for two *Scenedesmus* species if enriched with inorganic salts. Growth was always higher in nitrate enriched
media than with aminoacids (791). Vilicic (792) however,
found peptone and yeasts extracts as good growth substrates
for *S. quadricada* and *Chlorella* sp. Intermitent light greatly
reduced both cell number and dry matter of *S. obliquus,* with
a drop in chlorophyl and protein contents and an inhibitory
effect on phosphorus uptake (793). Growth of *S. quadricauda*
was tested under autotrophic, heterotrophic and mixotrophic
conditions; the latter gave the higher biomass concentration
7.6 g l^{-1} (794).

Despite the great daily and seasonal fluctuations in the
chemical composition of raw sewage flowing into the HROP, no
succession of species was observed by Abeliovich (795), as
long as water inflow was carefully controlled and as a result
the *S. obliquus* population was very stable during three years
of monitoring. Changes in algal shape were described by Oron
et al. (796) for *S. dimorphus* which were caused by operational
variables as well as the nutritional fertility of the medium.
In order to maximize algal productivities it was recommended
to maintain a high free carbon dioxide concentrations to avoid
carbon limitations (797). Toerien and Grobelaar report of
HROP operation during winter conditions with settled sewage at
the NIWR in Pretoria (798), *Scenedesmus* sp was the dominant
algae, with production rates of 9.0 g m^{-2} day at a residence
time of six days and about 3.5 g m^{-2} day at a residence time

of 16 days. Biological evaluation of *S. obliquus* and *S. Platensis* employing rats has been reported by von der Decken (799) and Walz and Brune (800) recommended a 10% *Scenedesmus* powder in formulas for chicks.

Brune and Novak (801), Markl (802), and Panikov (803) have modelled the growth of *Chlorella vulgaris* under carbon and light limitation and under double substrate (urea and phosphage) limitations. The growth responses of many algae in mass culture may best be represented as a Monod fit of the specific growth rate and carbon dioxide concentration. Lowenstean and Bachofen (804) found that in *Chlorella fusca* carbon dioxide assimilation and ATP synthesis were strongly dependent on light intensity. Double substrate limitation has been accounted by the model of Panikov and Pirt (805). The growth yield of *C. vulgaris,* defined as g cells harvested per kJ of light energy absorbed by the cells, was assessed under light limited conditions and its value ranged from $3 \times 1 \times 10^{-3}$ to 5.0×10^{-3}; there was also a growth inhibition due to oxygen (806,807). Pirt et al. (808) have reported values of 15.3×10^{-3} for the Sorokin *Chlorella* strain 211/8 h and 20.6 g kJ^{-1} for a newly selected mixed culture which consisted of an algae and three species of heterotrophic bacteria. The specific maintenance of the latter was in the range of $0 - 0.0066$ h^{-1} Lorenzen (809) from laboratory experiences has shown that aereation with air/CO_2 or pure CO_2 in intervals during the light phase is as effective in terms of productivity as continuous aereation; moreover 0.01 g 1^{-1}glucose may be the optimum concentration to increase the crop without risk of overgrowing with bacteria or other heterotrophic organisms. The form of inorganic carbon utilized for photosynthesis in *Chlorella vulgaris* is free CO_2 and biocarbonate cannot be utilized (810). Carbonic anhydrase enhanced photosynthesis under CO_2 limiting conditions but it was inhibitory above certain CO_2 concentration (811). However, controversy still exists and direct bicarbonate uptake cannot be ruled out (812). The main factor controlling productivity has been shown to be by Goldman et al. (813) the product of gas flow rate and CO_2 concentration in the gas phase; biocarbonate was an excellent source of inorganic carbon below specific pH levels and with bubbled CO_2 it was possible to achieve light limitation. Ogawa and Aiba (814) found that the specific growth rate of *C. vulgaris* under mixotrophic conditions was always larger than under autotrophic conditions when the light intensity was less than 10k lux; a totally opposite behavior was shown by *S. acutus.* Barroso and Nonato (815) found glucose a much better substrate for *Chlorella* than saccharose and acetate. Brueckner and Huefner (816) studied the effect of iron and

Berenshlein and Kostlan (817) the effect of aldehydes upon
Chlorella growth. Toledo et al. (817a) found that humic
acid increased copper tolerance for growth. De Franca and
Yuan (818) and El-Ayouty et al. (819) characterized nitrate
as a source of nitrogen. Active transport of orthophosphate
in *C. ellipsoidea* has been studied (820) as also the glucose
uptake, which mediated by the proton motive potential difference across the membrane (821,822). Acid pH was the most important factor in creating toxicity levels of sulphite and
sulfide upon *C. vulgaris* (823). Numerous reports deal with
the capacity of *Chlorella* cells to selectively take up various
heavy metal ions from aqueous systems and their distribution
and chemical state within *Chlorella* cells, special reference
made to the publications from the Miyazaki Medical College in
Japan (824-827).

In Asia many large scale *Chlorella ellipsoidea* and *C.
pyrenoidosa* production factories are operating and producing
purely cultivated micro-algae mainly for health foods (828).
Two methods of cultivation are practices, autotrophic and
mixotrophic employing acetic acid as carbon source. HROPs are
usually employed although closed fermentors have been tested
when glucose is used as carbon source (829). *C. vulgaris* was
used to decrease the BOD of fermented sludge from palm oil
mills (830). *C. pyrenoidosa* growth in extracts of digested
sludge produced cells with less protein content than those
grown on activated sludge (831). The specific growth rate
of *C. homosphaera* in heterotrophic growth was temperature
independent but the biomass yield had a maximum at 30°C (832).
Chicken and pig manures were suitable substrates for *C. pyrenoidosa* growth (833). Lincoln and Hill (834) described the
operation of a HROP in Florida treating swine manure in which
the culture population was dominated throughout much of the
year by several species of *Chlorella*. Net productivity was in
the range of 30 g m^{-2} day. Chemical flocculation and solar
drying were employed as recuperation and preservation operat--
ions. Indeed one major technical difficulty of employing unicellular algae for biomass production is harvesting in a cheap
way (834a). Malis-Arad et al. (835,836) have studied *Chlorella* flocculation induced by alkaline pH and Regan and Shicko
(837) have induced flocculation by acid pH and *Lactobacillus
casei*. Frasinel et al. (838) give the fatty acid and sterol
composition of three *Chlorella* species. Wright et al. (839)
have shown that the amount of fat and sterols varies if the
algae are grown under autotrophic, photoheterotrophic and
heterotrophic conditions. In six algae species sterol composition was higher in autotrophic growth; in five species
linolenic acid was also increased. Picard et al. (840) and

de la Nove et al. (841) have employed a two state system to
grow *Oocystis* as a tertiary treatment of wastewaters. *Ulva
lactuca* grows well in full strength sewage mixed with 0.1%
sludge extract (842); however, with concentrations above 2%
of sludge extracts chlorosis and plasmolysis occurred (843).
Selesnatrum capriocornutum is widely used in the algal assay
to study biological responses to water quality, hence its
growth under continuous nitrogen limited conditions is well
known (844) as also its cell chemical composition as a
function of growth conditions (845). *Euglena gracilis* was
shown to grow heterotrophically on demineralized-hydrolyzed
whey; although glucose consumption occurred no galactose was
utilized (846). Light affected negatively glucose consumption
and showed growth on glucose containing medium; these effects
were not related to photosynthetic activity per se but could
be explained by modifications in the permeability of the cellular membrane (847). Phenylalanine inhibited growth of mid-
exponential phase *E. gracilis* cells (848) causing a shortage
of intracellular tryptophan (849). Respiratory oxygen uptake
and photosynthetic oxygen evolution in the light were strongly
inhibited by Zn, Cd and Hg (850). Sludge extracts from activated sludge were better than digested sludge for cultivation
of *E. gracilis;* two percent sludge extracts on *Euglena* medium
maximized growth rates (851). Polyamines have been detected
in *E. gracilis* that also have been found in thermophilic
microorganisms (852). Under special conditions the salt tolerant algae species *Dunaliella* accumulates large quantities
of glycerol (853). Reports have been published on systems
description (854) and process development (855) to produce
this particular microbial biomass product. Miura et al. (856,
857) have studied the algae *Chlamydomonas reinhardtii* as a
hydrogen production biosystem in an alternating light/dark
cycle. Badger et al. (858) have studied the CO_2 uptake
mechanism in this algae which is correlated with a bicarbonate
influx pump which conforms with the postulate that algae are
capable of dirctly using biocarbonate from the external medium for photosynthesis. Berdykulov and Nurieva (859) were
able to obtain 4.6 g 1^{-1} of *V. reinhardtii* biomass and Florencio and Vega (860) have studied nitrate and ammonia assimilation.

Mokady et al. (861) tested sewage grown, drum dried
Oocystis, Scenedesmus, Micractinium and *Chlorella-Euglena*
as partial and total soybean replacements in feeding formulas
for rats and chicks. With rats total replacement was possible without impairing PER values, however, the best chickens'
growth performance was obtained when the algal protein only
substituted for 25% of the soybean protein. Lipstein and

Hurwitz (862) used drum dried, sewage-grown *Chlorella* in
poultry diets. The algae and sorghum substituted fish meal,
soya bean meal and corn. No deleterious results were observed as long as the algae level in the diet did not exceed
15%. Sandbank and Hepner (863) employed *Oocystis*, *Scenedes-*
mus, *Euglena*, and *Anikstrodesmus* in the diets of fish (*Cyp-*
rinus caprio and *Tilapia*). There was no harmful effect of
the algae, or the aluminum used in algae harvesting, on
health, growth and rate of survival of the fish. Ponds where
fish were fed an algae-meal diet containing only 3% fish-meal
gave better yields than those run on a commercial diet with
15% fish-meal. Secondary toxicity tests on rats raised on
diets containing meat from chickens or fish, fed algae
rations were reported by Yannai et al. (864). They found no
differences from rats maintained on diets containing chickens
or fish fed control rations.

   Ogawa et al. (865) have discussed a reassessment of the
algal yield in terms of the kcal of light energy absorbed by
the cells. Lee and Pirt (866) have discussed these data and
found them in the low side as compared to their own data
(867,868), discrepancies that could be explained with maintenance energy requirements and a possible growth rate dependent maintenance factors. *Chlorogonium elongatum* can
grow in acetate at rates among the highest reported for green
algae (869). A growth model has been developed for *S. acutus*
(870). Bannister (871) describes a model for nutrient-
satured algal growth. Brown et al. (872,873) discuss phosphate uptake and phosphate limited growth kinetics of *S.*
capricornutum. Under phosphorus limitation protein was the
major fraction in which excess nitrogen accumulated intracellularly (874). Mixotrophic growth has been studied by
Follman et al. (875). Growth rates equations have been discussed in detail by Stephanopoulos and Fredrickson (876).
Carbon limited growth kinetics for *S. quadricauda*, *C. salina*,
and *C. vulgaris* are presented by Brune and Novak (877). The
ability of 27 algae belonging to 11 taxonomic divisions to
growth at the expense of organic nitrogen in axenic cultures
was reported by Neilson and Larson (878). Humphrey (879)
found for several algae a higher photosynthetic rate and a
higher photosynthesis: respiration ratio in oscillating
light-dark regimes than when grown under constant illuminat-
ion. Cadmium toxicity was tested for *Coelastrum proboscideum*
(880); tannins at 0.005% concentration completely inhibited
C. reinhardtii and *Cryptomonas* sp (881). Nitrogen fixation
has been suggested in a hot spring green algae (882). *Cyani-*
dium caldarium and *Protococcus sulphurarius*, two species of
unicellular, *eukaryotic*, *thermoacidophilic* algae have been

found in acid hot springs in North and Central America (883).
The classification of the former is still being discussed,
but Nagashima and Fukuda (884) suggests that it was a primi-
tive red algae. The green colonial alga *Botryococcus brauni*
has unusually high levels of hydrocarbons and its biosynthesis
has been recently discussed (885). Lipid and pigment pattern
variations have been presented for *Fritschiella tuberosa*
(886).

III. MB PRODUCTION BY MIXED CULTURES

A "structured" mixed culture concept denotes a stable and
defined mixed population under specific conditions (887).
Samuelov (888) states that preliminary observations demonstrate that carefully designed mixed cultures are more stable,
easily maintained and less frequently contaminated, and hence
these important advantages make them attractive for further
investigations.

In Section II in various opportunities mixed culture
systems in which one specie was stimulated to predominate were
described, i.e.: low pH submerged culture fungal fermentations, solid substrate fungal growth, photosynthetic bacteria,
green algae and aerobic bacteria in HROPs. In this section
a more complete description is given of several mixed culture
systems that have been the subject of research and development
activities.

1.   Photosynthetic Systems

The algal-bacteria interactions for MB production and/or
water treatment systems have been studied recently by Yanagimoto et al. (889); Lee (890; Lee and Pirt (891) and Escher
and Characklis (892). Fritz et al. (893) have presented a
growth dynamic model which was solved numerically, simulated
and the results compared with experimental data from a wastewater stabilization pond. Wastewater treatment lagoons operated in New Zealand, South Africa and Israel have clearly
shown how complex the interactions among species are and their
effects in causing biological instabilities (894). Bollman
and Robinson (895) reported that three *Chlorophyta*, including
Chlorella, competed successfully with bacteria for organic
acids as a source of carbon and energy in axenic culture.
Hydroxylamine release by *Arthrobacter* sp inhibited *Chlorella*
growth (896). In simulated diurnal experiments with *Chlorella*
pyrenoidosa and *Pseudomonas fluorescens, the bacteria grew
mainly during the daylight period at the time of glycolate*

excretion by the algae (897,897a). These data support the
hypothesis that bacterial growth can depend directly, in
starvation eco-systems like lakes, on algal photosynthesis
and extracellular release (898). In CO_2 limited continuous
culture experiments of *Anabaena* sp and *Zooglea* sp, a dominant heterotroph (899), dissolved organic matter was found
to substitute for CO_2 in order to maximize growth rate and
biomass production by phototrophs when heterotrophic bacteria
were present (900). The physical association between algae
and bacteria emphasizes the importance of the symbiotic association in these organisms where bacteria are harbored in the
gelatinous sheaten of blue-green algae (899,901). Under
natural conditions bacteria provided CO_2 and (or) cofactors
for *A. variabilis, A. nidulans, Ch. pyrenoidosa,* and *S. capricornutum* photosynthesis which in turn supplied oxygen for
bacterial respiration (902). Many planktonic blue-green algae
produce metal chelating substances active at pH's above 7
(903,904).

Anabaena and *Macrocystis* inhibited *Chlorella* growth (905).
Anabaena and *Microcystis* also inhibited *Chlorella* growth, and
Microcystis inhibited in turn *Anabaena* (906). The potent inhibitory effect of *Microcystis* was probably due to the production of extracellular products. The pH of the eco-system
might favor one algae over the other if there is carbon limitation, although this conditions might not be present in
natural waters (907); moreover, several authors have observed
in these eco-systems species shifts with changing nitrogen/
phosphorus ratios (908). Nitrogen fixing cyanobacteria play
a key role in water reservoirs in the southwestern USA and
aquatic actinomycetes follow its bloom influencing the reduction and perhaps degradation of the blue-green algae (909).

In HROP's treating solids free pig manure an unstable
algae population developed, *Coelastrum sphaericum* first predominated followed by *Scenedesmus acutus* and *Chlorella* (910).
Balloni et al. (911) operated a two stage in series photosynthetic bacterial system, in which the first contained non-
sulphur purple bacteria (*Rhodopseudomonas gelatinosa, R.
palustris, Rh. acidophila*) or green filamentous bacteria
(*Chloroflexus*-type), followed by a second stage in which
heterocystous nitrogen-fixing blue-green bacteria predominated
(*Anabaena cylindrica*) or non-heterocystous filamentous cyanobacteria (*Oscillatoria spendida* and *Spirulina platensis*).

Fujii et al. (912) have isolated a mixed culture of
photosynthetic bacteria and algae (*Chlorella* sp) capable of
growing under anaerobic-light conditions on 0.1% methanol

and 0.1% bicarbonate at 30°C. Maximum growth rate was 0.092 h^{-1} and a cell yield of 0.87 based on methanol used, although dry cell weight was less than 1 gl^{-1}.

2.   Non-Photosynthetic Microorganisms on Soluble Substrates

A. Yeast. Mostecky et al. (913) have employed a mixture of *C. utilis* and *Cryptococcus diffluents* in SWL and stillage. Rychtera (914) describes in more detail and process and laboratory data and has included a mixture of *C. utilis, C. diffluens* and *C. tropicalis.* In all cases ethanol was added as a carbon source. Lines (915) has used a mixture of *E. fibuligera* and *C. utilis* in potato processing wastes.

B. Fungi. A 1 m^3 h^{-1} pilot plant with a packed bed biotower was used by Wheatley et al. (916) to treat wastes from cheese, butter and cream production. *Fusarium* and *Geotrichum* grew together as a filamentous biomass on the surface of the supports. The plant was run at pH 4-5 and at between 5 and 10 kg of BOD day^{-1} m^{-3}. BOD removal was between 30 and 50% and biomass production between 0.1 and 0.5 kg of dry biomass day^{-1}.

C. Yeast and fungi. Fedurina (917) grew *Spicaria* and *Trichosporon astaneum* on the mash from yeasts fermentations. Ali et al. (918) showed that a mixture of *C. utilis* and *T. viride* was superior in terms of protein enrichment of wastes from date palm extraction processes than the yeast alone. Opoku and Adoga (919) found the same result with cassava as substrate employing *S. cerevisiae* and *T. viride*. A mixed culture of *G. candidum, Candida krusei* and *Hansenula anomala* was employed by Barker et al. (920) for treating whiskey distillery spent wash. The mixed culture did not give superior yields or productivities, but did give proportionally greater COD reduction and would be expected to be more stable and resistant to contamination than a monoculture. At a dilution rate of 0.1 h^{-1} the mixed culture gave a biomass yield of 4.8 gl^{-1} with a COD reduction of 31.5%, while in mixed batch culture a biomass yield of 12.5 gl^{-1} and a COD reduction of 54.9% were obtained. Illanes and Schaffeld (921) have developed a system to treat leached beet cossettes by acid pretreatment. *T. viride* growth in suspension in the solid phase and *Candida marina* ATCC 22974 growth in the hydrolyzate.

D.   Yeasts and bacteria. Moebus and Teuber (922) describe in detail their lactic acid fermentation followed by *S. cerevisiae* growth on mixtures of whole or deproteinized whey and hydrolyzed starch. Weber et al. (923) describes a two step process for the bio-oxidation of calcium bisulphite liquor; in the first step *Candida* sp is employed and the second is a mixed culture of 2 *Trichosporon* yeasts, 2 *Arthrobacter* strains (*A. terregens, A. citreus*), *Pseudomonas putida* and *Chromobacterium vidaceum.* Operating details of the process have been recently presented and discussed (924,925). Medium containing 2% of prehydrolyzed cassava starch was used to grow batchwise *Bacillus subtilis* and *C. utilis.* The final biomass product at the end of 80 h contained 52% protein and *B. subtilis* amounted to only 2-5% of the dry matter (926). Kargi and Shuler (927) describe a mixed bacteria - *C. utilis* process to convert poultry waste mixed with molasses into a higher protein feed-stuff. Poznanski et al. (927a) grew *Escherichia coli* R-12 and *S. cerevisiae* in a mixture of liquid pig waste plus beet molasses and obtained an increase of 17.3 gl^{-1} in dry matter; 6.7 gl^{-1} in total nitrogen, 87.5% sugar utilization, 32.9% true protein in the biomass.

E.   Aerobic bacteria and bacterial-fungal mixtures. A mixed bacterial culture was grown on butadiene-acrylonitrile rubber and various nitrogen sources like urea, glutamic acid or fertilizers (928). Mixed culture of *Methylomonas flagellata,* a methane utilizing bacterium, and *Alcaligenes eutrophus,* a hydrogen-utilizing bacterium, was attempted in order to use the carbon dioxide produced from methane by the former bacterium as the carbon source for the autotrophic growth of the latter (929). However, growth of the mixed culture was poor compared to growth in pure culture. *A. eutrophus* seemed to inhibit the growth of *M. flagellata* by its preferential uptake of carbox dioxide and by production of a growth inhibitor. Hsu and Thayer (930) grew a nine-member mixed, cellulolytic culture on the alkali extracts of mesquite wood. Kargi et al. (931) found the optimal conditions to treat poultry waste as follows: temperature 25°C, pH 7.5 and 1.5% of initial solids. The poultry waste was enriched with molasses and the microbial biomass product was recovered by flocculation with calcium hydroxide (932). Heddle (933) found that approximately 70% of the feed COD from slaughter house wastes could be recovered as aerobic biomass having a crude protein content of 50-55%. Evans et al. (934) found correlations between substrate utilization (measured in terms of COD, solids and BODs) and mean residence time in aerobic reactors, applicable to whole slurries of excretas from beef cattle, laying hens and piggery wastes. Dried swine wastes

were mixed with *Escherichia coli* and *Rhizopus oligosporus*
biomass previously grown on deproteinized whey and and resulting biomass was mixed with potato flakes and given as
feed for steers (935). Brown et al. (936) describe a system
where piggery waste is diluted with water, an additional
carbon source added and the mixture used to grow a strain of
Aspergillus niger. Garraway (937) presented results of a
three aerobic reactors in series for eliminating odor of
piggery slurries; in the system considerable nitrogen losses
took place. The effluent of an anaerobic digestor was excellent substrate for further aerobic treatment when compared
to a physico-chemical pretreatment methods (938). Hissett
et al. (939) described the relationship between microbial respiration and temperature in aerobic tanks treating pig
wastes. Mixed culture interactions in aerobic waste treatment have been discussed by Somerville (940). The identification and role of filamentous bacteria within the overall
framework of the activated sludge ecology were discussed by
Dallas-Newton and Forster (941). Continuously fed aerobic
systems repeatedly resulted in the development of filamentous bacteria and bulking of the sludge; on the contrary
intermitently fed reactors did form good settling sludges.
They did form flocs as a result of higher substrate uptake
rates followed by a starvation phase of maintenance (942,943).
Cometabolism exists in dealing with small concentrations of
herbicides by sewage aerobic bacteria (944). Deep aeration
tanks of 10 m were described by Shimiza and Odawara (945),
operated at a BOD loading of 15.3 kg m^{-3} day, a BOD removal
of 93% employing 0.49 kWh/kg BOD removed. Non-linear Monod
models were described by Vavilin and Visiliev (946). An
anaerobic-aerobic system for potato processing wastewaters
was the most economical among eleven alternatives evaluated
(947).

3.   Non-Phothsynthetic Microorganisms on Insoluble Substrates

   A. Aerobic systems. Microorganisms are naturally involved in the recycling of waste nutrients from agricultural
and animal wastes. When both substrates are brought together
an upgrading and enhancement of their biological and economical values can be achieved hence defining biodegradation
processes. Seal and Eggins (948) studied the thermophilic
mixed fungal flora developed in aerated pig slurry and straw
mixtures. A pilot scale, solid substrate packed tower fermentor is briefly described. Thayer and David (949) were
successful in developing a mixed culture that exceeded pure
cultures in mesquite wood hydrolysis. They were later employed (950), in protein upgrading alkali treated rice hulls.

Wu et al. (951) used a mixture of *Cellulomonas flavigena* and *Pseudomonas putida* in submerged culture on paper pulp; 98% decomposition was obtained after five days and a protein of 0.17 g/g of pulp used was reached. After adjusting the final broth to pH 10 and 60°C the cellular RNA degradation to monocucleotides was enhanced. Sugarcane bagasse was alkali and steam treated and inoculated with *Cellulomonas* and *Alcaligenes faecalis* as a possible source of β glucosidase in order to enhance cellobiose uptake (952). Chang et al. (953) employed a mixed culture of seven strains of bacteria to grow in an optimized growth medium containing 4% alkali-treated treated rice hulls and mineral salts. A batch total productivity of 0.9 gl^{-1} h^{-1} and a 42% substrate hydrolysis were obtained. An isolated mixed culture consisting of *Cellulomonas* sp and *Bacillus subtilis* was grown on milled alkali pretreated sugarcane bagasse pith. A maximum cell biomass yield per g of hydrolyzed pith of 0.5 was obtained when the original slurry had 30 gl^{-1} of pith plus urea, yeast extracts and minerals. The yields were higher when pith from unburnt sugarcane was used as compared to those obtained with in the field of burnt sugarcane. The presence of a chemical inhibitor in the latter was discarded by the authors, hence only a substrate modification could be the reason for the differences on the substrate's enzymatic susceptibility (954,955). An enzyme preparation from a mutant strain of *Cellulomonas* CS1-17 acted synergistically with low levels of *Trichoderma reesei* cellulase in degrading alkali-pretreated sugarcane bagasse (956,957). Viesturs et al. (958) developed a solid-state fermentation process for converting wheat straw into protein-enriched ruminant feed employing a mixed culture of *Chaetomium cellulolyticum* or *Trichoderma lignorum* and *Candida lipolytica*. The process was done in horizontal stirred fermentors at 30°C for 7 days. Cellulose degradation was 33% and the product contained 16-18% protein. *In vitro* dry matter digestibility of the resulting product was around 51%. Ringpfeil et al. (959) have used the solid-free liquids from piggery waste supplemented with methanol as substrate for growth of an isolated acidophilic bacteria MB58. The resulting mixed culture was comprised of about 80% of the inoculated microorganism. de la Torre (960,961) has isolated a mixed culture of *Cellulomonas flavigena* and *Xanthomonas* sp which grew quite well on alkali pretreated sugarcane bagasse. The experiments were done fed batch achieving a substrate concentration of 25 gl^{-1}; biomass productivity was 0.7 gl^{-1} h^{-1} and a cell yield based on utilized substrate of 0.5. The resulting biomass was separated and the solid residue had a crude protein content of 12% and an *in vitro* dry matter digestibility of 60%. Sarhar and Trabhu (962) grew a

mixed culture of *Cellulomonas* sp B_1 and *Trichoderma* sp in
alkali pretreated sugarcane bagasse and obtained a 95% decomposition of the substrate in a 1 day shake culture of 10
gl^{-1} slurry. The nutritional requirements and growth responses of a *Cellulomonas* sp on cellulosic substrates were
studied by Han (963). Growth was enhanced by adding yeast
extracts and ammonium salts were the best nitrogen sources.
Mixed cultivation with *A. faecalis* which utilizes cellobiose,
increased the cell yield almost six-fold. *Candida utilis*
QM8240 was grown on acid hydrolyzed mesquite wood followed
24 h later with a *Pseudomonas* JM127 culture (964). Thouvenot
et al. (965) employed a self heating compost system to treat
lignocellulosic residues plus porcine manure. In it *Humicola
insolvens* and *Humicola grisea* var *thermoidea* were predominant
and 25% of original nitrogen was transformed into fungal
protein.

General reviews of composting have appeared (966-970) and
specific applications include: feathers (971), cotton gin-
trash (972), municipal wastes (973), refuse and sewage sludge
(974), refuse and bark (975), grain dust (976), cottonseed
wastes (977), leaves (978), gelatin waste (979), bagasse and
vinasses (980), and food processing waste sludges (981).

Coupling of heat and mass transfer effects in aerobic
solid substrate fermentation and heat production rates have
been studied by Rathbun and Shuler (982) and by Mote and
Griffis (983). The heat output/temperature relationship
has been proposed as a process control parameters (984). This
is directly affected by the degree of ventilation as air not
only evaporates moisture but cools the pile (985,986). de
Bertoldi et al. (987) provide a comparison of three windrow
compost systems and two publications describe bench scale
systems (988,989). A kinetic model of the composting process
has been discussed by Whang and Meenaghan (990) and Psavianos
et al. (991) give a methodology for designing uniform air
distribution systems.

Composting of lignocellulosic residues like straw mixed
with animal manures, like swine liquid manure were studied
by van Fraassen and van Dijk (992). Only small increases in
protein resulted in the final product as nitrogen losses
occurred as ammonia. It was concluded that there was little
incentive to use this technique to prepare a roughage, however, as mentioned before other authors have obtained better
conversions to microbial protein. This technique is usually
followed to prepare material in which mushroom spawn is
planted in order to develop fruiting bodies. Although Smith

and Spencer (993) have prepared with a more rapid technique
adequate composts without animal manures. Spent mushroom
compost could be eventually used as an animal feed if it
proves to be free of pathogenic microorganisms and the protein
content reaches adequate levels. Fermor et al. (994,995)
have studied the predominant microflora on composts; for
composts without animal manures there was initially a build-
up of thermophilic bacteria (*Bacillus subtilis*) and actinomycetes (*Thermoactinomyces vulgaris*) appeared much earlier
at the expenses of undesirable mesophilic fungi (*Aspergillus
fumigatus*), although *Chaetomium thermophile* was the most heat
tolerant species and was the fungus most commonly isolated
during composting. Kleyn and Wetzler (996) in common manure
spent compost found the following: a) the most common bacterial isolate was *Bacillus licheniformis,* b) the most common
actinomycetes isolates were *Streptomyces diastaticus* and
Thermoactinomyces vulgaris, and c) the most common fungal
isolates were *Aspergillus fumigatus* and *Humicola grisea* var
thermoidea. Steaming for 14 h with live steam resulted in
70-76% reduction in microbial numbers. Küster and in der
Schmitten (997) found inhibitory substances of abiotic origin
towards *Bacillus megaterium* in municipal garbage which decreased substantially during composting. There were no water
soluble antibiotics produced during the microbial processes.
Millner (998) has studied the thermophilic and thermo-toler-
and actinomycestes in sewage-sludge compost. Streptomyces
species were predominant in cured composts and its prevalence
was due to their degradative and antagonistic capacities as
well as for some, their property of melanoid pigment production, compounds which are similar to humic acids and are part
of the important soil-constituent humus. Concentrations of
thermophilic actinomycetes, especially *Thermo actinomyces*
spp and *Micropolyspora* spp, at the worksites and at downwind
locations were much lower than those found in association
with reported episodes of farmer's lung from moldy hay. This
disabling disease which can cause fatalities is called hyper-
sensitivity pneumonitis or extrinsic allergic alveolitis,
caused by one mechanism involving activation of the alternate
pathway of complement (APC) by glycoproteins of certain
thermoactinomycetes. Recently Stutzenberger and Bowder (999)
have found activation of the APC by extracellular products of
Thermomonospora curvata, microorganism with potential for
cellulose bioconversions. Hence, a certain amount of caution
should be employed in assessing thermophilic-solid substrate
processes, specially since *Thermoactinomyces* sp have proven
to be the most effective to date, with *T. reesei,* in attacking natural cellulose (1000). Cellulolytic actinomycetes
were grown together with fungi on insoluble cellulose or

cellulose derivatives, resulting in a more extensive utilization of the carbon source and a faster and higher production and cellulase synthesis (1001). Also when yeasts or bacteria were added, the reducing sugar content was kept low such that repression of cellulase formation was prevented.

The decomposition rate of organic matter in soil is a slow process and is initially dependent on the amount of microbial biomass which has invaded it and the conditions prevailing for its growth and activity (1002,1003). Fungi are undoubtedly the most important decomposers of plant material and it has been found in nature that no single fungal species is able to use all components and it is well established that a succession of different groups of fungi will appear on different substrates (1004). Interactions between bacteria and fungi are common (1005) and many of the former are active hollocellulose degraders (1006,1007). Detailed studies on coconut fiber decay have been discussed (1008). In this case basidiomycetes were the only microbes capable of complete biodegradation of coir fibres, specifically *Trametes versicolor* and *Stereum rugosum*. There seems to be a strong competition and inhibition between basidiomycetes and other soil fungi. Aerobic mixed culture systems as discussed previously aim at speeding up what normally proceeds slower in nature, however, a definite and specific microbial inoculum is preferred over the uncontrolled succession of microflora which pertains in nature (biodegradation) or in accelerated process (composting).

The microbiology of lignin biodegradation has been substantially clarified in recent years due to the amount of fundamental research which has reached a high level and current progress is indeed rapid. Amer and Drew (1009) in this series have recently reviewed the microbiology of lignin biodegradation. This phenomena is totally different to those schemes hydrolyzing holocellulose (1010). The complex polymer is completely metabolized by pure cultures of various higher fungi, primarily basidiomycetes and is substantially degraded by some ascomycetes and imperfect fungi. All of them eucaryotic microbes. In anaerobic environments lignin and humus are recalcitrant and require oxygen for effective depolymerization and solubilization (1011,1011a). This microbial conversion is of importance in the production of feed and food from lignocellulosics (1012).

Lacasse is the major enzyme produced by *A. bisporus* during mycelium growth and it is associated with lignin biodegradation (1013-1015). The significance of specific oxidations of lignin with radical species produced by lacasse/oxygen or peroxidase/hydrogen peroxide has been investigated in white rot fungi (1016,1017). Leonowicz and Grzywnowicz (1018) have discussed recently the methodology of lacasse evaluation. A screening of tropical and temperate white rots (1019) has shown no relationship between specificity for lignin, decay and growth rates and enzymatic activities. Petroski et al (1020) characterized the lacasse production by *Polyporus anceps*. Superoxide radicals have been studied in *Coriolus versicolor* (1021). This fungi was a superior lignin biodegrader than the brown-rot *Poria placenta* (1022) and *Streptomyces viridosporus* and *S. setonii* (1023). In *Fomes annosus* there are lacasse induced and lacasse constitutive strains (1024). Wicklow et al. (1025) studied the microflora involved in lignin and cellulose modifications in ruminant dung. *Sporotrichum pulverulentum* was found to be the best biodegrader of modifice lignins under nitrogen limited conditions (1026). Nutritional regulation of lignin degradation by the white rot *Phanerochaete chrysosporium* has been extensively studied (1027-1032). *Trichoderma* strains have been isolated recently with lacasse activities (1033). Lignin by-products produced by *S. viridosporus* have been studied by Crawford (1034). Direct delignification of bark chips by a mixture of *Bacillus* and *Cellulomonas* was demonstrated by Deschamps et al. (1035). Glanser et al. (1036) studied the biodegradation of peat by a mixed culture of bacteria and yeasts. Recently, Leisola and co-workers (1036a) at the ETH in Zurich have shown that lignin degradation by *Phanerochaete chrysosporium* is preceded by secondary growth where cell bound polysaccharides created conditions where cells probably suffered from oxygen limitation.

The chemistry of lignin biodegradation is a complex problem to characterize and study (1037-1039). Modified or labeled lignins have been employed in establishing mechanisms (1040-1048). Bacterial degradations have been studied by the group at Compiegne (1049-1052), Munchen (1053,1054), INRA (1055), Julich (1056), and Zurich (1057).

Interest has also been shown for the use as feed of waste activated sludge, composed mainly of aerobic bacteria, from industrial effluents like those from pulp mills (1057a) as those from municipal waters (1057b).

B. Anaerobic systems. Recycling of animal waste as a
livestock feed has been looked upon as a method of lowering
the disposable waste load and at the same time of employing
a potentially useful source of feed nutrients including the
bacterial biomass present. Cophrophagy indeed has been cus-
tomary in many wild and domestic animals. Various alternatives have been explored: a) ensilage of the waste, b) ensilage of a mixture of the waste and crop residues or grains,
c) total anaerobic digestion (methane production) of the
waste and feeding of the slurry or the recuperated microbial
biomass and mixing it with grains.   .

Caswell studied the optimum water levels for ensiling
wood-shaving broiler litter for feeding wethers. During
ensilage at 40% moisture, coliforms were eliminated and the
product had an acceptable digestion coefficient (]058).
D'Urso et al. (1059) found complete coliform disappearance
after 32 days of ensilage of poultry manure mixed with maize
and molasses. Jayal et al. (1060) have concluded that the
ensilage of poultry excreta with maize and molasses has a
satisfactory potential for supplying dietary nitrogen, calcium and phosphorus to ruminants. Mixtures of poultry excreta
and oat forage (1061) and sugarcane tops (1062) and molasses
have been ensiled causing in the process an improvement of
feed flavor and a significant decrease in the number of pathogenic microorganisms. Weiner (1063) studied the microflora
during ensilage of corn and swine waste. Malossini et al.
(1064-1067) have evaluated the ensilage of cattle manure with
mixtures of straw and corn from the chemical, microbiological,
toxicological and nutritional point of views and found it adequate for feeding young cattle. Good fermentation in terms
of acid production was found when cattle waste was ensiled
with young rye straw; maximum fermentation and destruction of
pathogenic organisms were complete after one week of ensilage
(1067a).

Prior and Hashimoto (1068) have reported the feeding of
sheep and cattle with centrifuged biomass from anaerobic
digestor effluent. Between 10-20% of the material could be
incorporated in the diet. Marchaim and Criden (1069) have
fed digested manure to fish ponds (common and silver carp and
tilapia) substituting up to 50% of the pelleted feed and also
have fed wet digested slurry to calves, cows and sheep.
Lizdas et al. (1070) fed liquid slurry and centrifuged cake
to cattle and found that the animals had performed as well as
or better than the production cattle at the feedlot.

On these anaerobic acidification processes of lignocellulosic substrates one of the first reactions to take place
is the hydrolysis of the major structural polymers cellulose,
hemicellulose, starch, protein and pectin. Experimental evidence indicates that lignin is not biodegraded under anaerobic
conditions (1071,1072) but many of the lignin-derived molecules are (1073). Although recently Akin (1073a) has shown
the existance of a filamentous facultative anaerobic non-
cellulolytic microbe which under anaerobiosis was capable of
attacking lignified tissue. The anaerobic hydrolyzers have
been studied recently especially the rumen microflora. A
very strong interaction between the microbes and the fibers
has been found which controls the biodegradation. Either
there is a total attachment of the microbe to the surface or
the microbe is within striking distance from the fiber surface imbedded in a gel matrix. Hydrolytic enzymes which are
true extracellular enzymes or surface bound are within the
gel and usually within outer membrane packages or subcellular
visicles. In this way they are conserved and used where and
when needed. The enzyme-fiber adsorption described frequently in the literature might very well be instead a capsule-
fibre interaction where enzymes are located within the cap-
sule membranes. These hypothesis have come from truly fascinating electron microscopy studies (1074). What has been
difficult however, is to correlate one type of microbe with
enzymes capable of attacking highly ordered cellulose, where
rod-like organisms seem to predominate (1075). Different
cell wall types are usually digested at different rates, as
also the layers within them. Rumen protozoa are also involved in the cell wall hydrolysis of forages (1076) as also
phycomycete fungi (1076a). Harbers and Thouvenelle (1077)
found that mesophyll and phloem from ensiled corn and sorghum leaves were readily attacked by rumen microflora, but
the remaining vascular tissue, sclerenchyma and most of the
epidermis, which are highly lignified tissues, were undigested. The same behavior for several grasses was informed
by Akin and Burdick (1077a). Morris and van Gylswyk (1078)
found definite fiber attachment for all ruminococci and
Bacteroides succinogenes. No attachment was seen for
Fusobacterium saccharolyticum and *Buryrivibrio fibrisolvens*.
Most of these bacteria practically used the hexoses and pentoses produced but relatively little of the uronic acids.
Differences in the proportions of the principal fiber at-
taching bacteria and variations in availability of cell wall
constituents could be the major causes affecting overall
biodegradability (1079). Cheng et al. (1080) described the
events in the digestion of fresh legume leaves and found
bacterial growth in micro-colonies composed of single or

mixed morphological types. Most of these micro-colonies
were enclosed in an extensive fibrous exo-polysaccharide
glycocalyx which could convey an essential measure of protection within this rich and competitive environment (1081).

    The rumen per se is an ideal fermentation site. The
microflora is dense and diverse: about 200 species of bacteria and 20 of protozoa containing 10^{10}-10^{11}bacterial and
10^6 protozoa per ml. Feedstuffs are biodegraded to short
chain fatty acids and microbial biomass and these products
serve as sources of energy and protein. The microbial population also degrades a considerable portion of the dietary
protein to simple nitrogenous compounds for use in cell synthesis. Dietary lipids are hydrolyzed and the unsaturated
fatty acids are hydrogenated (1082-1084). The rumen can be
considered a continuous culture microbial biomass fermenter
with an average particle residence time of 1 to 3 days
(1084a). Numerous studies have shown microbial cell protein
to have a relatively constant amino acid composition, regardless of the feed, with an overall digestibility in the range
of 70-90% (1085). Many of the bacteria excreted amino acids
during active growth but ceased shortly after growth became
limited (1086). The bacterial interactions in the rumen are
extremely complex. Chesson et al. (1087) found that the
hydroxycinnamic acids, ferulic and p-coumaric, suppressed the
growth of *R. albus, R. flavefaciens,* and *B. succinogenes* and
the cellulose digestion of the last two. In pure culture
these acids were partially hydrogenated, however, in the
rumen fluid a short-fall in the recovery of added phenolic
acids was noted. *R. flaveciens* was more effective than *B.*
succinogenes in degrading total dry matter, however, the
latter was better in degrading the hemicellulose and hemicellulose sugars (1088). *Bacteroides ruminicola* had adequate
cell yields on xylose, arabinose and rhamnose and were fermented mainly to 1,2 propanediol, succinate and acetate
(1089). Steward et al. (1090) report data that confirms the
fact that there is a limit to the amount of starch that can
be present without detriment to the digestion of roughage.
A blcok diagram description of ruminal nitrogen metabolism
has been presented and discussed by Baldwin and Denham (1091).
Decreasing the extent of protein degradation in the rumen
can be achieved by including in the diet protein with low
solubility or by formaldehyde treatment; however, this may
cause nitrogen limitation for microbial growth which can be
overcome by providing non-protein nitrogen like urea (1092).
However, it seems that protein solubility is not by itself
an indication of the protein's resistance or susceptibility
to hydrolysis by rumen bacterial proteases, instead

structural characteristics like the presence of crosslinking
disulfide bonds (1093). Brock et al. (1094) have studied recently the proteolytic activity of rumen microorganisms and
have concluded that rumen bacteria process primarily, serine,
cysteine and metalloproteinases. Amino acid deamination
occurs in the rumen as is the major ammonia producing process
in the rumen (1095). Ammonia is incorporated rapidly into
rumen bacteria and used for protein synthesis as only protein
and nucleic acids form the bulk of bacterial nitrogen compounds entering the small intestine where they are absorbed
(1096). Rumen ammonia concentration had no apparent effect
on either concentration or distribution of volatile fatty
acids; also no systematic trends in digestibility were observed (1097). Functional electron transport chains are
operable in many rumen bacteria with coupled energy production which critically control metabolic activities (1098).

Besides nutrient protection the addition of chemical
agents, like propionate enhancers, methane and diaminase inhibitors and decreasing the acetate/propionate ratio, can
modify the rumen fermentation (1099). The use of propionate
enhancers like monensin and lasalocid (1100-1106) have demonstrated the increased propionate production which in turn
suppresses methane formation. Fiber digestion was reported
to be inhibited (1101) but dry matter digestibility and nitrogen retention were not altered (1104) and energy digestibility remained largely unchanged (1105). Lactate production
was also reduced (1106). The resulting depression of methane
production is accompanied by an accumulation of H_2, which is
usually used by rumen bacteria like *Bacteroides ruminicola,
Anaerovibrio lipolytica* and *Selenomonas ruminantium* as a
source of reducing power (1107). When the dilution rate is
increased in the rumen the contrary effect is produced: more
acetate if produced (1108). However, *in vitro* chemostat
experiments with a population of mixed rumen bacteria have
shifted the results to more propionic acid and less methane
production (1109). Branched chain fatty acids are required
by the cellulolytic anaerobes (1109a) and these are usually
produced by non-cellulolytic rumen bacteria like *Bacteroides
amylophilus* and *Megasphaera elsdenii* (1110). A saccharolytic
spirochete that associated and interacted with *B. succinogenes* was isolated from the rumen fluid; the spirochete enhanced cellulose breakdown by the bacteria probably by im-
parting to it a passive motility towards the cellulose fibers
(1111). When starch is fed, *Streptococcus bovis* numbers
within the rumen increase which produce more lactic acid and
the pH drops affecting the growth of other rumen bacteria
(1112,1113). Intermicrobial hydrogen transfer has been

demonstrated between a rumen anaerobic fungus and rumen methanogens (1114) and this co-culture has been shown to biodegrade pure cellulose at a rate of 3 gl^{-1} day and attack
cellulose in lignocellulosic by-products (1115). Peptides,
acetate and other volatile fatty acids improved significantly
the growth of *B. ruminicola* and were used for maintenance;
little ammonia was produced during growth but after it ceased
there was a linear increase in ammonia (1116).

In anaerobic digesters, the microbial interactions are
also extremely complex (1117). Two types of holocellulose
breakdown were reported by Lane (1118): one involving the
close association of bacteria and fibers and the other typical
of truly extracellular enzymes. The conversion of cellulose
to acids, to acids and ethanol and to methane and carbon dioxide by specific symbiotic cultures have been studied by
Khan and co-workers at the NRC in Ottawa, Canada (1119-1125).
Recently they have found that at increased cellulose feed
rates, the number of cellulolytic bacteria, because of their
faster growth rate, exceeded the number of propionate and
acetate using bacteria and hence, the pH was lowered which
in turn inhibited further cellulose biodegradation and acid
formation (1126). They suggested a two phase reactor: the
lower one mildly agitated where cellulose was in suspension
and a packed section above it where the acetate converters
and methanogens were immobilized. Rates of 5 gl^{-1} day of
pure cellulose degradation were achieved as compared to rates
in the neighborhood of 1.5 gl^{-1} day typical of anaerobic
systems (1126). Researchers at Dynatech, Boston published
their conversion data of several lignocellulosic residues
to organic acids in a packed bed, external circulation loop
upflow fermentor (1127,1128). Untreated materials like
bagasse were poorly converted (16%) as compared to marine
algae biomass (90%). A wide spectrum of organic acids was
produced besides acetic, distribution which was dependent of
pH and temperature. This effect was reported in chemostats
fermenting sucrose where dilution rate (1129), pH (1130,1131),
and nitrogen source (1132) influenced the accumulated products. Datta (1133-1135) has obtained very good conversion
of corn stover to acids, about 84% including hot water solu-
bles, hemicellulose, cellulose and lignin in alkaline pretreated material. Playne also reports the acidogenic conversion of alfalfa (1136), Clausen et al. (1137,1138) or orchard
grass and Nishio et al. (1139) for mandarin orange peel. In
all these systems the acidified product could eventually be

used as an animal feed or the acids could be employed as raw
materials for microbial biomass production, serving only as
intermediate products between lignocellulose and microbial
cells.

Bekers et al. (1140) have proposed a process to convert
the juices from alfalfa, grasses and beet tops into microbial biomass by subjecting it to an anaerobic fermentation,
recuperating the biomass by centrifugation, granulating it
and drying.

Maceration of non-woody annual crops is another anaerobic
process where under neutral or acidic conditions pectin found
in the primary plant tissue is enzymatically degraded. Such
liquors can be a source of chemicals for microbial biomass
production. A recent review of such processes has been published (1141). Recently, Yoshihara and Kobayashi (1142,1143)
have employed alkalophilic bacteria with high macerating
activity in the production of fibers. Under alkaline conditions the fibers swell improving substrate accessibility
by pectinolytic enzymes; pectin is chemically demethylated
and depolymerized by a pectate transeliminase with an alkaline
pH optimum. Hence, it is expected that research activity in
this field will become more common.

Finally ensilage as a fermentative preservation method
(1144,1145) is also the result of anaerobic acid production
mainly by lactic bacteria on soluble carbohydrates. Microbial
biomass has been added as a nitrogenous supplement (1146)
and ensiled products of these types seem attractive as an
alternative to drying, however, careful pH control is required in order to avoid excessive protein losses (1147-1152).
Ensiling of whole sugarcane is interesting as adequate amounts
of ethanol are also produced (1153-1156). Ensilage of alkali treated crops (1157,1158) and by-products like bagasse
(1159) produces a neutralized feed as discussed previously
(1).

IV. FINAL COMMENT

The length of this review confirms our early statement:
microbial biomass as a source of nutrients is receiving worldwide attention. It is obvious from the previous sections that
in some cases there is repetitive research work that does not
produce new insights to process development, however, there
have been outstanding new ideas brought to the general attention which undoubtedly have increased the number of process

alternatives available. We have not included here a discussion about product processing alternatives or end product use as an animal feed, its chemical, biological and
nutritional evaluation as such subjects merit themselves
a separate analysis. It is encouraging to see so much activity in a field where biotechnology can play a unique and important role: the procurement of adequate supplies of food
utilizing local raw materials and technology. No doubt a
first priority item for any country in the world.

V.   ACKNOWLEDGEMENT

This review was done in part under contract from the
Scientific Department of the Organization of American States
(OAS) and within the activities of the Guatemala-Biotech-
nology MIRCEN sponsored by UNEP-UNESCO-ICRO.

VI.   REFERENCES

1.   Rolz, C. and Humphrey, A. A. *Adv. Biochem. Eng. 21*, 1
      (1982).
1a.   Ratledge, C. *Prog. Ind. Microbiol. 16*, 119 (1982).
2.   Cousin, M. A. *Ann. Reports Ferm. Processes 4*, 31 (1980).
2a.   Litchfield, J. H. *BioScience 30*, 387 (1980).
2b.   Waslien, C. I. and Steinkraus, K. H. *BioScience 30*,
      397 (1980).
2c.   Weissermel, K. *Umsch. Wiss. Tech. 80*, 551 (1980).
2d.   Dunlap, C. E. and Chiang, L. C. In "Utilization and
      Recycle of Agricultural Wastes and Residues," (Shuler,
      M. L., ed.), CRC Press, Boca Raton, FL, p. 19 (1980).
2e.   Shuler, M. L. In "Utilization and Recycle of Agricul-
      tural Wastes and Residues," (Shuler, M. L., ed.), CRC
      Press, Boca Raton, FL, p. 67 (1980).
2f.   Anderson, A. W. and Anderson, J. F. In "Utilization
      and Recycle of Agricultural Wastes and Residues,"
      (Shuler, M. L., ed.), CRC Press, Boca Raton, FL, p. 237
      (1980).
2g.   Mateles, R. I. In "Microbial Technology, Society for
      General Microbiology Symposium 29," (A. T. Bull; Ell-
      wood, D. C. and Ratledge, C., eds.), Society for General
      Microbiology, Cambridge, UK, p. 29 (1979).
2h.   Cooney, C. L., Rha, Ch., and Tannenbaum, S. R. *Adv.
      Food Res. 26*, 1 (1980).
3.   Gray, P. and Berry, D. R. In "Handbook of Organic Waste
      Conversion," (Bewick, M.W.M., ed.), van Nostrand-Rein-
      hold, New York, p. 339 (1980).

3a.   Smith, J.E. in "Handbook of Organic Waste Conversion"
      (M.W.M. Bewick, ed.) p. 209, van Nostrand-Reinhold,
      New York (1980).
3b.   Soeder, C. J. *Hydrobiol. 72,* 197 (1980).
4.   Drews, G. *Naturwiss. Rundsch. 32*(1) 11 (1979).
5.   Faust, U. and Prave, P. *Process Biochem. 14* (11) 28 (1979).
6.   Hamer, G. and Hamdan, I.Y. *Chem. Soc. Rev. 8,*143 (1979).
7.   Rose, A.H. in "Microbial Biomass" (A.H. Rose, ed.) p. 1,
      Academic Press, London (1979).
8.   Hang, Y.D. in "Food Processing Waste Management" (J.H.
      Green and A. Kramer, eds.) p. 442, AVI Publishing Co.,
      Westport, CT (1979).
9.   Senez, J. *Proc. GIAM V,* 307 (1979).
9a.   Piret, T. *Ann. de Gembloux 85,* 83 (1979).
9b.   Dilov, Kh. *Khidrobiol. 10,* 3-11 (1979).
10.   Forage, A.J. and Righelato, R.C. *Prog. Ind. Microbiol. 14,*
      59 (1978)..
11.   Litchfield, J.H.*Chem. Tech. 8,* 218 (1978).
12.   Prave, P. and Faust, U. *Chem. Ber. 14,*552 (1978).
13.   Rogers, P.L. *Food Technol. Australia 30* (3) 109 (1978).
14.   Tannenbaum, S.R., Cooney, C.L., Demain, A.M. and Haverberg,
      L., in "Protein Resources and Technology: Status and
      Research Needs" (M. Milner, N. Scrimshaw, D.I.C. Wang, eds.)
      p. 502, AVI Publishing Co., Westport, CT (1978).
15.   Vetterlein, G. and Kauruff, W. *Lebensm. Ind. 25,* 437 (1978).
16.   Waslien, C, Myers, J., Bessel, K. and Oswald, W. in
      "Protein Resources and Technology: Status and Research
      Needs" (M. Milner, N. Scrimshaw, D.I.C. Wang, eds.) p. 522,
      AVI Publishing Co., Westport, CT (1978).
17.   Worgan, J.T. in "Plant Proteins" (G. Norton, ed.) p. 191,
      Butterworths, London (1978).
18.   Abou-Zeid, A. and Farid, M.A.M. *Zentralbl. Bakteriol. Para-
      sitenkd. Infektionskr. Hyg. Abt. 2: Microbiol. Landwirtsch.
      Technol. Umwelt-Schutzes 133* (7/8) 657 (1978).
18a. El-Nawawy, A.S. in "GIAM V, Global Impacts of Applied
      Microbiology. GIAM and its relevance to developing countries"
      (W.R. Stanton, J.E. DaSilva, eds.)p. 139, UNEP/ICRO,
      University of Malaya Press (1978).
19.   Laskin, A.I. *Ann. Reports Ferm. Processes 1,* 151 (1977).
20.   Litchfield, J.H. *Adv. Appl. Microbiol. 22,*267 (1977).
21.   Litchfield, J.H. *Food Technol. 31* (5) 175 (1977).
21a. Litchfield, J.H. *Science 219,* 740 (1983).
22.   Moo-Young, M. *Process Biochem. 12*(4) 6 (1977).
23.   Skryabin, G.K. and Eroshin, V.K.*Microbiol. 46,* 657 (1977).
24.   Tannenbaum, S.R. in "Food Proteins" (J.R. Whitaker, S.R.
      Tannenbaum, eds.) p. 315, AVI Publishing Co., Westport, CT
      (1977).
24a. Senez, J.C. in "Microbial Conversion Systems for Food and
      Fodder Production and Waste Management" (T.G. Overmire, ed.)
      p. 83, KISR, Kuwait, (1977).

24b. Searty, S.A. and Masson, J.C., in "Microbial Conversion
     Systems for Food and Fodder Production and Waste
     Management," (T.G. Overmire, ed), 145, KISR, Kuwait (1977).
24c. Srinivasan, M.C., *Chem. Ind. Develop.*, *11*(6), 26 (1977).
25. Rehm, H.J., *Ernaehr. Umsch.*, *23*, 307 (1976).
26. Tannenbaum, S.R. and Pace, G.W., in "Food from Waste,"
     (G.G. Birch; K.J. Parker, J.T. Worgan, eds.) 8, Applied
     Science Publishers, London (1976).
26a. Senez, J., *Folia vet. lat. 6*, (Suppl.) 97 (1976).
27. Ratladge, C., *Chem. Ind.*, *21*, 918 (1975).
28. Dimmling, W., *Dechema Monogr.*, *83*, 1 (1979).
29. Mateles, R.I., *Proc. GIAM, V.*, 315 (1979).
30. Blenke, H., *Adv. Biochem. Eng.*, *13*, 121 (1979).
31. Blanch, H., *Ann. Reports Ferm. Processes*, *3*, 47 (1979).
32. Goldman, J.C., *Water Research.*, *13*, 1 (1979); *13*, 119
     (1979).
33. Benemann, J.R., Weissman, J.C. and Oswald, W.J., in
     "Microbial Biomass," (A.H. Rose, ed.), 177, Academic
     Press, London (1979).
34. Oswald, W.J. and Benemann, J.R., in "Biochemical and
     Photosynthetic Aspects of Energy Production," (A. San
     Pedro, ed.) 59, Academic Press, New York (1980).
35. Richmond, A. and Preiss, K., *Interdiscipl. Science Rev.*,
     5(1), 60 (1980).
36. Bhattachanya, A.N. and Taylor, J.C., *J. Anim. Sci.*, *41*(5),
     1438 (1975).
37. Tagari, H., *Refuah. vet.*, *35*(3), 123 (1978).
38. Arndt, D.L., Day, D.L. and Hatfield, E.F., *J. Anim. Sci.*,
     48, 157 (1979).
39. Calvert, C.C., *J. Anim. Sci.*, *48*, 178 (1979).
40. Wilson, P.N., Brigstocke, T.D.A. and Williams, D.R.,
     Process Biochem., *15*(7), 36 (1980).
41. Fleming, H.P. and McFeeten, R.F., *Food Technol.*, *35*(1),
     84 (1981).
42. Speck, M.L., *Food Technol.*, *35*(1), 71 (1981).
43. Wang, H.L. and Hesseltine, C.W., *Food Technol.*, *35*(1),
     79 (1981).
44. Hesseltine, C.W., *Process Biochem.*, *16*(3), 2 (1981).
45. Baun, J.N. and Brown, W.L., *Food Technol.*, *35*(1), 74
     (1981).
46. Ralph, B.J., *Food Technol. Australia*, *28*(7), 247 (1976).
47. Hesseltine, C.W., *Process Biochem.*, *12*(6), 24 (1977);
     12(9), 29 (1977).
48. Nagai, S., *Proc. GIAM V*, 413 (1979).
49. Cannel, E. and Moo-Young, M., *Process Biochem.*, *15*(2),
     (1980); *15*(6), 24 (1980).

50.   Knapp, J.S. and Howell, J.A., *Topics Enzyme Ferment. Biotechnol.*, *4*, 8- (1980).

50a. Aidou, K.E., Hendry, R. and Wood, B.J.B., *Adv. Appl. Microbiol.*, *28*, 201 (1982).

51.   Spencer-Martins, I. and van Uden, N., *Europ. J. Appl. Microbiol. Biotecnol.*, *4*, 29 (1977).

52.   Spencer-Martins, I. and van Uden, N., *Europ. J. Appl. Microbiol. Biotechnol.*, *6*, 241 (1979).

53.   van Uden, N., Cabeca-Silva, C., Madeira Lopen, A., Spencer-Martins, I., *Biotechnol. Bioeng.*, *22*, 651 (1980).

54.   Azoulay, E., Jouanneau, F., Bertrand, J.C., Raphael, A., Janssens, J., Lebeault, J.M., *Appl. Environ. Microbiol.*, *39*, 41 (1980).

54a. Azoulay, E., Janssens, J., Raphael, A., Bertrand, J.C., in "Advances in Biotechnology," (M. Moo-Young and C.W. Robinson, eds.) 33, Vol. II, Pergamon Press, Toronto (1981).

55.   Ng, B.H., Eswaran, S., Tan, E.L., Lee, T.K., *Malays. Appl. Biol.*, *6*(2), 123 (1977).

56.   Lemmel, S.A., Heimsch, R.C., Korus, R.A., *Appl. Environ. Microbiol.*, *39*, 387 (1980).

57.   Admassu, W., Korus, R., Heimsch, R., Lemmel, S.A., *Biotechnol. Bioeng.*, *23*, 2361 (1981).

58.   Sa-Correia, I. and van Uden, N., *Europ. J. Appl. Microbiol. Biotechnol.*, *13*, 24 (1981).

59.   Yoshizawa, K., Tanno, K., Suzuki, O., Koto, K., *Nihon Juzo Kyokai Zasshi*, *73*, 959 (1978).

59a. Dreimame, M., Tenisone, R., Luka, V., Sel. Kul'tiv Prod. Aminokislot Fermentov, p. 90-95, Zinatne: Riga, USSR (1979); CA *93*: 68626j.

59b. Wilson, J.J., Khachatourians, G.G. and Ingledew, W.M., *Biotechnol. Letters*, *4*, 333 (1982).

59c. Oten-Gyang, K., Moulin, G. and Galzy, P., *Acta Microbiol. Acad. Sci., Hung*, *27*, 155 (1980).

59d. Touzi, A., Prebois, J.P., Moulin, G., Deschamps, F., and Galzy, P., *Eur. J. Appl. Microbiol. Biotechnol.*, *15*, 232 (1982).

60.   Hang, Y.D., *Appl. Environ. Microbiol.*, *39*, 470 (1980).

61.   Hang, Y.D. and Woodams, E.E., *J. Food Science*, *46*, 1498 (1981).

61a. Hang, Y.D., in "Current Developments in Yeast Research," (G.G. Stewart and I. Russell, eds.) p. 41, Pergamon Press, Toronto (1981).

62.   Oliva, R.U. and Hang, Y.D., *Appl. Environ. Microbiol.*, *38*, 1027 (1979).

63.   Prior, B.A., Lategan, P.M., Botes, P.J., Potgieter, H.J., *S. Afr. Food. Rev.*, *7*, (1 Suppl.) 120 (1980).

63a. Prior, B.A., Botha, M., Custers, M., Casaleggio, C., in
"Advances in Biotechnology," (M. Moo-Young and C.W.
Robinson, eds.), p. 337, Vol. II, Pergamon Press,
Toronto (1981).

64. Taguchi, H., Yoshida, T., Ocampo, T.A., Sawada, H.,
*Ann. Reports Int. Center Coop. Research Dev. Microbiol.
Eng., 1,* 1 (1978), see also, *Proc. GIAM V, 466* (1979).

65. Braun, R., Meyrath, J., Stuparek, W. and Zerlauth, G.,
Process Biochem., 14(1), 16 (1979).

65a. Meyrath, J. and Braun, R., in "Current Developments in
Yeast Research," (G.G. Stewart and I. Rossell, eds.)
p. 25, Pergamon Press, Toronto (1981).

66. Kamel, B.S., *Process Biochem., 14*(6), 12 (1979).

66a. Aligedi, Kh., Bechkov, M., Roshkova, Z., *Nauchni Tr.
Vissch, Inst. Khranit., Vkusova Prum-st. Plovdiv., 24*(1),
275 (1977).

66b. Aligedi, Kh., Bechkov, M., *Nauchni Tr. Vissch. Inst.
Khranit. Vkusova Prum-st. Plovdiv., 23*(1), 253 (1977).

67. Guiraud, J.P., Devouge, C., Galzy, P., *Biotechnol. Lett.,
1,* 461 (1979).

68. Kajs, T.M. and Vanderzant, C., *Dev. Ind. Microbiol., 21,*
481 (1980).

69. Zarnescu, A., Sterpu, I., Stroia, I. and Stancu, M.,
Lucr. Cercet. Inst. Cercet. Ind. Chim. Aliment., 13, 131,
(1978).

69a. Martinet, F., Ratomahenina, R ., Gracille, J., Galzy, P.,
Biotechnol. Letters 4(1), 9-12 (1982).

69b. Nakahara, T., Sasaki, K., Tabuchi, T., *J. Ferment.
Technol., 60,* 89 (1982).

70. Mudgett, R.E., Rajagopalan, K. and Rosenau, J.R., *Trans.
Am. Assoc. Agr. Eng., 23,* 1590 (1980).

71. Okada, N., Ohta, T., Ebine, H., *J. Jap. Soc. Food Sci.
Technol., 27,* 213 (1980).

71a. Mouchet, R., *Ind. Alim. Agric., 96* (9/10) 1003 (1979).

71b. Lui Sui Fong, J.C., *Int. Sugar J., 84,* 5 (1982).

71c. Alian, A., El-Gindy, M., El-Amry, Z., *Pakistan J. Science
26,* (1/6) 15 (1974).

71d. Ramirez, M. and Gonzalez, I.M., *J. Agr. U. Puerto Rico,
64,* 148 (1980).

71e. Davy, C.A.E., Wilson, D., Lyon, J.C.M., in "Advances in
Biotechnology," (M. Moo-Young and C.W. Robinson, eds.)
p. 343, Vol. II, Pergamon Press, Toronto (1981).

71f. Forage, A.J., *Process Biochem., 13*(1), 8 (1978).

71g. Miskiewicz, T., Oleszkiewicz, J.A., Kosinska, K.,
Koziarski, S., Kramarz, M., and Ziobrowski, J., *Agr.
Wastes, 4,* 3 (1982).

71h. Oleszkiewicz, J.A., Kosinska, K. and Koziarski, S.,
     Env. Protect. Eng., *5*, 155 (1979).
71i. Oleszkiewicz, R., Ryznar, G., *Proceed. Symp. Livestock
     Wastes* (Amarillo, TX) 350 (1981).
71j. Karthigesan, J. and Brown, B.S., *J. Chem. Tech.
     Biotechnol.*, *31*, 55 (1981).
71k. Tauk, S.M., *Cientifica, 7*, (Brasil) 173 (1979).
71l. Klappach, G., Weichert, D., Meinhold, I., Feiler, E.,
     Becker, U. and Schneider, J., German Democratic Pat.
     142 452 (1980), FSTA *13* 3G, 192.
71m. Moresi, M. and Marchionni, G., *Eur. J. Appl. Microbiol.
     Biotechnol.*, *16*, 204 (1982).
71n. Tauk, S.M., *Eur. J. Appl. Microbiol. Biotechnol.*, *16*,
     223 (1982).
71o. Ercoli, E. and Ertola, R., *Biotechnol. Lett.* 5, 457
     (1983).
72. Kadlec, K., Pelechova, J., Krumphanzi, V. and Sokolov, T.,
     Kvasny Prum, *25*(4), 82 (1979); *Microbiol Abstracts*
     831-A15.
73. Krumphanzl, V., Sokolov, T., Pelekhova, Ya., Kadlec, K.
     and Cholev, I., *Gidroliz, Lesokhim. Prum-st.*, (6), 28
     (1978); CA *89*, 178 063y.
74. Volfova, O. and Kyslikova, E., *Folia Microbiol.*, *24*, 157
     (1979); CA *93*, 24 414j.
75. Volfova, O., Kyslikova, E., Kikyta, B. and Panos, J.,
     Folia Microbiol., *24*, 163 (1979); CA *93*, 24 415k.
75a. Volfora, O., Kyslikova, E., Sikyta, B. and Zalabak, V.,
     Biol. Chem. Vet., *14*, 237-242 (1978).
75b. Sikyta, B., Volfova, O. and Kylikova, E., *Wissenschaft
     Z. Ernst-Moritz-Arndt Universitat Greisfswald Math.
     Naturwissenschaft Reihe*, *29*, 145-146 (1980).
76. Semenov, V.F. and Podgorski, V.S., *Microbiol. Zh.*, *40*,
     293 (1978); CA *89*, 88 814q.
77. Malinovskaya, N.M., Repka, V.P., Kulik, A.P. and
     Kosenko, V.A., *Vopr. Khim Tekhnol.*, *51*, 66 (1978); CA
     90, 166 548t.
78. Eklund, E., Hatakka, A., Mustranta, A. and Nybergh, P.,
     Europ. J. Appl. Microbiol., *2*, 143 (1976).
79. Araujo Neto, J.S. and Panek, A.D., *Rev. Microbiol.*, *6*,
     59 (1975); CA *87*, 18 654x.
80. Cho, H.O., Rhee, C.O. and Chae, S.K., *Korean J. Food
     Sci. Technol.*, *5*, 101 (1973); FSTA 7, 3G 146.
81a. Ghonaim, S.A., Abou-Zeid, A.A., Abd El-Fattah, A.F.,
     Farid, M.A., *Zentralbl. Bakteriol. Parasitenkd.
     Infektionskr. Hyg. II.*, *135*, 82 (1980).
82. Moo-Young, M., Moreira, A.R., Daugulis, A.J. and
     Robinson, C.W., *Biotechnol. Bioeng. Symp.*, *8*, 205 (1978).
83. Gonzalez, A. and Moo-Young, M., *Biotechnol. Letters 3*,
     143 (1981).

84. LeDuy, A., *Process Biochem.*, *14*(3), 5 (1979).
85. Quierzy, P., Therien, N., LeDuy, A., *Biotechnol. Bioeng.*, *21*, 1175 (1979).
86. Beja da Costa, M., *Ciencia Biol.*, *5*(1), 61 (1980); CA *92*, 162 151b.
86a. Chaudry, M.Y., Shah, M.A., Shah, F.H., *Pak. J. Biochem.*, *11*(1-2) 12 (1978).
86b. Pozmogova, I.N., Semoschina, T.N., Perevezentsev, V.P., *Appl. Biochim. Microbiol.*, *17*, 196 (1981).
86c. Bobleter, O., Niesner, R. and Rohr, M., *J. Appl. Polym. Sci.*, *20*, 2083 (1976).
86d. Khandelwal, G.D., Apte, K.P., Chatterjee, M.K., Moolik, S.K.R., *Indian Pulp Pap.*, *34*(6), 3 (1980).
86e. Wilson, D.U. and Thayer, D.W., *Am. Inst. Chem. Eng. Symp. Series*, *74*(172), 136 (1978).
86f. Mavluni, M.I. and Ismailov, A.U., *Uzb. Biol. Zh.* (6), 8 (1980); CA *94*, 154 935m.
86g. Gonzalez, I. and Knackmuss, K.P., *Acta Biotechnol.*, *2*, 377 (1982).
86h. Davey, G. and Bruce, J., *Biotechnol. Bioeng.*, *24*, 2753 (1982).
86i. Balasubramanya, R.H. and Bathawdekar, S.P., *Indiana J. Agric. Sci.*, *50*, 965-970 (1980).
87. Teuber, M., in "Proc. COST-Workshop on Production and Feeding of SCP," p. 8, KFA, Julich, Federal Republic of Germany (1980).
88. Gueriviere, J.F., ed la, *Ind. Alim. Agric.*, *98*, 395 (1981).
89. Moresi, M., Nacca, C., Nardi, R. and Palleschi, C., *Eur. J. Appl. Microbiol. Biotechnol.*, *8*, 49 (1979).
90. Moresi, M. and Sebastiani, E., *Eur. J. Appl. Microbiol. Biotechnol.*, *8*, 63 (1979).
91. Moresi, M., Colicchio, A. and Sansovani, F., *Eur. J. Appl. Microbiol. Biotechnol.*, *9*, 173 (1980).
92. Moresi, M., Colicchio, A., Sansovani, F. and Sebastiani, E., *Eur. J. Appl. Microbiol. Biotechnol.*, *9*, 261 (1980).
93. Blakebrough, N. and Moresi, M., *Eur. J. Appl. Microbiol. Biotechnol.*, *12*, 173 (1981).
94. Blakebrough, N. and Moresi, M., *Eur. J. Appl. Microbiol. Biotechnol.*, *13*, 1 (1981).
95. Burgess, K.J., *Irish J. Food Science Technol.*, *1*, 107 (1977).
96. Castillo, F.J. and de Sanchez, S.B., *Acta Cientifica Venezolana*, *29*, 113 (1978).
97. Castillo, F.J., de Sanchez, S.B. and Goncalves, J.A., *Acta Cientifica Venezolana*, *30*, 588 (1979).

98. de Sanchez, B. and Castillo, F.J., *Acta Cientifica Venezolana, 31,* 24 (1980).

98a. Sanchez, L. and Castillo, F.J., *Acta Cientifica Venezolana, 31,* 154 (1980).

98b. de Bales, S.A. and Castillo, F.J., *Appl. Environ. Microbiol., 37,* 1201 (1979).

98c. Pedrique, M. and Castillo, F.J., *Appl. Environ. Microbiol. 43,* 303 (1982).

98d. Gomez, A. and Castillo, F.J., *Biotechnol. Bioeng., 25,* 1341 (1983).

99. Edelsten, D., Nielsen, E.W. and Refstrup. E., in "Brief Comm. 20th Int. Dairy Congress," p. 965, Paris (1978).

99a. Edelsen, D., Ebbesen, F. and Hertel, J., *Milchwissenschaft, 34,* 733 (1979).

100. Yaziciouglu, T., Celikkol, E., Ocal, S., Aran, N. and Umeroglu, S., "Some Trials on the Utilization of Whey, Blacwater of Olive and Vinasse for Production of SCP in Turkey," Publ. No. 42, Marmara Scientific and Industrial Research Institute (1980).

101. Gilliland, S.E., *J. Dairy Sci., 62,* 882 (1979).

102. Gilliland, S.E. and Stewart, C.F., *J. Dairy Sci., 63,* 989 (1980).

103. Terra, N.N., *Rev. Farm. Bioquim. Univ. Sao Paulo, 14*(1), 55 (1976); CA *87,* 199 453z.

104. Knight, S., Smith, W. and Mickle, J.B., *Cultured Dairy Products, J., 7*(2), 17 (1972).

105. Osovik, A.N., Gridina, L.E., Marichenko, A.B. and Bashirova, R.S., *Appl. Biochem. Microbiol., 11,* 423 (1975).

106. Foda, M.S., *Milchwissenschaft, 31,* 138 (1976).

107. Harju, M., Heikonen, M. and Kreula, M., *Milchwissenschaft, 31,* 530 (1976).

108. Zalashko, M.V., Shamgin, V.K., Zalashko, M.V., Obraztsova, N.V. and Chaika, L.L., *20th Int. Dairy Congress E.,* 936 (1978); FSTA *10,* 10G 66.

108a. Chaika, L.L., Zalashko, L.S., Skums, L.D. and Vinokurova, Z.P., *Mikrobiol. Zh.* (Kiev) 39, 494 (1977); CA *87,* 132 294e.

108b. Zalashko, M.V., Oleshko, V.S., Zalashko, L.S. and Chaika, L.L., *Vestsi Akad. Navuk BSSR, Ser. Biyal Navuk.* (3) 74 (1980); CA *93,* 130 520a.

109. Giec, A. and Kosikowski, F.V., *J. Dairy Science, 61,* (Suppl. 1), 104 (1978).

109a. Giec, A. and Kosikowski, F.V., *J. Food Science 48,* 1892 (1982).

110. Hernandez, E., Meza, E. and Lazano, N., *20th Int. Dairy Congress E,* 935 (1978); FSTA *10,* 12G 791.

111. Barraquio, V. L., Silverio, L. G., Revilleza, R. P., and
     Fernández, W. L. *NSDB Technol. J. 4*(4), 68 (1979).
112. Andrzejewski, W., Surazynski, A., Chojnowski, W., and
     Koziak, W. *Prem. Ferment. Rolny 18*(6), 1 (1974); FSTA
     7, 8G, 523.
113. Vrignaud, Y. *Oesterreichische Milchwirtschaft 31*(2),
     405 (1976).
114. Vrignaud, Y. *Technique Laitiere* 912, 31 (1977).
115. El-Akher, M. A., Abd, M. A., Alian, A. M., and El-Shimy,
     N. M. *Chem. Mikrobiol. Technol. Lebensm. 3*, 81-84
     (1974).
116. Moulin, G., Ratomahemina, R., and Galzy, P. *Lait 56*
     (553/556), 135 (1976).
117. Mohandas, D. V. and Srinivasan, R. A. *Indian J. Micro-
     biol. 17*(3), 133 (1977).
118. Moon, N. J., Hammond, E. G., and Glatz, B. A. *J. Dairy
     Science 60* (Suppl. 1), 41 (1977).
119. Moon, N. J., Hammond, E. G., and Glatz, B. A. *J. Dairy
     Science 61*, 1537 (1978).
120. Moon, N. J. and Hammond, E. G. *J. Am. Oil Chem. Soc.
     55*, 683-688 (1978).
121. Dluzewski, M., Kowalczyk, R., Abram, M., and Dordzik, B.
     *Zesz. Nauk. Szk. Gl. Gospod. Wiejsk. Warszawue Technol.
     RolnoSpozyw.* (8), 109 (1973); CA *80*, 25 901j.
122. Lane, A. G. *J. Appl. Chem. Biotechnol. 27*, 165 (1977).
123. Abreu, L.E.V., Terra, N. N., Meller, A. C., Mussoi, E.,
     Brum, M.A.R. de, and Almeida, J. de P. *Rev. Centro
     Ciencias Rurais 6*(3), 275-280 (1976).
124. Vananuvat, P. and Kinsella, J. E. *J. Food Science 40*,
     336 (1975).
125. Vananuvat, P. and Kinsella, J. E. *J. Food Science 40*,
     823 (1975).
126. Anon. Deutsche *Malkerei-Zeitung 96*(50), 1656 (1975);
     FSTA *8*, 5P 831.
127. Coton, S. G. In "Food from Waste" (G. G. Birch, K. J.
     Parker, and J. T. Worgan, eds.), p. 221, Applied
     Science Publishers, London (1976).
128. Damerow, G. and Gerhold, E. *Deutsche Milchwirtschaft
     27*(25), 796 (1976); FSTA *8*, 11P 2013.
129. Anon. *Food Eng. Int. 2*(7), 42 (1977).
130. Knipschildt, M. E. *Tejipar 27*(1), 14 (1978); FSTA *10*,
     7P 1009.
131. Pace, G. W. and Goldstein, D. J. In "Single Cell Pro-
     tein II" (S. R. Tannenbaum and D.I.C. Wang, eds.), p.
     330, MIT Press, Cambridge (1975).
132. Joux, J.L.F. French Pat. Appl. 2 292 039 (1976).
133. Muller, H., Muller, F., Hermann, P., and Bowden, A.
     Swiss Pat. 575 722 (1976).

134. Muller, H. US Pat. 3 968 257 (1976).
135. Ringpfeil, M., Meyer, D., Beck, D., and Iske, U. German
      Dem. Rep. Pat. 124 811 (1977).
136. Beck, D., Iske, U., Meyer, D., Ringpfeil, M., and Mein-
      hold, I. German Dem. Rep. Pat. 132 352 (1978).
137. Anon. Jap. Pat. 5 331 946 (1978).
138. Muller, H. Brit. Pat. 1 511 975 (1978).
139. DuChaffaut, J. A., Magnox, C. R., and Oberto, P.L.C.
      US Pat. 4 165 389 (1979).
140. Anon. US Pat. 4 183 966 (1980).
140a Kappeli, O., Halter, N., and Puhan, Z. In "Advances in
      Biotechnology" (M. Moo-Young and C. W. Robinson, eds.),
      p. 351, Vol. II, Pergamon Press, Tronto (1981).
140b Gholson, J. H. and Gough, R. H. *La. Agriculture* *23*(3),
      19 (1980).
140c Poncet, S. and Jacob, F. H. In "Current Developments
      in Yeast Research" (G. G. Stewart and F. H. Jacob,
      eds.), p. 93, Pergamon Press, Toronto (1980).
140d González, J. B. and Berry, D. R. *Biotechnol. Letters 4*,
      369 (1982).
140e Sandhu, D. K. and Waraich, M. K. *Biotechnol. Bioeng.*
      25, 797 (1983).
141. Chang, C. T. and Yang, W. L. *Taiwan Sugar 20*(5), 193
      (1973).
142. Chung, M. *Taiwan Sugar 23*(6), 241 (1976).
143. Madi, A., El-Nagar, A., and Shehata, Y. M. *Indian J.
      Microbiol. 15*, 27 (1975); *Pakistan J. Biochem. 8*,
      49 (1975).
144. Mahmoud, S. A., Abdel-Hafez, A. M., El-Sawy, M., and
      Ramadan, E. M. *Zentralbl. Bakteriol. Parasitenkd.
      Infektionskr. Hyg. Abt. 2 132*, 326 (1977).
145. El-Sawy, M., Mahmoud, S. A., Abdel-Hafez, A. M., and
      Ramadan, E. M. *Zentralbl. Bakteriol. Parasitenkd.
      Infektionskr. Hyg. Abt. 2 132*, 641 (1977).
146. El-Sawy, M., Mahmoud, S. A., Abdel-Hafez, A. M., and
      Ramadan, E. M. *Zentralbl. Bakteriol. Parasitenkd.
      Infektionskr. Hyg. Abt. 2 132*, 648 (1977).
147. Estevez, R. and Almazán, O. *Rev. Derivados de Caña de
      Azúcar* (Cuba) *7*(3), 60 (1973).
148. Estevez, R. and Almazán, O. *Rev. Derivados de Caña de
      Azúcar* (Cuba) *8*(2), 51 (1974).
149. Hernández, P. and Aguila, M. *Cent. Azucar* (Cuba) *2*(1),
      89 (1975).
150. Amazán, O. and Klibansky, M. *Boletin de Tecnologia,*
      GEPLACEA (Mexico) No. 11 (1979).
151. Cabello, A. *Boletin de Tecnología,* GEPLACEA (Mexico)
      No. 11 (1979).

151a   Klibansky, M. and Peña, M. A. *Cuba Azucar* (Oct/Dec),
       37 (1979).

152.   Moreira, A., de Menezes, T.J.B., and Arakaki, T. *Col.
       Inst. Tecnol. Alim.* (Campinas) 7(1), 97 (1976).

153.   Mian, F. A., Ajdary, A., and Fazeli, A. *J. Ferment.
       Technol. 54*, 76 (1976).

153a   Rickard, P. A. and Hewetson, J. W. *Biotechnol. Bioeng.
       21*, 2337 (1979).

154.   Peschard, E. and Viniegra, G. *Biotechnol. Bioeng. Symp.
       7*, 119 (1977).

155.   Konstantinova, R. and Lippert, E. *Zentralbl. Bakteriol.
       Parasitenkd. Infektionskr Hyg. Abt. 2 135*, 185 (1980);
       135, 402 (1980).

156.   Heisel, M. *Linde Ber. Tech. Wiss. 45*, 84 (1979); *Chie-
       mie Ing. Technik. 52*, 60 (1980).

156a   Heisel, M. *Eng. Costs Prod. Economics 5*(1), 21 (1980).

157.   El Sheik Idris, E.T.A. and Berry, D. R. *Biotechnol.
       Lett. 2*, 61 (1980).

157a   El Sheik Idris, E.T.A. and Berry, D. R. *Biotechnol.
       Lett. 2*, 105 (1980).

158.   Almazán, O., Klibansky, M., and Otero, M. A. *Biotechnol.
       Lett. 3*, 663 (1981).

159.   Ichibashi, M., Mori, H., and Mizukoshi, S. Jap. Pat.
       78 27782, 10 August 1978; CA *89*, 178 113q.

159a   Chahal, D. S., Singla, M., and Gupta, R. P. *Proc.
       Indian Natl. Sci. Acad.* (B) *45*, 18 (1979).

159b   Choi, S. Y., Ryu, D.D.Y., and Rhee, J. S. *Biotechnol.
       Bioeng. 24*, 1165 (1982).

159c   Hsiao, H. Y., Chiang, L. C., Ueng, P. P., and Tsao, G.
       T. *Appl. Environ. Microbiol. 43*, 840 (1982).

159d   Moebus, O., Teuber, M., and Reuter, H. *Kieler Milch-
       wirtschaftl. Forsch. 31*, 297 (1979).

159e   Moebus, O. and Teuber, M. *Eur. J. Appl. Microbiol. 15*,
       194 (1982).

159f   Mishra, I. M., El-Temtamy, S. A., and Schugerl, K.
       Eur. J. Appl. Microbiol. 16, 197 (1982).

160.   Anderson, R., Weisbum, R. B., and Robe, K. *Food Pro-
       cessing 35*(7), 58 (1974).

161.   Nakano, J. *Appl. Polym. Symp. 28*, 85 (1975).

162.   McKee, L. A. and Quicke, G. V. *South African J. Science
       73*(12), 379 (1977).

163.   Andersen, R. A. *Pulp & Paper Canada 80*(4), 43 (1979).

164.   Barta, J. *Antonie van Leeuwenkoek J. Microbiol. Serol.
       35* (Suppl.: Yeast Symp.), J7 (1969).

165.   Chaudry, M. Y., Shah, M. A., and Shah, F. H. *Pak. J.
       Biochem. 10*(2), 39-43 (1977).

166. Simark, R. E. and Cameron, A. *Pulp Paper Mag. Can. 75* (C), 107 (1974).

167. Zhukova, G. M., Chikileva, E. A., Nikolaeva, N. D., and Ashikmina, G. F. *Tr. Vses. Nauchno-Pruizvod Ob'edin. Bum. Prom. sti. Permsk. Fil. 2I,* 83 (1976); CA *89,* 4571x.

167a. Rychtera, M., Panuschka, G., Fischer, H., and Vernerova, J. *Kusany Prum. 25,* 270 (1979).

167b Setzermann, U. and Zimmer, G. *Wiss Beitr. Ingenieur-hochsch. Koethen* (1), 86 (1978).

168. Ignat'eva, O. I., Yakovleva, I. V., and Sapotnitskii, S. A. *Bum. Prom. st.* (11), 18 (1979); CA *92,* 56793c.

169. Pineault, G., Pruden, B. B., and Loutfi, H. *Can. J. Chem. Eng. 55,* 333 (1977).

170. Sestakova, M. *Folia Microbiol. 24,* 318 (1979).

171. Revah-Moiseev, S. and Carroad, A. *Biotechnol. Bioeng. 23,* 1067 (1981).

171a Carroad, P. A. and Tom, R. A. *J. Food Science 43,* 1158 (1978).

171b Cosio, I. G., Fisher, R. A., and Carroad, P. A. *J. Food Science 47,* 901 (1982).

172. Abe, S. and Tagaki, M. Jap. Kokai Tokkyo Koho 79 98 389, 3 August 1979; CA *91,* 173 385m.

173. Solomon, B. D., Erickson, L. E., and Hess, J. L. *Biotechnol. Bioeng. 23,* 2333 (1981).

174. Araujo, A. and D'Souza, J. *J. Ferment. Technol. 58,* 399 (1980).

175. Nishio, N. and Nagai, S. *Eur. J. Appl. Microbiol. Biotechnol. 11,* 156 (1981).

176. Anon. Fr. Demande 2 395 311, 19 Jan. 1979; CA *91,* 138 868z.

177. Kolarova, N. and Farkas, V. *Eur. J. Appl. Microbiol. Biotechnol. 13,* 184 (1981).

178. Beja da Costa, M. *Ciencia Biol.* (Coimbra) *5,* 61 (1980).

179. Wickstrom, L. and Leisola, M. In "Advances in Biotechnology" (M. Moo-Young and C. W. Robinson, eds.), p. 21, Vol. II, Pergamon Press, Toronto (1981).

179a Nojiri, M. *Proc. GIAM V,* 320 (1979).

179b Kamikubu, T., Tanaka, M., Taniguchi, M., Morita, T., and Matsuno, R. In "Advances in Biotechnology" (M. Moo-Young and C. W. Robinson, eds.), p. 311, Vol. II, Pergamon Press, Toronto (1981).

179c Musenge, H. M., Anderson, J. G., and Holdom, R. S. *Biotechnol. Lett. 2,* 35 (1980).

179d Moo-Young, M., Chahal, D. S., and Vlach, D. *Anim. Feed Science Technol. 5,* 175 (1980).

179e   Taniguchi, M., Kumetani, Y., Tanaka, M., Matsuno, R.,
       and Kamikubo, T. *Eur. J. Appl. Microbiol. Biotechnol.*
       14, 74 (1982).

179f   Kaytan, M. and Kolankaya, N. *Hacettepe Bull. Nat. Sci.*
       Eng. 9, 19 (1980).

179g   Morrison, S. M., Elmond, G. K., Grant, D. W., and Smith,
       V. J. *Dev. Ind. Microbiol. 18*, 145 (1977).

180   Barford, J. P. and Hall, R. J. *Biotechnol. Bioeng. 23*,
       1735 (1981); *23*, 1763 (1981).

181.   Toda, K., Yabe, I., and Yamagata, T. *Biotechnol.*
       Bioeng. 22, 1805 (1980).

182.   de Kok, H. E. and Roels, J. A. In "Advances in Bio-
       technology" (M. Moo-Young and C. W. Robinson, eds.),
       p. 271, Vol. I, Pergamon Press, Toronto (1981).

183.   Geurts, Th.G.E. and de Kok, H. E. *Biotechnol. Bioeng.*
       22, 2031 (1980).

184.   Dekkers, J.G.J., de Kok, H. E., and Roels, J. A. *Bio-*
       technol. Bioeng. 23, 1023 (1981).

185.   Nelligan, I. and Callam, C. T. *Biotechnol. Lett. 2*,
       531 (1980).

186.   Woehrer, W. and Roehr, M. *Biotechnol. Bioeng. 23*, 567
       (1981).

187.   Woehrer, W., Hampel, W., and Roehr, M. In "Advances in
       Biotechnology" (M. Moo-Young and C. W. Robinson, eds.),
       p. 419, Vol. I, Pergamon Press, Toronto (1981).

188.   Okada, W., Fukuda, H., and Marikawa, H. *J. Ferment.*
       Technol. 59, 103 (1981).

189.   Nanba, A., Hirota, F., and Nagai, S. *J. Ferment.*
       Technol. 59, 383 (1981).

190.   Miskiewicz, T. *J. Ferment. Technol. 59*, 411 (1981).

191.   Daikaru, K., Yamasahi, Y., Kuki, K., Shioya, S., and
       Takamatsu, T. *Biotechnol. Bioeng. 23*, 2069 (1981).

192.   Ramirez, A., Durand, A., and Blachere, H. T. *Bio-*
       technol. Letters 3, 555 (1981).

192a   Daikaru, K., Yamasaki, Y., Morikawa, H., Shioya, S., and
       Takamatsu, T. *J. Ferment. Technol. 60*, 67 (1982).

192b   Wasungu, K. M. and Sinard, R. E. *Biotechnol. Bioeng.*
       24, 1125 (1982).

192c   Josic, D. *Lebensm. Wiss. Technol. 15*, 5 (1982).

193.   Haraldson, A. and Bjorling, T. *Eur. J. Appl. Microbiol.*
       Biotechnol. 13, 34 (1981).

194.   Rolz, C. *Interciencia 7*, 153 (1981).

195.   dos Anjos Magalhaes, M. M., Vairo, M.L.R., and Borzani,
       W. *J. Ferment. Technol. 57*, 534 (1979).

196.   Pekur, G. N., Bur'gan, N. I., and Pvalenko, N. M.
       Appl. Biochim. Microbiol. 17, 248 (1981).

197. Tanner, R. D., Richmond, L. D., Wei, C. J., and Woodward, J. *J. Chem. Tech. Biotechnol. 31*, 290 (1981).

198. Balagopal, C. and Maini, S. B. *J. Root Crops 2*, 49 (1976).

199. Balagopal, C. and Maini, S. B. *J. Root Crops 3*, 33 (1977).

199a Muindi, P. J. and Hanssen, J. F. *J. Sci. Food Agric. 32*, 655 (1981).

200. Worgan, J. T. In "Food from Waste" (G. G. Birch, K. J. Parker, and J. T. Worgan, eds.), p. 23, Applied Science Publishers, Ltd. London (1976).

201. Imrie, F.K.E. and Righelato, R. C. In "Food from Waste" (G. G. Birch, K. J. Parker, and J. T. Worgan, eds.), p. 79, Applied Science Publishers, Ltd. London (1976).

202. Fuska, J. and Kollarova, A. *Biologia 32*, 973 (1977).

203. Deshpande, K. S. and Joshi, R. N. *Nycopath. Mycol. Appl. 45*, 151 (1977).

204. Ghewande, M. P. and Bansode, D. K. *Science Culture 43*, 345 (1977).

205. Gewaily, E. M. *Ann. Agr. Science*, Moshtohor 7, 31 (1977).

206. Stakheev, I. V. and Babitskaya, V. G. *Mikol. Fitopatol. 12*, 490 (1978).

207. Gupta, R. P., Singh, A., Kalra, M. S., Sharma, V. K., and Gupta, S. K. *Indian J. Nutr. Diet. 14*, 302 (1977).

208. Zeteleki-Horvath, K. *Acta Aliment. Acad. Sci. Hung. 7*, 225 (1978).

209. Zeteleki-Horvath, K. and Vas, K. *Acta Aliment. Acad. Sci. Hung. 6*, 241 (1977).

210. Zeteleki-Horvath, K., Abd-El Bakey, M., and Vas, K. *Acta Aliment. Acad. Sci. Hung. 7*, 167 (1978).

210a Zeteleki-Horvath, K. and Vas, K. *Acta Aliment. Acad. Sci. Hung. 9*, 191 (1980).

210b Zeteleki-Horvath, K. and Vas, K. *Acta Aliment. Acad. Sci. Hung. 9*, 209 (1980).

211. Rosenberg, H., Obrist, J., and Stobs, S. J. *Econ. Botany 32*, 413 (1978).

211a Israilides, C. J., Grant, G. A., and Anderson, A. W. *Agr. Wastes 4*, 323 (1982).

212. Jones, J. A. and Bu'Lock, J. D. *Biotechnol. Letters 1*, 81 (1979).

213. Bhatia, I. S. and Arneja, J. S. *J. Sci. Fd. Agric. 29*, 619 (1978).

214. Ghai, S. K., Kahlon, S. S., Chahal, D. S. *Indian J. Exp. Biol. 17*, 789 (1979).

215. Abraham, M. J. and Srinivasan, R. A. *J. Food Science Technol. 16,* 11 (1979).

216. Blain, J. A., Anderson, J. G., Todd, J. R., and Divers, M. *Biotechnol. Letters 1,* 269 (1979).

217. Anderson, J. G., Blain, J. A., Marchetti, P., and Todd, J. R. *Biotechnol. Letters 3,* 451 (1981).

218. de González, I. M. and de Murphy, N. F. *J. Agric. Univ. Puerto Rico 63,* 325 (1979).

219. de González, I. M. and de Murphy, N. F. *J. Agric. Univ. Puerto Rico 63,* 330 (1979).

220. de González, I. M. and de Murphy, N. F. *J. Agric. Univ. Puerto Rico 64,* 138 (1980).

221. Quinn, J. P. and Marchant, R. *Water Research 14,* 545 (1980).

222. Ek, M. and Eriksson, K. E. *Biotechnol. Bioeng. 22,* 2273 (1980).

223. Malfait, J. L., Wilcox, D. J., Mercer D. G., and Barker, L. D. *Biotechnol. Bioeng. 23,* 863 (1981).

224. Barker, T. W. and Worgan, J. T. *Eur. J. Appl. Microbiol. Biotechnol. 11,* 234 (1981).

225. Gibriel, A. Y., Mahmoud, R. M., Goma, M. and Abou-Zeid, M. *Agr. Wastes 3,* 229 (1981).

226. Garg, S. K. and Neelakantan, S. *Biotechnol. Bioeng. 23,* 1653 (1981).

226a Garg, S. K. and Neelakantan, S. *J. Food Science Technol. 18,* 64 (1981).

226b Garg, S. K. and Neelakantan, S. *J. Fd. Technol. 17,* 271 (1982).

226c Garg, S. K., and Neelakantan, S. *Biotechnol. Bioeng. 24,* 2407 (1982).

227. Vered, Y., Milstein, O., Flowers, H. M., and Gressel, J. *Eur. J. Appl. Microbiol. Biotechnol. 12,* 183 (1981).

227a Miller, T. F. and Srinivasan, V. R. *Biotechnol. Bioeng. 25,* 1509 (1983).

228. Schmidt, O., Puls, J., Sinner, M., and Dietrichs, H. H. *Holzforsch. 33,* 192 (1979).

229. Milstein, O., Vered, Y., Gressel, J., and Flowers, H. M. *Eur. J. Appl. Microbiol. Biotechnol. 13,* 117 (1981).

230. Kim, J. H., Iibuchi, S., and Lebeault, J. M. *Eur. J. Appl. Microbiol. Biotechnol. 13,* 208 (1981).

231. Kim, J. H. and Lebeault, J. M. *Eur. J. Appl. Microbiol. Biotechnol. 13,* 151 (1981).

232. Foda, M. S. *J. Sci. Food Agr. 32,* 1109 (1981).

233. Martin, A. M. *Biotechnol. Letters 4,* 13 (1982).

234. Gregory, K. F., Reade, A. E., Santos-Nuñez, J., Alexander, J. C., Smith, R. E., and MacLean, S. J. *Anim. Feed Sci. Technol. 2,* 7 (1977).

234a Gregory, K. F. *Conservation Recycling 5,* 33 (1982).

235. Mikami, Y., Gregory, K. F., Levadoux, W. L., Balagolapan, C., and Whitwill, S. T. *Appl. Environ. Microbiol. 43,* 403 (1982).

236. Ingman, M. *Wochenbl. Papierfabr. 108,* 193 (1980).

237. Amailagic, M., Nadazdin, M., Dzinic, M., and Pavlovic, D. *Veterinaria* (Sarajevo) *29,* 101 (1980).

237a Betucci, L. and Perez, C. *Rev. lat-amer. Microbiol. 23,* 151 (1981).

237b Eaton, D. C., Chang, H. M., Joyce, T. W., Jeffries, T. W., and Kirk, T. K. *TAPPI 65,* 89 (1982).

237c Saguido, M.A.P., Cayabyab, V. A., and Uyenco, F. R. Paper presented at the Workshop on Bioconversion of Lignocellulosic and Carbohydrate Residues in Rural Communities, Bali (1979).

238. Ghosh, A. K. and Sengupta, S. *J. Fd. Sci. Technol. 14,* 6 (1977).

239. Ghosh, A. K. and Sengupta, S. *J. Fd. Sci. Technol. 15,* 237 (1978).

240. Ghosh, A. K. and Sengupta, S. *Indian J. Microbiol. 19,* 217 (1979).

241. Manachini, R. L. *Tecnol. Alimentari 11*(5), 17 (1979); FSTA *12,* 4G 220.

242. Sakamoto, R., Niimi, T., and Takahashi, S. *J. Agr. Chem. Soc. Japan 52,* 75 (1978).

243. Sakamoto, R., Niimi, T., and Takahashi, S. *J. Agr. Chem. Soc. Japan 52,* 83 (1978).

244. Labaneiah, M. E., Abou-Donia, S. A., Esmat, S. A., El-Zalaki, M., and Mohame, M. S. *Alex. J. Agr Res. 25,* 71 (1977).

245. Labaneiah, M. E., Abou-Donia, S. A., Mohamed, M. S., and El-Zalaki, E. M. *Alex. J. Agr. Res. 25,* 281 (1977).

245a Labaneiah, M. E., Abou-Donia, S. A., Mohamed, M. S., and El-Zalaki, E. M. *J. Food Technol. 14,* 95 (1979).

246. Carels, M. and Shepherd, D. *Can. J. Microbiol. 23,* 1360 (1977).

247. Carroad, P. A. and Wilke, C. R. *Appl. Environ. Microbiol. 33,* 871 (1977).

248. Duvnjak, Z., Erick, M., and Tamburasev, G. *Mljekarstvo 28,* 38 (1978); FSTA *11,* 2P 254.

248a Duvnjak, Z., Spinderk, I., and Tamburasev, G. *Mljekarstvo 29,* 161 (1979); FSTA *13,* 1G 18.

249. Kurtzman, R. H. *Mycologia 70,* 179 (1978).
250. Jauhri, K. S., Kumari, M. L., and Sen, A. *Zbl. Bakt. II 133,* 588 (1978).
251. Jauhri, K. S., Kumari, M. L., and Sen, A. *Zbl. Bakt. II 133,* 597 (1978).
252. Jauhri, K. S. and Sen, A. *Zbl. Bakt. II 133,* 604 (1978).
253. Jauhri, K. S. and Sen, A. *Zbl. Bakt. II 133,* 609 (1978).
254. Jauhri, K. S. and Sen, A. *Zbl. Bakt. II 133,* 614 (1978).
255. Vecher, A. S., Solomko, E. F., Shachov, E. N., Dudka, I. A., Bukhalo, A. S., Paromchik, I. I., and Pchelintseva, R. K. *Dokl. Akad. Nauk. 23,* 855 (1979); CA *91,* 171 407w.
256. Ghosh, A. K. and Sengupta, S. *J. Fd. Sci. Technol. 19,* 57 (1982).
257. Gupta, J. K., Kumar, L., Vadehra, D. V., and Gupta, Y. P. *J. Sci. Ind. Res. 35,* 325 (1976).
258. Sadana, J. C., Lachke, A. H., and Shewale, J. G. *J. Sci. Ind. Res. 38,* 442 (1979).
259. Griffin, H. L., Kaneshiro, T., Kelson, B. F., and Sloneker, J. In "Symposium on Enzymatic Hydrolysis of Cellulose" (M. Bailey, T. M. Enari, and M. Linko, eds.), p. 419, Aulanko, Finland (1975).
260. Brown, D. E. and Fitzpatrick, S. W. In "Food from Waste" (G. G. Birch, K. J. Parker, and J. T. Worgan, eds.), p. 139, Applied Science Publishers, Ltd. London (1976).
261. De Menezes, T., Duchini, L. A., and Figueiredo, I. B. *Rev. Bras, Tecnol. 7,* 439 (1976).
262. Janus, J. M. *Proc. Bioconversion Symp.* (IIT) Delhi, 469 (1977).
263. Lerpido-Barraquio, V. C., Suga, K., Okumura, K., and Ichikawa, K. *Phillip. Agric. 61,* 31 (1977); CA *89,* 88 810k.
264. Babitskaya, V. G., Stakheev, I. V., and Plavskaya, A. I. *Mikol. Fitopatol. 13,* 118 (1979); CA *91,* 35 433z.
265. Peitersen, N. *Biotechnol. Bioeng. 19,* 337 (1977).
266. Peitersen, N. and Andersen, B. *Amer. Inst. Chem. Eng. Symp. Series 74*(172), 100 (1978).
267. Geethadevi, B. R., Sitaram, N., Kunhi, A.A.M., and Rao, T.N.R. *Indian J. Microbiol. 18,* 85 (1979).
267a Chahal, D. S., Moo-Young, M., and Dhillon, G. S. *Can. J. Microbiol. 25,* 793 (1979).
267b Sidhu, M. S. and Sandhu, D. K. *Biotechnol. Bioeng. 22,* 689 (1980).

268. Singh, A., Sethi, R.P., Kalra, M.S., and Chahal, D.S.,
      Indian J. Microbiol., 16, 37 (1976).
269. Sethi, R.P. and Sood, S.M., *Indian J. Mycol. Plant
      Pathol., 7,* 39 (1977).
270. Sitaram, N., Mohammad Kunhi, A.A., Geethadevi, B.R. and
      Ramachandra Rao, T.N., *Indian J. Microbiol., 18,* 90
      (1978).
271. Singh, A., Dhillon, G.S., Gupta, R.P. and Kalra, M.S.,
      Indian J. Exp. Biol., 16, 1317 (1978).
272. Dhillon, G.S., Singh, A. and Kalra, M.S., *J. Food
      Science Technol., 19,* 74 (1982).
273. Ghai, S.K. and Chahal, D.S., *J. Res., 16,* (Punjab Agric.
      U.), 60 (1979).
274. Ghai, S.K., Kahlon, S.S. and Sood, S.M., *J. Res., 16,*
      (Punjab Agric. U.), 64 (1979).
274a. Srinivasan, M.C., Rao, M., Deshpande, V., Mishra, C.,
      Kantham, B.C., Bastawade, K.B., Kulkarni, V.M., Phansalkar,
      S.B., Joglekar, A.V. and Jagannathan, V., *Hind. Antibiot.
      Bull. 19*(3-4), 31 (1977).
274b. Rao, M., Mishra, C., Seeta, R., Srinivasan, M.C.,
      Desphande, V.V., *Biotechnol. Lett., 5,* 301 (1983).
275. Ek, M. and Eriksson, K.E., *Proc. Bioconversion Symp.,*
      IIT (Delhi) 449 (1977).
276. Eriksson, K.E., Grunewald, A. and Vallander, L.,
      Biotechnol. Bioeng., 22, 363 (1980).
276a. Davey, G. and Bruce, J., *Biotechnol. Bioeng., 25,* 647
      (1983).
277. Daugulis, A.J. and Bone, D.H., *Eur. J. Appl. Microbiol.
      Biotechnol., 4,* 159 (1977).
278. Daugulis, A.J. and Bone, D.H., *Biotechnol. Bioeng., 20,*
      1639 (1978).
279. Bellamy, W.D., *Dev. Ind. Microbiol., 18,* 249 (1977).
280. Armiger, W.B., Moreira, A.R., Phillips, J.A. and
      Humphrey, A.E., *Amer. Inst. Chem. Eng. Symp. Series, 75*
      (184), 7 (1979).
281. Humphrey, A.E., *Adv. Chem. Series., 181,* 25 (1978).
282. Chahal, D.S. and Hawksworth, D.L., *Mycologia., 68,* 600
      (1976).
282a. Eriksen, J. and Goksoyr, J., *Arch. Microbiol., 110,* 233
      (1976).
283. Moo-Young, M., Chahal, D.S. and Vlach, D., *Proc.
      Bioconversion Symp.,* IIT (Delhi) 457 (1977).
284. Moo-Young, M., Chahal, D.S., Swan, J.E. and Robinson,
      C.W., *Biotechnol. Bioeng., 19,* 527 (1977).
285. Chahal, D.S. and Wang, D.I.C., *Mycologia, 70,* 160 (1978).
286. Chahal, D.S., Swan, J.E. and Moo-Young, M., *Dev. Ind.
      Microbiol., 18,* 433 (1977).

287. Pamment, N., Moo-Young, M., Hsieh, F.H. and Robinson,
      C.W., *Appl. Environ. Microbiol., 36,* 284 (1978).
288. Chahal, D.S. and Moo-Young, M., *Dev. Ind. Microbiol., 22,*
      143 (1981).
289. Pamment, N., Robinson, C.W. and Moo-Young, M., *Biotechnol.
      Bioeng. 21,* 561 (1979).
290. Chahal, D.S., Moo-Young, M. and Vlach, D., *Biotechnol.
      Bioeng., 23,* 2417 (1981).
291. Moo-Young, M. and Chahal, D.S., *Anim. Feed Science
      Technol., 4,* 199 (1979).
292. Moo-Young, M., Chahal, D.S. and Stickney, B., *Biotechnol.
      Bioeng., 23,* 2407 (1981).
293. Moo-Young, M., Daugulis, A.J., Chahal, D.S. and
      MacDonald, D.G., *Process Biochem., 14*(10), 38 (1979).
294. Moo-Young, M., in "Bioresources for Development,"
      (A. King and H. Cleveland, eds.) p. 155, Pergamon Press,
      New York (1978).
295. Moo-Young, M., McDonald, G. and Ling, A., *Biotechnol.
      Letters, 3,* 149 (1981).
296. Fahnrich, P. and Irrgang, K., *Biotechnol. Letters, 3,*
      201 (1981).
296a. Hecht, V., Schugerl, K. and Scheiding, W., *Eur. J. Appl.
      Microbiol. Biotechnol., 16,* 219 (1982).
296b. Gacera, S., Popov, S., Skrinjar, M. and Janhovits, I.,
      in "Proc. Microbiol. Conv. Raw Materials and Byproducts
      of Agriculture into Proteins, Alcohols and Other Products,"
      NoviSad, Yugoslavia, June 15-17th, p. 79 (1982).
296c. Karapinar, M. and Okugan, M., *J. Chem. Tech. Biotechnol.,
      32,* 1055 (1982).
297. Zadrazil, F. and Schliemann, J., paper presented at the
      IX International Congress of Edible Mushrooms, Taipei,
      Taiwan (1974).
298. Waslien, C.I. and Steinkraus, K.H., *BioScience, 28,* 397
      (1978).
299. Stanton, W.R., *Proc. GIAM V,* 180 (1978).
300. Hesseltine, C.W., *Dev. Ind. Microbiol., 22,* 1 (1981).
301. Hayes, W.A., *Process Biochem., 9*(10), 21 (1974).
302. Delmas, J., in "The Biology and Cultivation of Edible
      Mushrooms," (S.T. Chang and W. Hayes, eds.) p. 699,
      Academic Press, New York (1978).
303. Hayes, W.A. and Wright, S.A., in "Microbial Biomass,"
      (A.H. Rose, ed.) p. 142, Academic Press, New York (1979).
304. Chang, S.T., *Food Policy, 5*(1), 64 (1980).
305. Chang, S.T., *BioScience, 30,* 399 (1980).
306. Kurtzman, R.H., in "Protein Nutritional Quality of Foods
      and Feeds. Part 2," (M. Friedman, ed.) p. 305, Marcel
      Dekker, New York (1975).

307. Batra, L.R. and Millner, P.D., *Dev. Ind. Microbiol., 17,*
      117 (1976).
308. Steinkraus, K., in "Bioresources for Development,"
      (A. King and H. Cleveland, eds.), p. 135, Pergamon
      Press, New York (1978).
309. Djen, K.S. and Hesseltine, C.W., in "Microbial Biomass,"
      (A.H. Rose, ed.) p. 116, Academic Press, New York (1979).
310. Zadrazil, F., *Mushroom Science, 9,* 621 (1974).
311. Zadrazil, F., *Eur. J. Appl. Microbiol. Biotechnol., 9,*
      31 (1980).
312. Zadrazil, F., *Eur. J. Appl. Microbiol. Biotechnol., 4,*
      273 (1977).
313. Zadrazil, F., *Eur. J. Appl Microbiol. Biotechnol., 9,*
      243 (1980).
314. Zadrazil, F. and Brunnert, H., *Eur. J. Appl. Microbiol.
      Biotechnol., 9,* 37 (1980).
315. Zadrazil, F. and Brunnert, H., *Eur. J. Appl. Microbiol.
      Biotechnol., 11,* 183 (1981).
316. Zadrazil, F., in "The Biology and Cultivation of Edible
      Mushrooms," (S.T. Chang and W.A. Hayes, eds.) p. 521,
      Academic Press, New York (1978).
317. Zadrazil, F., in "Advances in Biotechnology," (M. Moo-
      Young and C.W. Robinson, eds.), p 369, Pergamon Press,
      Toronto (1980).
317a. Zadrazil, F., in "Straw Decay and its Effect on Disposal
      and Utilization," (E. Grossbard, ed.) p. 139, John Wiley
      & Sons, Chichester (1979).
318. Rajarathnam, S., Wankhede, D.B. and Patwardhan, M.V.,
      Eur. J. Appl. Microbiol. Biotechnol., 8, 125 (1979).
319. Platt, M.W., Chet, I. and Henis, Y., *Eur. J. Appl.
      Microbiol. Biotechnol., 13,* 194 (1981).
319a. Platt, M.W., Hadar, Y., Henis, Y. and Chet, I., *Eur. J.
      Appl. Microbiol. Biotechnol., 17,* 140 (1983).
320. Kaneshiro, T., *Dev. Ind. Microbiol., 18,* 591 (1977).
321. Virk, R.S., Sethi, R.P. and Garcha, H.S., *Indian J.
      Animal Sci., 50,* 293 (1980).
322. Danilyak, N.I., *Mikrobiol. Zh., 43,* (Kiev), 185 (1980);
      CA *92,* 211 709a.
323. Ginterova, A., Polster, M. and Janotkova, O., *Folia
      Microbiol., 25,* 332 (1980).
324. Eger, G., in "The Biology and Cultivation of Edible
      Mushrooms," (S.T. Chang and W.A. Hayes, eds.) p. 497,
      Academic Press, New York (1978).
325. Eger, G., Li, S.F. and Leal, H., *Mycologia, 71,* 577
      (1979).

326. Li, S.F. and Eger-Hummel, G., in "Advances in Biotechnology," (M. Moo-Young, C.W. Robinson and C. Vezina,
      eds) Vol. I, p. 101, Pergamon Press, Toronto (1981).
327. Singh, N.S. and Rajarathnam, S., *Curr. Science, 46,*
      617 (1977).
328. Ginterova, A. and Gallon, J.R., *Biochem. Soc. Trans., 7,*
      1293 (1979).
328a. Chang, S.T., Lau, O.W. and Cho, K.Y., *Eur. J. Appl.
      Microbiol. Biotechnol., 12,* 58 (1981).
328b. Lindenfelser, L.A., Detroy, R.W., Ramstack, J.M., Worden,
      K.A., *Dev. Ind. Microbiol., 20,* 541 (1979).
328c. Khanna, P. and Garcha, H.S., *Mushroom Newsletter Tropics,*
      2(3), 5 (1982).
328d. Singh, R.P., *Mushroom Newsletter Tropics, 3*(3), 8 (1983).
328e. Bano, Z. and Rajarathnam, S., *Mushroom Newsletter
      Tropics, 3*(3), 12 (1983).
328f. Chibenjaphol, S., *Mushroom Newsletter Tropics, 2*(3), 9
      (1982).
328g. Streeter, C.L., Conway, K.E. and Horn, G.W., *Mycol., 73,*
      1040 (1981).
328h. Streeter, C.L., Conway, K.E., Horn, G.W. and Mader, T.L.,
      J. Anim. Science., 54, 183 (1982).
329. Matteau, P.P. and Bone, D.H., *Biotechnol. Letters, 2,*
      127 (1980).
330. Adhikay, D.K., George, U. and Ghose, T.K., *Biotechnol.
      Letters, 4,* 197 (1982).
330a. Giovannuzzi-Sermanni, G., Badiani, M. and Luna, M.,
      Biotechnol. Letters, 4, 507 (1982).
330b. Gao, P.J., Ma, Q.R., Zhang, Y.Zh., Jiang, B.Y. and Wang,
      Z.N., in "GIAM VI Global Impacts of Applied Microbiology,"
      (S.O. Emejvaiwe, O. Ogunbu and S.O. Sanni, eds.) p. 253,
      Academic Press, London (1981).
331. Chang, S.T., in "The Biology and Cultivation of Edible
      Mushrooms," (S.T. Chang and W.A. Hayes, eds.) p. 573,
      Academic Press, New York (1978).
332. Samarawira, I., *Econ. Botany, 33,* 163 (1979).
333. Wang, C.W., Chew, M.Y. and Stanton, W.R., *Proc. GIAM V*
      349 (1979).
334. Chang, S.C. and Steinkraus, K.H., *Appl. Environ.
      Microbiol., 43,* 440 (1982).
334a. Chang, S.T. and Yau, C.K., in "GIAM VI Global Impacts
      of Applied Microbiology," (S.O. Emejuaiwe, O. Ogunbu
      and S.O. Sanni, eds.), p 647, Academic Press, London
      (1981).
334b. Santiago, C.M. and Peberdy, J.F., *Mushroom Newsletter
      for the Tropics, 3*(1), 6 (1982).

334c. Ibrahim, M.N.M. and Pearce, G.R., *Agr. Wastes, 2,* 199
      (1980).
334d. Hayes, W.A. and Lim, W.C., in "Straw Decay and its
      Effects on Disposal and Utilization," (E. Grossbard,
      ed.) p. 85, John Wiley, Chichester (1979).
334e. Latham, M.J., in "Straw Decay and its Effects on
      Disposal and Utilization," (E. Grossbard, ed.) p 131,
      John Wiley, Chichester (1979).
335. Gold, M.H. and Cheng, T.M., *Arch. Microbiol., 121,* 37
      (1979).
336. Zezula, V., Jilek, R., Vodickova, M., and Fuska, J.,
      Zb. Ved. Konf. SVST: Tvorba Ochr. Zivotn. Postredia,
      4th I, 184 (1979); CA *92,* 92 956s.
337. Kurtzman, R.H., in "The Biology and Cultivation of
      Edible Mushrooms," (S.T. Chang and Hayes, W.A., eds.)
      p 393, Academic Press, New York (1978).
338. Robert, J.C. and Durand, R., *Physiol. Plantarum, 46,*
      174 (1979).
339. Uno, I. and Ishikawa, T., *J. Bacteriol., 120,* 96 (1974).
340. Miyake, H., Takemaru, T. and Ishikawa, T., *Arch.
      Microbiol., 126,* 201 (1980).
341. Miyake, H., Tanaka, K. and Ishikawa, T., *Arch. Microbiol.
      126,* 207 (1980).
342. Uno, I., Nyunoya, H., Ishikawa, T., *J. Gen. Appl.
      Microbiol., 27,* 219 (1981).
343. Kamada, T., Kurita, R. and Takemaru, T., *Plant Cell
      Physiol., 19,* 263 (1978).
343a. Burrows, I., Seal, K.J. and Eggins, H.O.W., in "Straw
      Decay and its Effect on Disposal and Utilization,"
      (E. Grossbard, ed.) p 147, John Wiley & Sons, Chichester
      (1979).
344. Chen, P.C. and Hou, H.H., in "The Biology and
      Cultivation of Edible Mushrooms," *loc. cit,* p 629 (1978).
345. Cheng, S. and Tu, C.C., in "The Biology and Cultivation
      of Edible Mushrooms," *loc. cit,* p 606.
346. Szudyga, K., in "The Biology and Cultivation of Edible
      Mushrooms," *loc cit.* p 559.
347. Arita, I., in "The Biology and Cultivation of Edible
      Mushrooms," *loc. cit.* p 475.
348. Tonomura, H., in "The Biology and Cultivation of Edible
      Mushrooms," *loc. cit.* p 410.
348a. Cho, K.Y. and New, P.B., *Mushroom Newsletter Tropics,*
      3(2), 14 (1982).
349. Gramss, G., in "The Biology and Cultivation of Edible
      Mushrooms," *loc. cit.* p 424.

350. Tokimoto, K. and Komatsu, M., in "The Biology and Cultivation of Edible Mushrooms," *loc. cit.* p. 445.

351. Ito, T., in "The Biology and Cultivation of Edible Mushrooms," *loc cit.* p 461.

351a. Gyurko, P., *Karstenia, 18* (Suppl.) 70 (1978).

351b. Jalkanen, R. and Jalkanen, E., *Karstenia, 18* (Suppl.) 56 (1978).

351c. Roponen, I. and Kreula, M., *Karstenia, 18* (Suppl.) 58 (1978).

352. Tominaga, Y., in "The Biology and Cultivation of Edible Mushrooms," *loc. cit.* p 683.

353. Delmas, J., in "The Biology and Cultivation of Edible Mushrooms," *loc cit.* p 645.

353a. Reade, A.E. and McQueen, R.E., *Can. J. Microbiol., 29,* 457 (1983).

354. Toyama, N. and Ogawa, K., *Proc. Bioconversion Symp.,* IIT (Delhi) 305 (1976).

355. Vilela, L.C., Torillo, A.R., de Ocampo, A.T. and del Rosario, E.J., *Agric. Biol. Chem., 41,* 235 (1977).

356. Qadeer, M.A., Shah, F.H., Jabbar, A. and Ilahi, A., *Pak. J. Sci. Ind. Res., 20,* 120 (1977).

357. Jabbar, A. and Ilahi, A., *Agric. Biol. Chem., 45,* 1719 (1981).

358. Gulati, S.L., *Zbl. Bakt. II,* 135, 413 (1980).

359. Nishio, N., Kurisu, H. and Nagai, S., *J. Ferment. Technol., 59,* 407 (1981).

359a. Theja, K., Shamada, T.R., Sreekantiah, K.R. and Sreenivasa Mortly, V., *J. Food Science Technol., 20,* 84 (1983).

360. Sinha, S.N., Ghosh, B.L. and Ghose, S.N., *Can. J. Microbiol., 27,* 1334 (1981).

361. Chahal, D.S., Vlach, D. and Moo-Young, M., in "Advances in Biotechnology," (M. Moo-Young and C.W. Robinson, eds.) Vol. II, p. 327, Pergamon Press, Toronto (1981).

362. Ulmer, D.C., Tengerdy, R.P., Murphy, V.G. and Linden, J.C., *Dev. Ind. Microbiol., 21,* 425 (1980).

363. Ulmer, D.C., Tengerdy, R.P. and Murphy, V.G., *Biotechnol. Bioeng. Symp., 11,* 449 (1981).

364. Sekita, S., Yoshihira, K., Natori, S., Udagawa, S., Muroi, T., Sugiyama, Y., Kurata, H. and Umeda, M., *Can. J. Microbiol., 27,* 766 (1981).

365. Sethi, R.P. and Grainger, J.M., in "Advances in Biotechnology," (M. Moo-Young and C.W. Robinson, eds.) Vol. II, p. 319, Pergamon Press, Toronto (1981).

366. Bugdan, S.D. and Sivers, V.S., *Mikrobiol. Zh., 41,* 135 (1979); Microbial Abstracts 8616-A14; CA *91,* 4 121s.

367. Wicklow, D.T., Detroy, R.W. and Jessee, B.A., *Appl. Environ. Microbiol., 40,* 169 (1980).

367a. Zadrazil, F., Grinberge, J. and Gonzalez, A., *Eur. J. Appl. Microbiol. Biotechnol., 15,* 167 (1982).

367b. Wicklow, D.T., Detroy, R.W. and Adams, S., *Mycologia, 72,* 1065 (1980).

367c. Zadrazil, F. and Brunnert, H., *Eur. J. Appl. Microbiol. Biotechnol., 16,* 45 (1982).

368. Eriksson, K.E., Nilsson, T., Ander, P., Goodell, B. and Henningsson, B., Swedish Pat. 411 563, 14 January 1980; CA *93,* 68 688f.

368a. Pilon, L., Desrochers, M., Jurasek, L. and Neumann, P.J., *TAPPI 65,* 93 (1982).

368b. Pilon, L., Barbe, M.C., Desrochers, M., Jurasek, L. and Neumann, P.J., *Biotechnol. Bioeng., 24,* 2063 (1982).

268c. Bar-lev, S.S., Kirk, T.K. and Chang, H.M., *TAPPI 65,* 111 (1982).

369. Raimbault, M. and Alazard, D., *Eur. J. Appl. Microbiol. Biotechnol., 9,* 199 (1980).

370. Raimbault, M., Deschamps, F., Meyer, F. and Senez, J.C., *Proc. GIAM V,* 245 (1979).

371. Alazard, D. and Raimbault, M., *Eur. J. Appl. Microbiol. Biotechnol., 12,* 113 (1981).

372. Vezinhet, F., Roger, M., Oteng-Gyang, K. and Galzy, P., *Rev. Ferment. Ind. Aliment., 32*(3), 65 (1977).

373. Kokke, R., *J. Appl. Bacteriol., 43,* 303 (1977).

374. Plating, S.J. and Cherry, J.P., *J. Food Science, 44,* 1178 (1979).

375. Kamalakannan, V. and Motlag, D.B., *Can. J. Microbiol., 28,* 54 (1981).

376. Ogundana, S.K., *Lebensm. Wiss. Technol., 13,* 334 (1980).

377. Ogunsua, A.O., *Food Chem., 5,* 249 (1980).

378. Shieh, Y.S.C. and Beuchat, L.R., *J. Food Science., 47,* 518 (1982).

379. Sheih, Y.S.C., Worthington, R.E. and Phillips, R.D., *J. Food Science., 47,* 523 (1982).

380. Yokotsuka, T. and Sasaki, M., in "Advances in Biotechnology," (M. Moo-Young, and C.W. Robinson, eds.) Vol. II, p 461, Pergamon Press, Toronto (1981).

381. Lequerica, J.L., Vila, R. and Feria, A., *Rev. Agroquim. Technol. Aliment., 20,* 95 (1980).

382. Bajracharya, R. and Mudgett, R.E., *Biotechnol. Bioeng., 21,* 551 (1979).

383. Bajracharya, R. and Mudgett, R.E., *Biotechnol. Bioeng., 22,* 2219 (1980).

383a. Mudgett, R.E., Nash, J. and Rufner, R., *Dev. Ind. Microbiol.*, *23*, 397 (1982).

384. Gibb, E. and Walsh, J.H., *Trans. Br. Mycol. Soc.*, *74*, 111 (1980).

385. Carrizales, V., Rodriguez, H. and Sardina, I., *Biotechnol. Bioeng.*, *23*, 321 (1981).

386. Aidoo, K.E., Hendry, R. and Wood, B.J.B., *Eur. J. Appl. Microbiol. Biotechnol.*, *12*, 6 (1981).

387. Sugama, S. and Ohazaki, N., *J. Ferment. Technol.*, *57*, 408 (1979).

388. Ohazaki, N., Sugama, S. and Tanaka, T., *J. Ferment. Technol.*, *58*, 471 (1980).

388a. Narahara, H., Koyama, Y., Yoshida, T., Pichangkura, S., Ueda, R. and Taguchi, H., *J. Ferment. Technol.*, *60*, 311 (1982).

389. Smith, D.A., Kuhn, P.J., Bailey, J.A. and Burden, R.S., *Phytochem.*, *19*, 1673 (1980).

389a. Yadav, K.S., Mishra, M.M., Kapoor, K.K., *Agr. Wastes*, *4*, 329 (1982).

389b. Gaur, A.C., Sadasivam, K.V., Mathur, R.S. and Magu, S.P., *Agr. Wastes*, *4*, 453 (1982).

389c. Kundu, A.B., Ghosh, B.S., Ghosh, B.L. and Ghose, S.N., *J. Ferment. Technol.*, *61*, 185 (1983).

390. Clayton, R.K. and Sistrom, W.R., "The photosynthetic bacteria," Plenum Press, New York (1978).

391. Pfenning, N., in "The Photosynthetic Bacteria," (R.K. Clayton and W.R. Sistrom, eds.) p 3, Plenum Press, New York (1978).

392. Aiba, S., Koizumi, J. and Nishizawa, Y., *J. Chem. Technol. Biotechnol.*, *29*, 311 (1979).

393. Gobel, F., in "The Photosynthetic Bacteria," *loc. cit.* p 907 (1978).

394. Erikson, L.E., *J. Ferment. Technol.*, *58*, 53 (1980).

395. Aiking, H. and Sojka, G., *J. Bacteriol.*, *139*(2), 530 (1979).

396. Aretz, W., Kospari, H., Klemme, J.H., *FEMS Microbiol. Letters*, *4*, 249 (1978).

397. Kim, J.S., Ito, K. and Takahashi, H., *Agric. Biol. Chem.*, *44*, 827 (1980).

398. Watanabe, K., Kim, J.S., Ito, K., Buranakarl, L., Kampee, T. and Takahashi, H., *Agric. Biol. Chem.*, *45*, 217 (1981).

399. Kim, J.S., Ito, K. and Takahashi, H., *J. Ferment. Technol.*, *59*, 185 (1981).

400. Kim, J.S., Ito, K. and Takahashi, H., *Agric. Biol. Chem.* *46*, 937 (1982).

400a. Kim, J.S., Yamauchi, H., Ito, K. and Takahashi, H., *Agr. Biol. Chem.*, *46*, 1469 (1982).

401. Miyake, J., Tomizuka, N. and Kamibayashi, A., *Ferment. Technol., 60,* 199 (1982).

401a. Vrati, S. and Verma, J., *J. Ferment. Technol., 61,* 157 (1983).

402. Pirie, N.W., in "Food Protein Sources," (N.W. Pirie, eds) p. 33, Cambridge University Press, Cambridge (1975).

403. Tiedemann, A.R., Lopushinsky, W. and Larsen, H.J., *Soil Biol. Biochem., 12,* 471 (1980).

404. Venkataram, G.S., in "GIAM V," (W.R. Stanton and E.J. DaSilva, eds.) p 129 (1978).

405. Rodgers, G.A., Bergman, B., Henriksson, E. and Udris, M., *Plant & Soil, 52,* 99 (1979).

406. Loftis, S.G. and Kurtz, E.B., *Soil Science, 129,* 150 (1980).

407. Gorham, P.R. and Carmichael, W.W., *Pure Appl. Chem., 52,* 165 (1979).

408. Gorham, P.R. and Carmichael, W.W., *Prog. Water Technol., 12,* 189 (1980).

409. Alam, A., Shimizu, Y., Ikawa, M. and Sasner, J.J., *Env. Science Health, 13,* 493 (1978).

410. de la Fuente, G., Flores, A., Molina, M., Almengor, L., and Bressani, R., *Appl. Environ. Microbiol., 33,* 6 (1977).

411. Santillana, C., *Experientia, 38,* 40 (1982).

411a. Durand-Chastel, H., in "Algae Biomass," (G. Shelef and C.J. Soeder, eds.) p 51, Elsevier/North-Holland Biomedical Press, Amsterdam (1980).

412. Clement, C., Lonchamp, D., Rebeller, M. and Vanlendeghem, H., *Chem. Eng. Science, 35,* 119 (1980).

413. Faucher, U., Coupal, B. and Leduy, A., *Can. J. Microbiol., 25,* 752 (1979).

413a. Ayala, F.A. and Bravo, R., *Eur. J. Appl. Microbiol., 15,* 198 (1982).

414a. Richmond, A. and Preiss, K., *Interdis. Sci. Rev., 5,* 60 (1980).

414b. Richmond, A. and Vonshak, A., *Ergebnisse Limnol., 11,* 274 (1978).

414c. Richmond, A., Vonshak, A. and Arad, Sh., in "Albae Biomass," (G. Shelef and C.J. Soeder, eds.) p 63, Elsevier/North-Holland Biomedical Press, Amsterdam (1980).

414d. Vonshak, A., Boussiba, S., Abeliovich, A. and Richmond, A., *Biotechnol. Bioeng., 25,* 341 (1983).

415. Contreras, S., del Rio, A. and Soto, M.A., *Biotechnol. Bioeng., 18,* 1479 (1976).

416. Oron, G., Shelef, G. and Levi, A., *Biotechnol. Bioeng., 21,* 2169 (1979).

416a. Shelef, G., Azov, Y., Moraine, R. and Oron, G., in
      "Albae Biomass," (G. Shelef and C.J. Soeder, eds.)
      p 163, Elsevier/North-Holland Biomedical Press,
      Amsterdam (1980).

417. Nigam, B.P., Ramanathan, P.K. and Venkataram, L.V.,
      Biotechnol. Letters, 3, 619 (1981).

418. Venkataram, L.V., Madhari Devi, K., Mahadevaswamy, M.,
      Mohammed Kunhi, A.A., *Agr. Wastes, 4,* 117 (1982).

418a. Venkataram, L.V., Nigam, B.P. and Ramanathan, P.K.,
      in "Albae Biomass," (G. Shelef and C.J. Soeder, eds.)
      p 81, Elsevier/North-Holland Biomedical Press, Amsterdam
      (1980).

418b. Soong, P., in "Algae Biomass" (G. Shelef and C.J.
      Soeder, eds.) p 97, Elsevier/North-Holland Biomedical
      Press, Amsterdam (1980).

419. Juttner, F., *Biotechnol. Bioeng., 19,* 1679 (1977).

420. Juttner, F., *Process Biochem., 17*(2), 2 (1982).

421. Leduy, A. and Therien, N., *Can. J. Chem. Eng., 57,* 489
      (1979).

422. Leduy, A. and Therien, N., *Biotechnol. Bioeng., 19,* 1219,
      (1977).

423. Ogawa, T. and Aiba, S., *J. Appl. Chem. Biotechnol., 28,*
      515 (1978).

424. van Eykelenburg, C., *Antoine van Leeuwenhoek., 45,* 369
      (1979); *46,* 113 (1980).

425. Karagouni, A.D. and Slater, J.H., *FEMS Microbiol. Lett.,*
      4, 295 (1978).

426. Frischknecht, K. and Schneider, K., *Arch. Microbiol.,*
      120, 215 (1979).

427. Gavin McInnes, A., Walter, J.A. and Wright, J.L.C.,
      Lipids, 15 609 (1980).

428. Falkner, G. and Horner, F. and Simonis, W., *Planta, 149,*
      138 (1980).

428a. Parrot, L.M. and Slater, J.H., *Arch. Microbiol., 127,* 53
      (1980).

428b. Lee-Kaden, J. and Simonis, W., *Plant Cell Physiol., 20,*
      1179 (1979).

429. Contreras, S., del Rio., A., Pieber, M., Soto, M.A.,
      and Toha, J.C., *Biotechnol. Bioeng. Symp., 10,* 91 (1980).

430. Oren, A. and Shilo, M., *Arch. Microbiol., 122,* 77 (1979).

431. Friedberg, D., Fine, M. and Oren, A., *Arch. Microbiol.,*
      123, 311 (1979).

432. Padan, E., *Adv. Microbial Ecology, 3,* 1 (1979).

433. Yamaoka, T., Satoh, K. and Kotoh, S., *Plant Cell Physiol.,*
      19, 943 (1978).

434. van Liere, L., Zevenboom, W. and Mur, L.R., *Prog. Water*
      Technol., 8, 301 (1977).

435. Zevenboom, W., de Groot, G.J. and Mur, L.R., *Arch. Microbiol., 125,* 59 (1980).
436. Zevenboom, W. and Mur, L.R., *FEMS Microbiol. Lett., 6,* 209 (1979).
437. van Liere, L. and Mur, L.R., *J. Gen. Microbiol., 115,* 153 (1979).
438. van Liere, L., Mur, L.R., Gibson, C.E. and Herdman, M., *Arch. Microbiol., 123,* 315 (1979).
439. Ahlgren, G., *Oikos, 29,* 209 (1977).
440. Ahlgren, G., *Mitt. Int. Ver. Theor. Angew. Limnol., 21,* 88 (1978).
441. Ahlgren, G., *Arch. Hydrobiol., 89,* 43 (1980).
441a. Yanagimoto, M. and Saitoh, H., *J. Ferment. Technol., 60,* 305 (1982).
442. Materassi, R., Narese Filastro, M., Balloni, W., Paoletti, C. and Florenzano, G., *Ann. Microbiol. Enzimol., 26,* 15 (1976).
443. Ohmori, M., Ohmori, K. and Strotmann, H., *Arch. Microbiol. 114,* 225 (1977).
444. Sahu, J.K., Adhikary, S.P. and Pattnaik, H., *Indian J. Exp. Biol., 17,* 1401 (1979); *Microbios Lett., 10,* 129, (1979); *J. Indian Bot. Soc., 58,* 358 (1979).
445. Gotto, J.W., Tabita, F.R. and van Baalen, C., *J. Bacteriol., 140,* 327 (1979).
446. Laube, V.M., McKenzie, C.N. and Kushner, D.J., *Can. J. Microbiol., 26,* 1300 (1980).
446a. Moede, A., Greene, R.W. and Spencer, D.F., *Environ. Exp. Bot., 20,* 207 (1980).
446b. Verstraete, D.R., Storch, T.A. and Dunham, V.L., *Physiol. Plantarum, 50,* 47 (1980).
447. Kruger, G.H.J. and Juttner, F., *J. Gen. Microbiol., 119,* 27 (1980).
448. Okada, M., Sudo, R. and Aiba, S., *Biotechnol. Bioeng., 24,* 143 (1982).
449. Lang, D.S. and Brown, E.J., *Appl. Environ. Microbiol., 42,* 1002 (1981).
450. Sicko-Goad, L., Jensen, T.E. and Ayala, R.P., *Can. J. Microbiol., 24,* 105 (1978).
451. Rogerson, A.C., *Nature, 284,* 563 (1980).
452. Coleman, J.R. and Colman, B., *Plant Physiol., 65,* 980 (1980).
453. Coleman, J.R. and Colman, B., *Plant Physiol., 67,* 917 (1981).
454. Miller, A.G. and Colman, B., *Plant Physiol., 65,* 397 (1980); *J. Bacteriol., 143,* 1253 (1980).
455. Jacobs, R. and Lind, O., *Microb. Ecology, 3,* 205 (1977).

456. Abreu, F.A., Billyard, T.C. and Walton, T.J., *Phytochem.*, *16*, 351 (1977).

457. Murry, M.A. and Benemann, J.R., *Plant Cell Physiol.*, *20*, 1391 (1979).

458. Stratton, G.W. and Corke, C.T., *Can. J. Microbiol.*, *25*, 1094 (1979).

459. Apte, S.K. and Thomas, J., *Curr. Microbiol.*, *3*, 291 (1980).

459a. Musgrave, S.C., Kerby, N.W., Codd, G.A. and Stewart, W.D.P., *Biotechnol. Letters*, *4*, 647 (1982).

460. Bothe, H., Tennigkeit, J. and Eisbrenner, G., *Arch. Microbiol.*, *114*, 43 (1977).

461. Daday, A., Platz, R.A. and Smith, G.D., *Appl. Environ. Microbiol.*, *34*, 478 (1977).

462. Strandberg, G.W., *Dev. Ind. Microbiol.*, *18*, 649 (1977).

463. Belkin, S. and Padan, E., *Arch. Microbiol.*, *116*, 109 (1978).

464. Peterson, R.B. and Wolk, C.P., *Plant Physiol.*, *61*, 688 (1978).

465. Jeffries, T.W., Timourian, H. and Ward, R.L., *Appl. Environ. Microbiol.*, *35*, 704 (1978).

466. Hallenbeck, P.C., Kochian, L.V., Weissman, J.C. and Benemann, J.R., *Biotechnol. Bioeng. Symp.*, *8*, 283 (1978).

467. Lambert, G.R., Daday, A. and Smith, G.D., *FEBS Letters*, *101*, 125 (1979).

468. Miyamoto, K., Hallenbeck, P.C. and Benemann, J.R., *J. Ferment. Technol.*, *57*, 287 (1979); *Appl. Environ. Microbiol.*, *37*, 454 (1979); *Biotechnol. Bioeng.*, *21*, 1855 (1979).

469. Asada, Y., Tonomura, K. and Nakayama, O., *J. Ferment. Technol.*, *57*, 280 (1979).

470. Antarikanonda, P., Berndt, H., Mayer, F. and Lorenzen, H., *Arch. Microbiol.*, *126*, 1 (1980).

471. Strandberg, G.W., *Eur. J. Appl. Microbiol. Biotechnol.*, *9*, 19 (1980).

472. Berchtold, M. and Bachofen, R., *Arch. Microbiol.*, *123*, 227 (1980).

472a. Miura, Y., Yokoyama, H., Takahara, K. and Miyamoto, K., *J. Ferment. Technol.*, *60*, 411 (1982).

473. Pearl, H.W., *Oecologia*, *38*, 275 (1979).

474. Reynaud, P.A. and Roger, P.A., *Compt. Rend. D.*, *288*, 999, (1979).

475. Schanz, F., Allen, E.D. and Gorham, P.R., *Can. J. Bot.*, *57*, 2443 (1979).

476. Padhy, S.N., *Hydrobiol.*, *70*(1-2), 37 (1980).

477. Rice, E.L., Lin, C.Y. and Huang, C.Y., *Bot. Bull. Acad. Sinica.*, *21*, 111 (1980).
478. Hable, M. and Alexander, M., *Appl. Environ. Microbiol.*, *29*, 342 (1980).
479. Kellar, P.E. and Pearl, H.W., *Appl. Environ. Micorbiol.*, *40*, 587 (1980).
480. Ito, O., Cabrera, D. and Watanabe, I., *Appl. Environ. Microbiol.*, *39*, 554 (1980).
481. Majid, F.Z. and Khatun, R., *Curr. Science*, *49*, 146 (1980).
482. Kapoor, K. and Sharma, V.K., *Z. Allg. Mikrobiol.*, *20*, 465 (1980).
483. Barraquio, W.L., de Guzman, M.R., Barrion, M. and Watanabe, I., *Appl. Environ. Microbiol.*, *43*, 124 (1982).
484. Bors, J., Kloss, M., Zelles, L. and Fendrik, J., *J. Gen. Appl. Microbiol.*, *28*, 111 (1982).
485. Collins, C.D. and Boylen, C.W., *Appl. Environ. Microbiol.*, *44*, 141 (1982).
485a. Al-Maadhidi, J. and Henriksson, E., *Oikos*, *35*, 115 (1980).
486. Thayer, D.W., Yang, S.P., Key, A.B., Yang, M.H. and Barker, J.W., *Dev. Ind. Microbiol.*, *16*, 465 (1975).
487. Fu, T.T. and Thayer, D.W., *Biotechnol. Bioeng.*, *17*, 1749 (1975).
488. Thayer, D.W., *Dev. Ind. Microbiol.*, *17*, 79 (1976).
489. Thayer, D.W., *Amer. Inst. Chem. Eng. Symp. Series.*, *74* (172) 126 (1978).
489a. Yang, H.H., Yang, S.P. and Thayer, D.W., *J. Food Science.*, *42*, 1247 (1977).
490. Wani, S.P. and Shinde, P.A., *Mysore, J. Agric. Sci.*, *12*, 388 (1978).
491. Bomar, M.T. and Schmidt, S., *Umschau Wiss. Technik.*, *78*, 25 (1978).
492. Bomar, M.T., *Ernahrungswirtschaft* No. 8, 29 (1978).
493. Bomar, M.T., *Landwirtschaftliche Forsch.*, *34*, 217 (1978).
494. Bomar, M.T., *Ernahrungswirtschaft* No. 9, 28 (1978).
495. Bomar, M.T., Schmidt, S. and Fliegl, E., *Berich. Bundesforsch. Ernahrung* No. 2, 192 (1980).
495a. ibid "Produktion von protein reichen substrat aus cellulose und cellulosehaltingen stoffen," BFE-Bericht 1980/2.
496. Srinivasan, V.R., Fleenor, M.B. and Summers, R.J., *Biotechnol. Bioeng.*, *19*, 153 (1977).
497. Summers, R.J., Boudreaux, D.P. and Srinivasan, V.R., *Appl. Environ. Microbiol.*, *38*, 66 (1979).
497a. Blancas, A., Alpizar, L., Larios, G., Saval, S. and Huitron, C., *Biotechnol. Bioeng. Symp.*, *12*, 171 (1982).

498. Ramasawy, K., Prahasam, K., Bevers, J. and Verachtert,
      H., *J. Appl. Bacteriol.*, *46*, 117 (1979); in "Straw
      Decay and its Effect on Disposal and Utilization,"
      (E. Grossbard, ed.) p 155, John Wiley & Sons, Chichester
      (1979).
499. Haggett, K.D., Gray, P.P. and Dunn, N.W., *Eur. J. Appl.
      Microbiol. Biotechnol.*, *8*, 183 (1979).
500. Lee, S.E. and Humphrey, A.E., *Biotechnol. Bioeng.*, *21*,
      1277 (1979).
501. Daigneault-Sylvestre, D. and Kluepfel, D., *Can. J.
      Microbiol.*, *25*, 858 (1979).
502. Lee, H.C., Chen, S.H., Chang, J.S. and Lee, S.C.,
      K'o Hsueh Fa Chan Yueh K'an., *8*, 839 (1980); CA *93*, 202
      682e.
503. Hitchner, E.V. and Leatherwood, J.M., *Appl. Environ.
      Microbiol.*, *39*, 382 (1980).
504. Stewart, B.J. and Leatherwood, J.M., *J. Bacteriol.*,
      128, 609 (1976).
505. Choudhury, N., Gray, P.P. and Dunn, N.W., *Eur. J. Appl.
      Microbiol. Biotechnol.*, *11*, 50 (1980).
506. Rickard, P.A.D. and Peiris, S.P., *Biotechnol. Letters*,
      3, 39 (1981).
507. Enriquez, A., *Biotechnol. Bioeng.*, *23*, 1423 (1981).
508. Enriquez, A., Montalvo, R. and Canales, M., *Biotechnol.
      Bioeng.*, *23*, 1431 (1981).
508a. Enriquez, A. and Rodriquez, H., *Biotechnol. Bioeng.*,
      25, 877 (1983).
509. Kim, B.H. and Winpenny, J.W.T., *J. Ferment. Technol.*,
      59, 275 (1981).
510. Kim, B.H. and Winpenny, J.W.T., *Can. J. Microbiol.*, *27*,
      1260 (1981).
511. Berg, B., *Can. J. Microbiol.*, *21*, 51 (1975).
512. Breuil, C. and Kushner, D.J., *Can. J. Microbiol.*, *22*,
      1776 (1976).
513. Chang, W.T.H. and Thayer, D.W., *Can. J. Microbiol.*, *23*,
      1285 (1977).
514. Tewari, H.K. and Chahal, D.S., *Indian J. Microbiol.*,
      17, 23 (1977).
515. Tewari, H.K. and Chahal, D.S., *Indian J. Microbiol.*,
      17, 88 (1977).
516. Ramasamy, K. and Verachtert, H., *J. Gen. Microbiol.*,
      117, 181 (1980).
517. Beguin, P., Eisen, H. and Roupas, A., *J. Gen. Microbiol.*,
      101, 191 (1977).
517a. Kolankaya, N., *Hacettepe Bull. Nat. Sci. Eng.*, *9*, 1
      (1980).

518. Stoppok, W., Rapp, P., and Wagner, F. *Appl. Environ. Microbiol. 44*, 44 (1982).

519. Choi, W. Y., Haggett, K. D., and Dunn, N. W. *Austr. J. Biol. Sciences 31*, 553 (1978).

520. Haggett, K. D., Choi, W. Y., and Dunn, N. W. *Eur. J. Appl. Microbiol. Biotechnol. 6*, 189 (1978).

521. Rickard, P.A.D. and Laughlin, T.A. *Biotechnol. Letters 2*, 363 (1980).

522. Choudhry, N., Gray, P. P., and Dunn, N. W. *Biotechnol. Letters 2*, 427 (1980).

523. Peiris, S. P., Rickard, P.A.D., and Dunn, N. W. *Eur. J. Appl. Microbiol. Biotechnol. 14*, 169 (1982).

524. Rickard, P.A.D., Rajoka, M. I., and Ide, J. A. *Biotechnol. Letters 3*, 487 (1981).

525. Rickard, P.A.D. *Proc. 5th Australian Biotechnol. Conf. 37* (1982).

526. Guerreiro, M. M. *Cienc. Biol.* (Coimbra) *5*, 63 (1980); CA *92*,193 295f.

526a Nakamura, K. and Kitamura, K. *J. Ferment. Technol. 60*, 343 (1982).

526b Stegmann, W.H.C., Rapp, P., and Wagner, F. In "Advances in Biotechnology" (M. Moo-Young and C. W. Robinson, eds.), Vol. II, p. 27, Pergamon Press (1981).

527. Hagerdal, B., Ferchak, J., Pye, E. K., and Forro, J. R. *Adv. Chem. Series 181*, 331 (1979).

528. Hagerdal, B., Harris, H., and Pye, E. K. *Biotechnol. Bioeng. 21*, 345 (1979).

528a Hagerdal, B., Ferchak, J. D., and Pye, E. K. *Biotechnol. Bioeng. 22*, 1515 (1980).

528b Ferchak, J. D., Hagerdal, B., and Pye, E. K. *Biotechnol. Bioeng. 22*, 1527 (1980).

528c Moreira, A. R., Phillips, J. A., and Humphrey, A. E. *Biotechnol. Bioeng. 23*, 1325 (1981).

528d Moreira, A. R., Phillips, J. A., and Humphrey, A. E. *Biotechnol. Bioeng. 23*, 1339 (1981).

528e Meyer, H. P. and Humphrey, A. E. *Biotechnol. Bioeng. 24*, 1901 (1982).

528f Meyer, H. P. and Charles, M. *Biotechnol. Bioeng 24*, 1905 (1982).

528g Fennington, G., Lupu, D., and Stutzenberger, F. *Biotechnol. Bioeng. 24*, 2487 (1982).

529. Skowronska, T. *Pol. J. Soil Sci. 10*, 41 (1977); *10*, 51 (1977); CA *89*, 159 845p; *89*, 159 846q.

530. Sreenath, H. K., Joseph, R., and Murthy, V. S. *Folia Microbiol. 23*, 299 (1978); CA *93*, 41 130f.

531. Kluepfel, D., Biron, L., and Ishaque, M. *Biotechnol. Letters 2,* 309 (1980).

532. Ishaque, M. and Kluepfel, D. *Can. J. Microbiol. 26,* 183 (1980).

532a Kluepfel, D. and Ishaque, M. *Dev. Ind. Microbiol. 23,* 389 (1982).

533. Loginova, L. G., Yusopova, I. Kh., Tashpulatov, Zh. *Appl. Biochem. Microbiol. 17,* 106 (1981).

534. Yasui, T., Nakinishi, H., and Kobayashi, T. *J. Soc. Ferment. Technol. 58,* 79 (1980).

535. Nakanishi, K. and Yasui, T. *Agric. Biol. Chem. 44,* 1885 (1980).

536. Nakanishi, K. and Yasui, T. *Agric. Biol. Chem. 44,* 2729 (1980).

537. Ishaque, M. and Kluepfel, D. *Biotechnol. Letters 3,* 481 (1981).

538. Uching, F. and Nakane, T. *Agric. Biol. Chem. 45,* 1121 (1981).

539. Vance, I., Topham, C. M., Blayden, S. L., and Tampion, J. *J. Gen. Microbiol. 117,* 235 (1980).

540. Vance, I., Blayden, S. L., and Tampion, J. *Biotechnol. Letters 3,* 45 (1981).

540a Esteban, R., Villanueva, J. R., and Villa T. G. *Can. J. Microbiol. 28,* 733 (1982).

540b Ohba, R. and Ueda, S. *Agr. Biol. Chem. 46,* 2425 (1982).

541. Bisaria, V. S. and Ghose, T. K. *Enzyme Microb Technol. 3,* 90 (1980).

542. Bailey, M. J. and Nevalainen, K.M.H. *Enzyme Microb. Technol. 3,* 153 (1980).

543. Cuskey, S. M., Schamhart, D.H.J., Chase, T. *Dev. Ind. Microbiol. 21,* 471 (1980).

544. Hsu, T. A., Gong, C. S., and Tsao, G. T. *Biotech. Bioeng. 22,* 2305 (1980).

545. Inglin, M., Feinberg, B. A., and Loewenberg, J. R. *Biotechm. J. 185*(2), 515 (1980).

546. Ladisch, M. R., Gong, C. S., and Tsao, G. T. *Biotech. Bioeng. 22*(6), 1107 (1980).

547. Montenecourt, B. S., Kelleher, T. J., and Eveleigh, D. E. *Biotech. Bioeng. Symp. 10,* 15 (1980).

548. Reese, E. T. and Mandels, M. *Biotech. Bioeng. 22*(2), 323 (1980).

549. Reese, E. T. and Ryu, D. Y. *Enzyme Microb. Technol. 2*(3), 239 (1980).

550. Vohra, R. M., Shirkot, C. K., Dhawan, S., and Gupta, K. G. *Biotech. Bioeng. 22,* 1497 (1980).

551. Weber, M., Fuglietti, M. J., Percheron, F. *J. Chromatogr. 188*(2), 377 (1980).

552. Allen, AL. and Mortensen, R. E. *Biotech, Bioeng.* *23*(11), 2641 (1981).

553. Kubicek, C. P. *Eur. J. Appl. Microbiol. Biotechnol.* *13*, 226 (1981).

554. Lahodova, I., Farkas, V., Bauer, S., Kolarova, N., and Branyik, A. *Eur. J. Appl. Microbiol. Biotechnol.* *12*, 16 (1981).

555. Mukhopadhyay, S. N. In "Advances in Biotechnology" (M. Moo-Young, C. Vezina, and K. Singh, eds.), Vol. III, p. 277, Pergamon Press Toronto (1981).

556. Mukhopadhyay, S. N. *J. Ferment. Technol.* *59*(4), 309 (1981).

557. Ng, T. K. and Zeikus, J. G. *Appl. Environ. Microbiol.* *42*(2), 231 (1981).

558. Ostrikova, N. A. and Konovalov, S. A. *Appl. Biochem. Microbiol.* *17*(6), 642 (1981).

559. Tangno, S. K., Blanch, H. W., and Wilke, C. R. *Biotech. Bioeng.* *23*, 1837 (1981).

560. Woodward, J., Whaley, K. S., Zachry, G. S., and Wohlpart, D. L. *Biotechnol. Bioeng. Symp.* *11*, 619 (1981).

561. Woodward, J. and Arnold, S. L. *Biotech. Bioeng.* *23*, 1553 (1981).

562. Frein, E. M., Montenecourt, B. S., and Eveleigh, D. E. *Biotechnol. Lett.* *4*(5), 287 (1982).

563. Gottvaldova, M., Kucera, J., and Podrazky, V. *Biotechnol. Lett* *4*, 229 (1982).

564. Ghosh, V. K., Ghose, T. K., and Gopalkrishman, K. S. *Biotechnol. Bioeng.* *24*, 241 (1982).

565. Ghosh, A., Al-Rabiai, S., Ghosh, B. K., Trimiño, H., Eveleigh, D. E., and Montenecourt, B. S. *Enzyme Microb. Technol.* *4*, 110 (1982).

566. Mishra, S., Gopalkrishnan, K. S., and Ghose, T. K. *Biotechnol. Bioeng.* *24*, 251 (1982).

567. Shirkot, C. K., Mann, D., Dhawan, S., Gupta, A. K., and Gupta, K. G. *Biotechnol. Bioeng.* *24*(6), 1233 (1982).

567a Ogawa, K., Toyama, H., and Toyama, N. *J. Ferment. Technol.* *60*, 349 (1982).

567b Chaudhary, K. and Tauro, P. *Eur. J. Appl. Microbiol.* *15*, 185 (1982).

567c Gottvaldova, M., Kucera, J., and Podrazky, V. *Biotechnol. Lett.* *4*, 645 (1982).

567d Lee, S. B., Shin, H. S., Ryu, D.D.Y., and Mandels, M. *Biotechnol. Bioeng.* *24*, 2137 (1982).

567e Nanda, M., Bisaria, V. S., and Ghose, T. K. *Biotechnol. Lett.* *4*, 633 (1982).

567f   Marsden, W. L., Gray, P. P., and Dunn, N. W. *Biotechnol. Lett. 4,* 589 (1982).

567g   Sternberg, D. and Mandels, G. R. *J. Bacteriol. 144,* 1197 (1980).

567h   Warzywoda, M., Vandecasteele, J. P., and Pourquie, J. *Biotechnol. Lett. 5,* 243 (1983).

567i   Dekker, R.F.H. *Biotechnol. Bioeng. 25,* 1127 (1983).

568.   Churilova, I. V., Maksimov, U. I., Klesov, A. A. *Biokhimiya 45*(4), 669 (1980).

569.   Ghai, S. K. *Indian J. Exp. Biol. 18*(7), 703 (1980).

570.   Zhu, Y. S., Wo, Y. Q., Chen, W., Tan, C., Gao, J. H., Fei, J. Y., and Shih, C. N. *Enzyme. Microb. Technol. 4,* 3 (1982).

571.   Idel'chik, M. S. Pavlovskaya, Zh. I., and Nezhinets, E. A. *Appl. Biochem. Microbiol. 17*(6), 639 (1981).

572.   Mukhopadhyay, S. N. and Malik, R. K. *Biotech. Bioeng. 22,* 2237 (1980).

573.   Matteau, P. P. and Saddler, J. N. *Biotechnol. Lett. 4,* 513 (1982).

573a   Mukhopadhyay, A. K. and Sikyta, B. *Zbl. Bakt. II Abt. 135,* 682 (1980).

573b   Mukhopadhyay, A. K. and Sikyta, B. *Zbl. Bakt. II Abt. 136,* 644 (1981).

573c   Kubicek, C. P. *Can. J. Microbiol. 29,* 163 (1983).

573d   Sarhar, Ch. and Prabhu, K. A. *Agr. Wastes 6,* 99 (1983).

574.   Attia, R. M. and Gamal, R. F. *Rev. Microbiologia* (Brazil) *11*(2), 64 (1980).

575.   Conrad, D. *Biotechnol. Lett 3*(7), 345 (1981).

576.   Oguntimein, G. B. and Reilly, P. J. *Biotechnol. Bioeng. 22*(6), 1127 (1980).

577.   Oguntimein, G. B. and Reilly, P. J. *Biotechnol. Bioeng. 22*(6), 1143 (1980).

578.   Woodward, J. and Wohlpart D. L. *J. Chem. Tech. Biotechnol. 32,* 547 (1982).

579.   Tavobilov, I. M., Gorbacheva, I. V., Rodionova, N. A., and Bezborodov, A. M. *Appl. Biochem. Microbiol. 17*(3), 320 (1981).

580.   Agabekyan, E. L., Gorkina, N. B., and Fikhte, B. A. *Appl. Biochem. Microbiol. 17*(6), 645 (1981).

581.   Ivanova, I. I., Gozhova, E. P., and Burdenko, L. G. *Appl. Biochem. Microbiol. 16*(1), 45 (1980).

582.   Gorg, S. K. and Neelakantan, S. *Biotechnol. Bioeng. 24,* 737 (1982).

583.   Gorg, S. K. and Neelakantan, S. *Biotechnol. Bioeng. 24,* 109 (1982).

584. Allen, A. and Stenberg, D. *Biotech. Bioeng. Symp. 10,*
      189 (1980).
585. Trivedi, L. and Rao, K. K. *Indian J. Exp. Biol. 18*(3),
      240 (1980); CA *93,* 41 204h.
585a Vandamme, E. J., Logghe, J. M., and Geeraerts, H.A.M.
      J. Chem. Tech. Biotechnol. 23, 968 (1982).
586. Jirku, V., Kraxnerova, B., and Krumphanzl, V. *Folia*
      Microbiol. 25(2), 24 (1980).
587. Kato, Y. and Matsoda, K. *Agric. Biol. Chem. 44*(8),
      1751 (1980).
588. Srivastava, S. K., Ramachandran, K. B., and Gopalkrish-
      nan, K. S. *Biotechnol. Lett. 3*(9), 477 (1981).
589. Astapovich, N. I., Kondrat'eva, L. V., Lobanok, A. G.,
      Golovneva, N. A., Markina, V. L., and Ryahaya, N. E.
      Appl. Biochem. Microbiol. 17(6), 649 (1981).
590. Malik, K. A., Kauser, F., and Azam, F. *Mycologia 72*
      (2), 322 (1980).
590a Ogundero, V. W. *Acta Biotecnol. 3,* 65 (1983).
591. Wood, T. M., McCrae, S. I., and MacFarlane, C. C. *Bio-*
      chem. J. 189(1), 51 (1980).
592. Rapp, P., Knobloch, U., and Wagner, F. In "Advances in
      Biotechnology" (M. Moo-Young, C. Vezina, and K. Singh,
      eds.), Vol. III, p. 289, Pergamon Press, Toronto (1981).
593. Rapp, P., Grote, E., and Wagner, F. *Appl. Environ.*
      Microbiol. 41(4), 857 (1981).
594. Szacks, G., Reczey, K., Hernadi, P., and Dobozi, M.
      Eur. J. Appl. Microbiol. Biotechnol. 11, 120 (1981).
595. Deleyn, F., Claeyssens, M., and DeBruyne, C. K. *Can.*
      J. Biochem. 58(1), 5 (1980).
596. Wang. C. W. *Mushroom Newsletter for the Tropics 1*(4),
      10 (1982).
597. Kanda, T., Wakabayoshi, K., and Nisizawa, K. *J. Bio-*
      chem. (Tokyo) *87*(6), 1625 (1980).
598. Kanda, T., Nakubo, Sh., Wakabayashi, K., and Nisizawa,
      K. *J. Biochem.* (Tokyo) *84*(5), 1217 (1978); CA *90,*
      35 491w.
599. Highley, T. L. *Appl. Environ. Microbiol. 40*(6), 1145
      (1980).
600. Yoshioka, H. and Hayashido, Sh. *Agric. Biol. Chem.*
      44(8), 1729 (1980).
601. Hayashida, Sh. and Yoshioka, H. *Agric. Biol. Chem.*
      44(8), 1721 (1980).
602. Sadana, J. C., Shewale, J. G., and Deshpande, M. V.
      Appl. Env. Microbiol. 39(4), 935 (1980).

603. Schmidt, O. and Liese, W. *Holzforschung* 34(2), 67
      (1980); CA *93*, 64 382d.
604. Desrochers, M., Jurasek, L., and Paice, M. G. *Appl.*
      Environ. Microbiol. 41(1), 222 (1981).
605. Desrochers, M., Jurasek, L., and Paice, M. G. *Devel.*
      Ind. Microbiol. 22, 675 (1981).
606. Fahnrich, P. and Irrgang, K. *Biotechnol. Lett. 4*, 519
      (1982); *4*, 775 (1982).
607. Singh, N. and Kunene, I. S. *Mycologia 72*(1), 182
      (1980); CA *92*, 211 574c.
608. Yoshioka, H., Chavanich, S., Nitobol, N., and Haya-
      shida, Sh. *Agric. Biol. Chem. 45*(3), 579 (1981).
609. Yoshioka, H., Nagato, N., Chavanich, S., Nilubol, N.,
      and Hayashida, Sh. *Agr. Biol. Chem. 45*(11), 2425
      (1981).
610. Meyer, H. P. and Canevascini, G. *Appl. Environ. Micro-*
      biol. 41(4), 924 (1981).
611. Margaritis, A. and Creese, E. *Biotechnol. Lett. 3*(9),
      471 (1981).
611a Margaritis, A., Merchant, R., and Yaguchi, M. *Bio-*
      technol. Lett. 5, 265 (1983).
612. Taniguchi, M., Tanaka, M., Matsuno, R., and Kamikubo,
      T. *J. Ferment. Technol. 58*(2), 143 (1980).
613. Tanaka, M., Taniguchi, M., Morinaga, T., Matsuno, R.,
      and Kamikubo, T. *J. Ferment. Technol. 58*(2), 149
      (1980).
614. Tanaka, M., Taniguchi, M., Matsuno, R., and Kamikubo,
      T. *J. Ferment. Technol. 59*(3), 177 (1981).
615. Trivedi, L. S. and Rao, K. K. *Biotechnol. Lett. 3*(6),
      281 (1981).
616. Desai, J. D., Desai, A. J., and Patel, N. P. *J. Fer-*
      ment. Technol. 60(2), 117 (1982).
617. Tiunova, N. A., Rodionova, N. A. and Martinovich, L.
      I. *Appl. Biochem. Microbiol. 16*(1), 30 (1980).
618. Tiunova, N. A., Rodionova, N. A., Martinovich, L. I.,
      and Gogolev, M. N. *Appl. Biochem. Microbiol. 16*(1),
      137 (1980).
619. Kozlovskaya, L. N., Rodionova, N. A., Akparov, V. Kh.,
      Paseshnichenko, V. A., and Bezborodov, A. M. *Appl.*
      Biochem. Microbiol. 16(1), 34 (1980).
620. Kozlovskaya, L. N., Rodionova, N. A., and Bezborodov,
      A. M. *Appl. Biochem. Microbiol. 16*(4), 394 (1980).
621. Kubackova, M., Karacsonyi, S., and Varadi, J. *Folia*
      Microbiol. 20, 29 (1975).
622. Kubackova, M., Karacsonyi, S., and Toman, R. *Folia*
      Microbiol. 21, 28 (1976).

623. Kubackova, M., Karacsonyi, S., and Varadi, J. *Folia Microbiol. 20,* 29 (1975).

624. Kubackova, M., Karacsonyi, S., Bilisics, L., and Toman, R. *Folia Microbiol. 23*(3), 89 (1978); CA *93,* 643 520u.

625. Kubackova, M., Karacsonyi, S., Bilisics, L., and Toman, R. *Carbohydrate Res. 76,* 177 (1979).

625a Loureiro-Dias, M. C. *Appl. Environ. Microbiol. 44,* 744 (1982).

625b Kassim, B. A. *J. Ferment. Technol. 60,* 381 (1982).

625c Gunasekaran, M. *Mycologia 72,* 759 (1980).

625d Kubackova, M., Karacsonyi, S., and Varadi, J. *Folia Microbiol. 20,* 29 (1975).

625e Kubackova, M., Karacsonyi, S., and Toman, R. *Folia Microbiol. 21,* 28 (1976).

625f Kubackova, M., Karacsonyi, S., Bilisics, L., and Toman, R. *Folia Microbiol. 23,* 202 (1978).

625g Kubackova, M., Karacsonyi, S., Bilisics, L., and Toman, R. *Carbohyd. Res. 76,* 177 (1979).

625h Madan, M. and Sood, P. *Microbios Lett. 12,* 109 (1979).

625i Tong, C. C., Cole, A. L., and Shepherd, M. G. *Biochem. J. 191,* 83 (1980).

625j Kubicek, C. P. and Pitt, D. E. *Eur J. Appl. Microbiol. Biotechnol. 16,* 189 (1982).

625k Moloney, A. P., Considine, P. J., and Coughlan, M. P. *Biotechnol. Bioeng. 25,* 1169 (1983).

626. Biely, P., Kratky, Z., Kockova-Kratochvilova, A., and Bauer, S. *Folia Microbiol. 23*(5), 366 (1978); CA *93,* 43 900f.

627. Biely, P., Kratky, Z., Petrakova, E., and Bauer, S. *Folia Microbiol. 24,* 328 (1979).

628. Biely, P., Vrsanska, M., and Kratky, Z. *Eur. J. Biochem. 108*(1), 313 (1980); CA *93,* 65 534s.

629. Biely, P., Kratky, Z., Vrsanska, M., and Urmamicova, D. *Eur. J. Biochem. 108*(1), 323 (1980).

630. Biely, P., Ursanska, M., and Kratky, Z. *Eur. J. Biochem. 112*(2), 375 (1980); CA *94,* 12 173n.

631. Notario, V., Villa, T. G., and Villanueva, J. R. *J. Gen. Microbiol. 114*(2), 415 (1979).

632. Kratky, Z. and Biely, P. *Eur. J. Biochem. 112*(2), 367 (1980); CA *94,* 43 856s.

633. Stüttgen, E. and Sahm, H. *Eur. J. Appl. Microbiol. 15,* 93 (1982).

634. Aragno, M., Walter-Mauruschat, A., Mayer, F., and Schlegel, H. G. *Arch. Microbiol. 114,* 93 (1977); *114,* 101 (1977).

635.   Conrad, R. and Seiter, W.   *Soil Biol. Biochem. 13*,
       43 (1981).
636.   Conrad, R. and Seiter, W.   *FEMS Microbiol. Lett, 6*,
       143 (1979).
637.   Malik, K. A. and Schlegel, H. G.   *FEMS Microbiol. Lett.
       8*, 101 (1980).
638.   Aragno, M.   *FEMS Microbiol. Lett 3*, 13 (1978).
639.   Goto, E., Kodama, T., and Minoda, Y.   *Agric. Biol.
       Chem. 41*, 685 (1977).
640.   Goto, E., Kodama, T., and Minoda, Y.   *Agric. Biol.
       Chem. 41*, 1305 (1978).
641.   Kawasumi, T., Igarashi, Y., Kodama, T., and Minoda, Y.
       Agric. Biol. Chem. 44, 1985 (1980).
642.   Ohi, K., Matsumoto, H., Yamada, K., Okazaki, M., Kom-
       emushi, S., and Miura, Y.   *J. Ferment. Technol. 57*,
       195 (1979).
643.   Miura, Y., Okazaki, M., Ohi, K., and Nishimura, T.
       Agric. Biol. Chem. 45, 1181 (1981).
644.   Miura, Y., Okazaki, M., Ohi, K., Nishimura, T., and
       Komemushi, S.   *Biotechnol. Bioeng. 24*, 1173 (1982).
645.   Moringa, Y., Yamanaka, Sh , Ishizaki, A., and Hirose,
       Y.   *Agric. Biol. Chem. 42*, 439 (1978).
646.   Kester, T. G., Stasishina, G. N., Kasaeva, G. E., and
       Sid'ko, F. Ya.   *Mikrobiol. 47*, 17 (1978).
647.   Vollbrecht, D., El-Nawawy, M. A., and Schlegel, H. G.
       Eur. J. Appl. Microbiol. Biotechnol. 6, 145 (1978).
648.   Vollbrecht, D. and Schlegel, H. G.   *Eur J. Appl.
       Microbiol. Biotechnol. 6*, 157 (1978).
649.   Zeikus, J. G.   *Enzyme Microb. Technol. 1*, 243 (1979).
650.   Brock, T. D. and Freeze, H.   *J. Bacteriol. 98*, 289
       (1969).
651.   Oshima, T. and Imahori, K.   *Int. J. Syst. Bacteriol.
       24*, 102 (1974).
652.   Jackson, T. J., Ramaley, R. F., and Meinschein, W. G.
       Int. J. Syst. Bacteriol. 23, 28 (1973).
653.   Loginova, L. G., Userbaeva, G. B., and Egorova, L. A.
       Mikrobiol. 47, 881 (1978).
654.   Sasaki, T., Kimura, R., and Arima, K.   *Agr. Biol. Chem.
       36*, 2357 (1972).
655.   Ramaley, R. F. and Bitzinger, K.   *Appl. Microbiol. 30*,
       152 (1975).
656.   Cometta, S., Sonnleitner, B., and Fiechter, A.   *Eur.
       J. Appl. Microbiol. 15*, 69 (1982).
657.   Sonnleitner, B., Cometta S., and Fiechter, A.   *Eur.
       J. Appl. Microbiol. 15*, 75 (1982).

657a. Sonnleitner, B., Cometta, S. and Fiechter, A., *Biotechnol. Bioeng.*, *24*, 2597 (1982).

658. Yoshizaki, F. and Imahori, K., *Agr. Biol. Chem.*, *43*, 537 (1979).

659. Kuhn, H.J., Friederich, U. and Fiechter, A., *Eur. J. Appl. Microbiol.*, *6*, 341 (1979).

660. Kohn, H.J., Cometta, S. and Fiechter, A., *Eur. J. Appl. Microbiol.*, *10*, 303 (1980).

661. Djien, K.S., in "Fermented Foods. Economic Microbiology," (A.H. Rose, ed.), Vol. 7, p 15, Academic Press, London (1982).

661a. Steinkraus, K.H., *Acta Biotechnol.*, *3*, 3 (1983).

661b. Andres, C., *Food Processing, 39*(12), 67 (1978).

662. Mital, B.K. and Steinkraus, K.H., *J. Food Protection*, *42*, 895 (1979).

663. Patel, A.A., Waghmare, W.M. and Gupta, S.K., *Process Biochem.*, *15*(7), 9 (1980).

664. Pinthong, R., Macrae, R. and Rothwell, J., *J. Fd. Technol.*, *15*, 647 (1980); *15*, 653 (1980); *15*, 661 (1980).

665. Patel, A.A. and Gupta, S.K., *J. Food Protection*, *45*, 620 (1982).

666. Beuchat, L.R. and Nail, B.J., *J. Food Science*, *43*, 1109 (1978).

667. Bucker, E.R., Mitchell, J.H. and Johnson, M.G., *J. Food Science*, *44*, 1534 (1979).

667a. Stern, N.J., Hesseltine, C.W., Wang, H.L. and Konishi, F., *Can. Inst. Food Science Technol.*, *10*, 197 (1977).

667b. Kraidej, L., *Food*, *7*, 63 (1975).

668. Feilds, M.L., Hamad, A.M. and Smith, K.D., *J. Food Science*, *46*, 900 (1981).

669. Hamad, A.M. and Fields, M.L., *J. Food Science*, *44*, 456 (1979).

670. Umoh, V. and Fields, M.L., *J. Food Science*, *46*, 903 (1981).

671. Sathe, S.K. and Salunkhe, D.K., *J. Food Science*, *46*, 1374 (1981).

672. Robinson, R.K., *South African J. Dairy Technol.*, *10*, 139 (1978).

673. Cadena, M.A. and Robinson, R.K., *Ecol. Food Nutr.*, *8*, 169 (1979).

674. Anon, (to Nisshin Oil Mills Co. Ltd. Japan) Jap. 5, 432 078 (1979); FSTA *12*, 5H 846.

675. Anon. (to Japan Natural Food Co. Ltd.) Great Brit. Pat. 1539756 (1979); Microbiol. Abstracts 7518-A14.

675a. Ulloa, M., in "GIAM VI Global Impacts of Applied Microbiology," (S.O. Emejuaiwe, O. Ogunbi and S.O. Sanni, eds.) p 45, Academic Press, London (1981).

675b. Ulloa, M. and Herrera, T. in "Advances in Biotechnology,"
      (M. Moo-Young and C.W. Robinson, eds.) Vol. II, p 535,
      Pergamon Press, Toronto (1981).
675c. Nout, M.J.R., in "GIAM VI Global Impacts of Applied
      Microbiology," (S.O. Emejuaiwe, O. Ogunbo and S.O.
      Sanni, eds.) p 169, Academic Press, London (1981).
676a. Okafor, N., in "GIAM VI Global Impacts of Applied
      Microbiology," (S.O. Emejuaiwe, O. Ogunbi and S.O.
      Sanni, eds.) p 61, Academic Press, London (1981).
676b. Meuser, F. and Smolnik, H.D., *Starch, 32,* 116 (1980).
677. Nakamura, I.M. and Park, Y.K., *Starch, 27,* 295 (1975).
678. Ikediobi, C.D. and Onyike, E., *Agric. Biol. Chem., 46,*
      1667 (1982).
679. Okafor, N., *J. Appl. Bacteriol., 42,* 279 (1977).
680. Abe, M.O. and Lindsay, R.C., *J. Food Protect.* 44, 1570
      (1979).
681. Ngaba, P.R. and Lee, J.S., *J. Food Science, 44,* 1570
      (1979).
681a. Wood, B.J.B. in "Mixed Culture Fermentations," (M.E.
      Bushell and J.H. Slater, eds.) p 137, Academic Press,
      London (1981).
682. Laleye, L.C. and Simard, R.E., in "Advances in
      Biotechnology," (M. Moo-Young and C.W. Robinson, eds.)
      Vol. II, p 497, Pergamon Press, Toronto (1981).
683. Laleye, L.C. and Simard, R.E., *Dev. Ind Microbiol., 23,*
      601 (1982).
684. Kuzup, J. and Sistrunk, W.A., *J. Food Science, 47,*
      1001 (1982).
684a. Chen, K.H., McFeeters, R.F. and Fleming, H.P., *J. Food
      Science, 48,* 962 (1983); *48,* 967 (1983); *48,* 972 (1983);
      48, 975 (1983).
685. Maitra, R.K., Chaudhuri, D.R., Mitra, S.N. and Bose,
      A.N., *J. Inst. Chem., 51,* (India) 156 (1979); CA *92,*
      74 592r.
686. Aegeter, P. and Dunlap, C., *Appl. Environ. Microbiol.,*
      39, 937 (1980).
687. Katamine, S., Mamiya, Y., Sekimoto, K., Furukawa, N.
      and Yamanaka, Y., *J. Jap. Soc. Food Science Technol.,*
      25, 311 (1978); FSTA *12,* 3Q 47.
688. Moon, N.J., Beuchat, L.R., Kinkaid, D.T. and Hays, E.R.,
      J. Food Science, 47, 897 (1982).
688a. Rao, J., in "GIAM VI Global Impacts of Applied
      Microbiology," (S.O. Emejuaiwe, O. Ogunbi and S.O.
      Sanni, eds.) p 3, Academic Press, London (1981).
689. Sharpe, M.E., *J. Soc. Dairy Technol., 32*(1), 9 (1979).

690. Kilara, A. and Shahani, K.M., *J. Dairy Science, 61*, 1793 (1978).

691. Tamine, A.Y. and Deeth, H.C., *J. Food Protection, 43*, 939 (1980); *44*, 78 (1981).

692. de Haast, J., Lategan, P.M. and Novello, J.C., *S. Afr. J. Dairy Technol., 11*(1), 11 (1979).

692a. MacBean, R.D., Hall, R.J. and Linklater, P.M., *Biotechnol. Bioeng., 21*, 1517 (1979).

693. Reichart, O., *Acta Alim. Acad. Scient. Hungaricae, 8*, 373 (1979); Microbiol. Abstracts 2009-A16.

693a. Driessen, F.M., in "Mixed Culture Fermentations," (M.E. Bushell and J.H. Slater, eds.) p 99, Pergamon Press, London (1981).

694. Oberman, H. and Libudzisz, Z., *Acta Aliment. Pol., 4*, 201 (1978).

695. Zbikowski, Z., *Zezz. nauk ART Olszt.*, No. 16, 3 (1981).

696. Shahani, K.M. and Chandan, R.C., *J. Dairy Science, 62*, 1685 (1979).

697. Speck, M.L., *Dev. Ind. Microbiol., 19*, 95 (1978).

698. Speck, M.L., *J. Food Protection, 43*, 65 (1980).

699. Sozzi, T. and Smiley, M.B., *Appl. Environ. Microbiol., 40*, 862 (1980).

700. Yoo, J.Y., Min, B.Y., Suh, K.B. and Hah, D.M., *Korean J. Food Science Technol., 10*, 124 (1978); FSTA *11*, 2J 167.

701. Nes, I.F. and Skjelkvale, R., *J. Food Science, 47*, 1618 (1982).

702. Zaika, L.L. and Kissinger, J.C., *J. Food Science, 46*, 1205 (1981).

703. Thomas, T.D., *New Zealand J. Dairy Science Technol., 14*, 12 (1979).

704. Thomas, T.D., Ellwood, D.C. and Longyear, M.C., *J. Bacteriol., 138*, 109 (1979).

705. Thomas, T.D. and Turner, K.W., *Appl. Env. Microbiol., 41*, 1289 (1979).

706. Gotz, F., Sedewitz, B. and Elstner, E.F., *Arch. Microbiol., 125*, 209 (1980); *125*, 215 (1980).

707. Lloyd, G.T., Hillier, A.J., Banlow, I. and Jago, G.R., *Austr. J. Biol. Sciences, 31*, 565 (1978).

708. Zielke, H., Kneifel, H., Webb, L.E. and Soeder, C.J., *Eur. J. Appl. Microbiol. Biotechnol., 6*, 79 (1978).

709. Mikhlin, E.D. and Radina, V.P., *Appl. Biochem. Microbiol., 17*, 253 (1981).

710. Crickenberger, R. G., Henderson, H.E. and Reddy, C.A., *J. Anim. Science, 52*, 677 (1981).

711. Stieber, R.W. and Gerhardt, P., *J. Dairy Science, 62*, 1558 (1979).

712. Stieber, R.W. and Gerhardt, P., *J. Dairy Science, 63,* 722 (1980).

713. Stieber, R.W. and Gerhardt, P., *Biotechnol. Bioeng., 23,* 523 (1981).

714. Stieber, R.W. and Gerhardt, P., *Biotechnol. Bioeng., 23,* 535 (1981).

715. Forney, L.J. and Reddy, C.A., *Develop. Ind. Microbiol., 18,* 135 (1977).

716. Hietala, P.K., Westermarck, H.W. and Jaarma, M., *Nutr. Metab., 23,* 227 (1979).

717. Verbiscar, A.J., Banigan, T.F., Weber, C.W., Reid, B.L., Spencer Swingle, R., Trei, J.E. and Nelson, E.A., *J. Agric. Food Chem., 29,* 296 (1981).

718. Green, J.H., Goldmintz, D., Hale, M.B. and Flynn, D.J., *Develop. Ind. Microbiol., 14,* 310 (1973).

719. Moon, N.J., *J. Sci. Food Agric., 32,* 675 (1981).

720. Au, P.M. and Fields, M.L., *J. Food Science, 46,* 652 (1981).

721. Kazanas, N. and Fields, M.L., *J. Food Science, 46,* 819 (1981).

722. Anon. US Pat. 4,160,038, March 7, 1979, Microbiol. Abstracts 1027-A15.

723. Anon. G.B., Pat 1,549,196, July 25, 1979, Microbiol. Abstracts 1025-A15.

724. Bauchop, T and Clarke, R.T.J., *Appl. Environ. Microbiol., 32,* 417 (1976).

725. Bauchop, T., *Appl. Environ. Microbiol., 38,* 148 (1979).

726. Khan, A.W., Saddler, J.N., Patel, G.B., Calvin, J.R. and Martin, S.M., *FEMS Microbiol. Lett., 7,* 47 (1980).

727. Patel, G.B., Khan, A.W., Agnew, B.J. and Calvin, J.R., *Int. J. Syst. Bacteriol., 30,* 179 (1980).

728. Forsberg, C.W., Beveridge, T.J. and Hellstrom, A., *Appl. Environ. Microbiol., 42,* 886 (1981).

729. Forsberg, C.W. and Groleau, D., *Can. J. Microbiol., 28,* 144 (1982).

730. Groleau, D. and Forsberg, C.W., *Can. J. Microbiol., 27,* 517 (1981); *27,* 1132 (1981).

731. Yu, I. and Hungate, R.E., *Ann. Rech. Vet., 10,* 251 (1979).

732. Khan, A.W., *J. Gen. Microbiol., 121,* 499 (1980).

733. Saddler, J.N. and Khan, A.W., *Can. J. Microbiol., 25,* 1427 (1979); *26,* 760 (1980); *27,* 288 (1981).

734. McKenzie, C.R. and Bilous, D., *Can. J. Microbiol., 28,* 1158 (1982).

735. Gordon, J., Jimenez, M., Cooney, C.L. and Wang, D.I.C., *Amer. Inst. Chem. Eng. Symp. Series, 74*(181), 91 (1978).

736. Ait, N., Creuzet, N. and Forget. P., *J. Gen. Microbiol.*, *113*, 399 (1979).

737. Shinmyo, A., Garcia, D.V. and Demain, A.L., *J. Appl. Biochem.*, *1*, 202 (1979).

738. Garcia, D.V., Shinmyo, A., Madia, A. and Demain, A.L., *Eur. J. Appl. Microbiol. Biotechnol.*, *9*, 189 (1980).

739. Johnson, E.A., Madia, A. and Demain, A.L., *Appl. Environ. Microbiol.*, *41*, 1060 (1981).

740. Gordan, J. and Cooney, C.L., in "Advances in Biotechnology," (M. Moo-Young, and C.W. Robinson, eds.) p 15, Pergamon Press, London (1981).

741. Ng, T.K. and Zeikus, J.G., *Appl. Environ. Microbiol.*, *42*, 231 (1981).

742. Johnson, E.A., Sahajoh, M., Halliwell, G., Madia, A. and Demain, A.L., *Appl. Environ. Microbiol.*, *43*, 1125 (1982).

743a. Becker, E.W., *Process Biochem.*, *16*(5), 10 (1981).

743b. Goldman, J.C., in "Algae Biomass," (G. Shelef and J.C. Soeder, eds.) p 343, Elsevier/North Holland Biomedical Press, Amsterdam (1980).

744. Oswald, W.J., in "Microbial Conversion Systems for Food and Fodder Production and Waste Management," (T.G. Overmire, ed.) KISR, Kuwait (1977).

745. Taiganides, E.P., Chou, K.C. and Lee, B.Y., *Agr. Wastes*, *1*, 129 (1979).

746. Sekoulov, I.D., *Prog. Water Technol.*, *11*, 36 (1979).

747a. Shillinglaw, S.N. and Pieterse, A.J.H., *Water S.A.*, *3*, 183 (1977); *6*, 186 (1980).

747b. Shilo, M., in "Algae Biomass," (G. Shelef and C.J. Soeder, eds.) p 384, Elsevier/North Holland Biomedical Press, Amsterdam (1980).

748. Hill, D.T. and Lincoln, E.P., *Agr. Wastes, 3*, 43 (1981).

749. Essenberg, D.M., Koopman, B., Benemann, J.R. and Oswald, W.J., *Biotechnol. Bioeng. Symp.*, *111*, 429 (1981).

750. Benemann, J.R., Koopman, B., Weissman, J., Eisenberg, D. and Goebel, R., in "Algae Biomass," (G. Shelef and C.J. Soeder, eds.) p 457, Elsevier/North Holland Biomedical Press, Amsterdam (1980).

751a. Edwards, P., in "Algae Biomass," (G. Shelef and C.J. Soeder, eds.) p 191, Elsevier/North Holland Biomedical Press, Amsterdam (1980).

751b. Watson, A.S., "Aquaculture and algae culture. Process and products," Noyes Data Corp., Park Ridge, NJ (1979).

752. Dilov, Kh., *Khidrobiol.*, *10*, 3 (1979).

753. Soeder, C.J., *Hydrobiol.*, *72*, 197 (1980).

754a. Moraine, R., Shelef, G., Meydau, A. and Levi, A.,
      Biotechnol. Bioeng., 21, 1191 (1979).
754b. Azov, Y., Shelef, G., Moraine, R. and Levi, A., in
      "Algae Biomass," (G. Shelef and C.J. Soeder, eds.)
      p 245, Elsevier/North Holland Biomedical Press,
      Amsterdam (1980).
754c. Oron, G., Shelef, G. and Levi, A., *Hydrobiol., 77,*
      167 (1981).
755. Oron, G., Shelef, G., Levi, A., Meydau, A. and Azov,
      Y., *Applied Environ. Microbiol., 38,* 570 (1979).
756. Moraine, R., Shelef, G., Sandbank, E., Bar-Moshe, Z.
      and Shvantzbord, L., in "Algae Biomass," (G. Shelef
      and C.J. Soeder, eds.) p 531, Elsevier/North Holland
      Biomedical Press, Amsterdam (1980).
757. Mohn, F.H., in "Algae Biomass," (G. Shelef and C.J.
      Soeder, eds.) p 547, Elsevier/North Holland Biomedical
      Press, Amsterdam (1980).
758. Betzer, N., Argamon, Y. and Kott, Y., *Water Res., 14,*
      1003 (1980).
759a. Azov, Y., Shelef, G., Moraine, R. and Oron, G., in
      "Algae Biomass," (G. Shelef and C.J. Soeder, eds.)
      p 523, Elsevier/North Holland Biomedical Press,
      Amsterdam (1980); *J. Env. Eng. Div., 108* (EE4), 730
      (1982).
759b. Oron, G. and Shelef, G., in "Algae Biomass," (G. Shelef
      and C.J. Soeder, eds.) p 497, Elsevier/North Holland
      Biomedical Press, Amsterdam (1980).
760. Azov, Y., Shelef, G. and Narkis, N., *Appl. Environ.
      Microbiol., 43,* 491 (1982).
761. Azov, Y., Shelef, G. and Moraine, R., *Biotechnol. Bioeng.,
      24,* 579 (1982).
762. Dor, I. and Svi, B., in "Algae Biomass," (G. Shelef and
      C.J. Soeder, eds.) p 421, Elsevier/North Holland
      Biomedical Press, Amsterdam (1980).
762a. Azov, Y. and Goldman, J.C., *Appl. Environ. Microbiol.,
      43,* 735 (1982).
763. Persoone, G., Morales, J., Verlet, H. and de Pauw, N.,
      in "Algae Biomass," (G. Shelef and C.J. Soeder, eds.)
      p 505, Elsevier/North Holland Biomedical Press,
      Amsterdam (1980).
764. Miller, S.B. and Buhr, H.O., *Water, S.A., 7,* 8 (1981).
765. Berend, J., Simovitch, E. and Ollian, A., in "Algae
      Biomass," (G. Shelef and C.J. Soeder, eds.) p 799,
      Elsevier/North Holland Biomedical Press, Amsterdam
      (1980).

766. Dubinsky, Z., Aaronson, S. and Berner, T., in "Algae
     Biomass," (G. Shelef and C.J. Soeder, eds.) p 819,
     Elsevier/North Holland Biomedical Press, Amsterdam
     (1980).
767. Soeder, C.J., in "Algae Biomass," (G. Shelef and C.J.
     Soeder, eds.) p 9, Elsevier/North Holland Biomedical
     Press, Amsterdam (1980).
768. Nigam, B.P., Ramanathan, P.K. and Venkataram, L.V.,
     Biotechnol. Lett., *2*, 467 (1980).
769. Becker, E.W. and Venkataraman, L.V., in "Algae Biomass,"
     (G. Shelef and C.J. Soeder, eds.) p 34, Elsevier/North
     Holland Biomedical Press, Amsterdam (1980).
770. Sinchumpasak, O., in "Algae Biomass," (G. Shelef and
     C.J. Soeder, eds.) p 115, Elsevier/North Holland
     Biomedical Press, Amsterdam (1980).
771. Castillo, J., Merino, F. and Heussler, P., in "Algae
     Biomass," (G. Shelef and C.J. Soeder, eds.) p 123,
     Elsevier/North Holland Biomedical Press, Amsterdam
     (1980).
772. Heussler, P., Castillo, J., Merino, M. and Merino, M.F.,
     Ergebnisse der Limnol., *11*, 17 (1978).
773. Becker, E.W., in "Algae Biomass," (G. Shelef and C.J.
     Soeder, eds.) p 767, Elsevier/North Holland Biomedical
     Press, Amsterdam (1980).
774. Payer, H.D., Pithakpol, B., Nguitragool, M., Prabharaksa,
     C., Thananunkul, D. and Chavanna, S., *Ergebnisse der
     Limnol.*, *11*, 41 (1978).
775. Payer, H.D., Pabst, W. and Runkel, K.H., in "Algae
     Biomass," (G. Shelef and C.J. Soeder, eds.) p 787,
     Elsevier/North Holland Biomedical Press, Amsterdam
     (1980).
776. Shamala, T.R., Drawert, F. and Leupold, G., *Biotechnol.
     Bioeng.*, *24*, 287 (1982).
777. Shamala, T.R., Drawert, F. and Leupold, G., *Biotechnol.
     Bioeng.*, *24*, 1301 (1982).
778. Shamala, T.R., Drawert, F., Leupold, G. and Becker, W.,
     Biotechnol. Bioeng., *24*, 1183 (1982).
779. Gons, H.J. and Mur, L.R., *Arch. Microbiol.*, *125*, 9
     (1980).
780. Gons, H.J. and Mur, L.R. *Arch. Hydrobiol.*, *85*, 41 (1979).
781. Gavis, J., Chamberlin, C. and Lystad, L.D., *J. Phycol.*,
     15, 273 (1979).
782. Radmer, R. and Ollinger, O., *Plant Physiol.*, *65*, 723
     (1980).
783. Sandmann, G. and Boger, P., *Z. Pflanzen Physiol.*, *98*,
     53 (1980).

783a. Petersen, R., *Env. Science Technol.*, *16*, 443 (1982).

784. Goldman, J.C. and Graham, S.J., *Appl. Environ. Microbiol.*, *41*, 60 (1981).

785. Goldman, J.C., Dennett, M.R. and Riley, C.B., *Biotechnol. Bioeng.*, *24*, 619 (1979).

786. Livansky, K., Veber, K. and Prokes, B., *Acta Hydrochim. Hydrobiol.*, *7*, 369 (1979).

787. Dihoru, A., Cadar, D. and Vintila, M., *Rev. Roum Biologie Serie Biol. Vegetal.*, *24*, 27 (1979).

788. Georgiev, D., Dilov, Kh. and Avramova, S., *Khidrobiol.*, *7*, 14 (1978).

789. Bozhkova, M., Iordanova, S., Dilov, Kh., Karanova, S. and Runev, I., *Khidrobiol.*, *10*, 34 (1979).

790. Barna, A., Nagy-Toth, F., Bercea, V. and Stirban, M., *Stud. Univ. Babes-Bolyai Biol.*, *22*, 31 (1977); CA *88*, 101465c.

791. Dvorakova, J., *Biol. Plant*, *21*, 372 (1979); CA *91* 207162d.

792. Vilicic, D., *Acta Bot. Croat.*, *38*, 45 (1979); Microbiol. Abstracts 5110-C9.

793. Langowska, I. and Zawadzki, Z., *Pol. Arch. Hydrobiol.*, *26*, 257 (1979); Microbiol. Abstracts 6131-C9.

794. Guimarals, Robbs, P., Rosemberg, J.A. and de Oliveira Dias, G., *Rev. lat-amer. Microbiol.*, *21*, 181 (1979).

795. Abeliovich, A., in "Algae Biomass," (G. Shelef and C.J. Soeder, eds.) p 204, Elsevier/North Holland Biomedical Press, Amsterdam (1980).

796. Oron, G., Shelef, G. and Levi, A., *Biotechnol. Bioeng.*, *23*, 2185 (1981).

797. Azov, Y., *Appl. Environ. Microbiol.*, *43*, 1300 (1982).

789. Toerien, D.F. and Grobbelaar, J.V., in "Algae Biomass," (G. Shelef and C.J. Soeder, eds.) p 73, Elsevier/North Holland Biomedical Press, Amsterdam (1980).

799. von der Decken, A., in "Algae Biomass," (G. Shelef and C.J. Soeder, eds.) p 661, Elsevier/North Holland Biomedical Press, Amsterdam (1980).

800. Walz, O.P. and Brune, H., in "Algae Biomass," (G. Shelef and C.J. Soeder, eds.) p 733, Elsevier/North Holland Biomedical Press, Amsterdam (1980).

801. Brune, D.E. and Novak, J.T., in "Fuels from Biomass and Waste," (D.L. Klass and G.H. Emert, eds.) p 99, Ann Arbor Science, Ann Arbor (1981).

802. Markl, H., in "Algae Biomass," (G. Shelef and C.J. Soeder, eds.) p 360, Elsevier/North Holland Biomedical Press, Amsterdam (1980).

803. Panikov, N., *J. Chem. Tech. Biotechnol.*, *29*, 442 (1979).

804. Lowenstein, A. and Bachofen, R., *Arch. Microbiol.*, *116*, 169 (1978).

805. Panikov, N. and Pirt, S.J., *J. Gen. Microbiol.*, *108*, 295 (1978).

806. Ogawa, T., Fujii, T. and Aiba, S., *Arch. Microbiol.*, *127*, 25 (1980).

807. Shelp, B.J. and Canvin, D.T., *Plant Physiol.*, *65*, 780 (1980).

808. Pirt, S.J., Lee, Y.K., Richmond, A. and Watts Pirt, M., *J. Chem. Tech. Biotechnol.*, *30*, 25 (1980).

809. Lorenzen, H., in "Algae Biomass," (G. Shelef and C.J. Soeder, eds.) p 411, Elsevier/North Holland Biomedical Press, Amsterdam (1980).

810. Miyachi, Sh. and Shiraiwa, Y., *Plant Cell Physiol.*, *20*, 341 (1979).

811. Hogetsu, D. and Miyachi, D., *Plant Cell Physiol.*, *20*, 747 (1979).

812. Shelp. B.J. and Canvin, D.T., *Plant Physiol.*, *65*, 774 (1980).

813. Goldman, J.C., Dennett, M.R. and Riky, C.B., *Biotechnol. Bioeng.*, *23*, 995 (1981).

814. Ogawa, T. and Aiba, S., *Biotechnol. Bioeng.*, *23*, 1121 (1981).

815. Barroso, C.V. and Nonato, A.M. de J., *Rev. Microbiol.*, *9*, (Brasil), 84 (1978); CA *90*, 36 168h.

816. Brueckner, U. and Hoefner, W., *Z. Pflanzernernaehr. Bodenkd.*, *140*, 441 (1977); CA *87*, 130 221e.

817. Berenstein, A.F. and Kostlan, N.V., *Ferment. Spirtovaya Promyshlennost*, No. 2, 26 (1978); CA *88*, 150 572u.

817a. Toledo, A.P.P., Tundisi, J.G. and D'Aquino, V.A., *Hydrobiol.*, *71*, 261 (1980).

818. De Franca, F.P. and Yuan, M.C., *Rev. Microbiol.*, *9*, (Brasil), 88 (1978).

819. El-Ayouty, E.Y., Khalil, A.I. and El-Asadi Mohamed, A., *Egypt J. Bot.*, *22*, 111 (1979); CA *94*, 170 795w.

820. Aiba, S., Ohtaka, H. and Inoue, T., *Biotechnol. Lett.*, *1*, 245 (1979).

821. Komor, E., Schwab, W.G.W. and Tanner, W., *Biochim. Biophys. Acta.*, *555*, 524 (1979).

822. Komor, E., *Biochem. Soc. Trans.*, *8*, 861 (1980).

823. Spedding, D.J. Ziegler, I., Hampp, R. and Ziegler, H., *Z. Pflanzenphysiol.*, *97*, 205 (1980).

824. Nakajima, A., Horikoshi, T. and Sakaguchi, T., *Europ. J. Appl. Microbiol., Biotechnol.*, *12*, 76 (1981).

825. Nakajima, A., Horikoshi, T. and Sakaguchi, T., *Agr. Biol. Chem.*, *45*, 903 (1981).

826. Sakaguchi, T., Nakajima, A. and Horikoshi, T., *Europ. J. Appl. Microbiol. Biotechnol.*, *12*, 84 (1981).

827. Horikoshi, T., Nakajima, A. and Sakaguchi, T., *Agr. Biol. Chem.*, *45*, 781 (1981).

828. Kawaguchi, K., in "Algae Biomass," (G. Shelef and C.J. Soeder, eds.) p 24, Elsevier/North Holland Biomedical Press, Amsterdam (1982).

829. Hu, W.S., Jeng, H.P., Wang, H.H. and Lii, C.Y., *Bull. Inst. Chem. Acad. Sin.*, *25*, 1 (1978).

830. Sivalingam, P.M., Wuan, T.O., Fusetani, N. and Watanabe, T., *Prog. Water Technol.*, *11*, 69 (1979).

831. Wong, M.H., *Arch. Hydrobiol.*, *88*, 426 (1980).

832. de Franca, F.P., Gomes, M.P.D., Alvernaz, M.T.G. and SilvaFilho, M.S., *Rev. lat. Microbiol.*, *20*, 235 (1978).

833. Cheung, Y.A. and Wong, M.H., *Agr. Wastes*, *3*, 109 (1981).

834. Lincoln, E.P. and Hill, D.T., in "Algae Biomass," (G. Shelef and C.J. Soeder, eds.) p 229, Elsevier/ North Holland Biomedical Press, Amsterdam (1980).

834a. Koopman, B. and Lincoln, E.P., *Agr. Wastes*, *5*, 231 (1983).

835. Malis Arad, Sh., Friedlander, M., Benarie, R. and Richmond, A.E., *Plant Cell Physiol.*, *21*, 27 (1980).

836. Malis Arad, Sh., McGowan, R.E., Friedlander, M. and Richmond, A.E., in "Algae Biomass," (G. Shelef and C.J. Soeder, eds.) p 400, Elsevier/North Holland Biomedical Press, Amsterdam (1980).

837. Regan, D.L. and Skicko, J., *Eur. J. Appl. Microbiol. Biotechnol.*, *7*, 155 (1979).

838. Frasinel, C., Patterson, G.W. and Dutky, S.R., *Phytochem.*, *17*, 1567 (1978).

839. Wright, D.C., Berg, L.R. and Patterson, G.W., *Phytochem.*, *19*, 783 (1980).

840. Picard, G.A., de la Noue, J., Piette, J.P. and Kirouac, C., *Water Res.*, *13*, 1203 (1979).

841. de la Noue, J., Picard, G.A., Pitte, J.P. and Kirovac, C., *Water Res.*, *14*, 1125 (1980).

842. Chan, K.Y., Ting, Y.P., Chiu, Y.N. and Wong, P.K., Chemosphere, 8, 205 (1979).

843. Wong, M. and Lau, K., *Chemosphere*, *8*, 217 (1979).

844. Reynolds, J.H., Middlebrooks, E.J., Porcello, D.B., and Greeney, W.J., *J. Water Pollut. Control Fed.*, *47*, 2420 (1975).

845. Claesson, A., *Physiol. Plant 48*, 59 (1980).

846. Nicolas, P., Freyssinet, G. and Nigon, V., *Plant Physiol.*, *65*, 631 (1980).

847.   Freyssinet, G. and Nigan, V., *Eur. J. Appl. Microbiol. Biotechnol., 9*, 295 (1980).

848.   Oda, Y., Nakano, Y.and Kitaoka, S., *J. Gen. Microbiol. 128*, 853 (1982).

849.   Oda, Y., Nakano, Y. and Kitaoka, S., *Agric. Biol. Chem., 46*, 2153 (1982).

850.   de Filippis, Hampp, R. and Ziegler, H., *Arch. Microbiol., 128*, 407 (1981).

851.   Wong, M.H., *Env. Poll., 21*, 121 (1980).

852.   Villanueva, V.R., Adlakha, R.C. and Calvayrac, R., *Phytochem., 19*, 787 (1980); *19*, 962 (1980).

853.   Gibbs, N. and Duffus, C.M., *Appl. Env. Microbiol., 31*, 602 (1976).

854.   Williams, L.A., Foo, E.L., Foo, A.S., Kuhn, F. and Heden, C.G., *Biotechnol. Bioeng. Symp., 8*, 115 (1978).

855.   Chen, B.J. and Chi, C.H., *Biotechnol. Bioeng., 23*, 1267 (1981).

856.   Miura, Y., Yaki, K., Shoga, M. and Miyamoto, K., *Biotechnol. Bioeng., 24*, 1555 (1982).

857.   Miura, Y., Yagi, K., Nakano, Y. and Miyamoto, K., *J. Ferment. Technol., 59*, 441 (1981).

858.   Badger, M.R., Kaplan, A. and Berry, J.A., *Plant Physiol., 66*, 407 (1980).

859.   Berdykulov, Kh.A and Nurieva, D., *Uzb. Biol. Zh.*, (1) 34 (1980); CA *92*, 196 336f.

860.   Florencio, F.J. and Vega, J.M., *Phytochem., 21*, 1195 (1982).

861.   Mokady, S., Yannai, S., Einav, P. and Berk, Z., in "Algae Biomass," Biomedical Press, Amsterdam (1980).

862.   Lipstein, B. and Horwitz, S., in "Algae Biomass," (G. Shelef and C.J. Soeder, eds.) p 667, Elsevier/ North Holland Biomedical Press, Amsterdam (1980).

863.   Sandbank, E. and Hepher, B., in "Algae Biomass," (G. Shelef and C.J. Soeder, eds.) p 697, Elsevier/North Holland Biomedical Press, Amsterdam (1980).

864.   Yannai, S., Mokady, S., Sachs, K., Kantorowitz, B. and Berk, Z., in "Algae Biomass," (G. Shelef and C.J. Soeder, eds.) p 756, Elseiver/North Holland Biomedical Press, Amsterdam (1980).

865.   Ogawa, T., Fujii, T. and Aiba, S., *Biotechnol. Bioeng., 20*, 1493 (1978).

866.   Lee, Y.K. and Pirt, S.J., *Biotechnol. Bioeng., 24*, 507 (1982).

867.   Lee, Y.K. and Pirt S.J., *J. Gen. Microbiol., 124*, 43 (1981).

868.   Watts Pirt, M. and Pirt, S.J., *J. Gen. Microbiol., 119* 321 (1980).

869.   Kreuzberg, K. and Hempfling, W., in "Advances in
       Biotechnology," (M. Moo-Young, C.W. Robinson and C.
       Vezina, eds.) Vol. I, p 287, Pergamon Press, Toronto
       (1981).
870.   Ten Hoopen, H.J.G., Roels, I.A., van Gemert, J.M.,
       Novel, P.J. and Fuchs, A., in "Advances in Biotechnology,"
       (M. Moo-Young, C.W. Robinson and C. Vezina, eds.)
       Vol. I, p 315, Pergamon Press, Toronto (1981).
871.   Banniester, T.T., *Limnol. Oceanogr., 24,* 76 (1976).
872.   Brown, E.J., Harris, R.F. and Koonce, J.F., *Limnol.
       Oceanogr., 23,* 26 (1978).
873.   Brown, E.J. and Button, D.K., *J. Phycol., 15,* 305 (1979).
874.   Rhee, G.Y., *Limnol. Oceanogr., 23,* 10 (1978).
875.   Follmann, H., Mankl, H. and Vortmeyer, D., *Chem. Ing.
       Tech., 50,* 319 (1978).
876.   Stephanopoulos, G. and Fredrickson, A.G., *Bull. Math.
       Biol., 41,* 525 (1979).
877.   Brune, D.E. and Novak, J.T., *Eur. J. Appl. Microbiol.
       Biotechnol., 13,* 71 (1981).
878.   Neilsson, A.H. and Larsson, T., *Physiol. Plantarum, 48,*
       542 (1980).
879.   Humphrey, G.F., *J. Exp. Mar. Biol. Ecol., 40,* 63 (1979).
880.   Harnung, U., von Witsch, H. and Munz, A., in "Advances
       in Biotechnology," (M. Moo-Young and C.W. Robinson, eds.)
       Vol. II, p 663, Pergamon Press , Toronto (1981).
881.   Grant, W.D. and McMurtry, C.M., in "Microbiol Ecology,"
       (M.W. Loutit and J.A.R. Miles, eds.) p 427, Springer-
       Verlag, Berlin (1978).
882.   Yamada, T. and Sakaguchi, K., *Arch. Microbiol. 124,*
       161 (1980).
883.   De Luca, P., Gambardella, R. and Merola, A., *Bot. Gaz.,
       140,* 418 (1979).
884.   Nagashima, H. and Fukuda, I., *Phytochem., 20,* 439 (1981).
885.   Largeau, C., Casadevall, E. and Berhaloff, C., *Phytochem.
       19,* 1081 (1980).
886.   Wettern, M., *Phytochem., 19,* 513 (1980).
887.   Harrison, D.E.F., *Adv. Appl. Microbiol., 24,* 129 (1978).
888.   Samuelov, N.S., *Adv. Biotechnol. Process., 1,* 293 (1983).
889.   Yanagimoto, M., Tahara, H. and Saitoh, H., Report Nat.
       Food Res. Inst. No 36, 84 (1980).
890.   Lee, Y.K., in "Mixed Culture Fermentations," (M.E.
       Bushell and J.H. Slater, eds.) p 151, Academic Press,
       London (1981).
891.   Lee, Y.K. and Pirt, S.J., *J. Chem. Technol. Biotechnol.,
       31,* 295 (1981).

892. Escher, A. and Characklis, W.G., *Biotechnol. Bioeng.*, *24*, 2283 (1982).

893. Fritz, J.J., Middleton, A.C. and Meredith, D.D., *J. Wat. Poll. Control Fed.*, *51*, 2724 (1979).

894. Brockett, O.D., *Prog. Water Technol.*, *9*, 941 (1978).

895. Bollman, R.C. and Robinson, G.C., *J. Phycol.*, *13*, 1 (1977).

896. Berger, P.S., Rho, J. and Gunner, H.B., *Water Res.*, *13*, 267 (1979).

897. Nalewajko, C., Lee, K and Fay, P., *Microb. Ecol.*, *6*, 199 (1980).

897a. Nalewajko, C., in "Aquatic Microbial Communities," (J. Cairns, ed.) p 589, Garland Publishing Inc., New York (1977).

898. Chrost, R.J., *Acta Microbiol. Pol.*, *27*, 55 (1978).

898a. Meffert, M.E. and Overbeck, J., *Arch. Hydrobiol.*, *87*, 118 (1979).

899. Caldwell, D.E. and Caldwell, S.J., *Can. J. Microbiol.*, *24*, 922 (1978).

900. Schiefer, G.E. and Caldwell, D.E., *Appl. Environ. Microbiol.*, *44*, 84 (1982).

901. Lange, W., *Can. J. Microbiol.*, *22*, 1181 (1976).

902. Tison, D.L. and Lign, A.J., *Can. J. Microbiol.*, *25*, 1315 (1979).

903. Lange, W., *Can. J. Microbiol.* *20*, 1311 (1974).

904. Murphy, T.P., Lean, D.R.S. and Nalewajko, C., *Science*, *192*, 200 (1976).

905. Vincent, W.F. and Silvester, W.B., *Water Res.*, *13*, 711 (1979); *13* 717 (1979).

906. Lam, C.W.Y. and Silvester, W.B., *Hydrobiol.*, *63*, 135 (1979).

907. Goldman, J.C. and Shapiro, J., *Science, 182,* 306 (1973).

908. Schindler, D.W., *Science, 195,* 260 (1977).

909. Silvey, J.K.G. and Wyatt, J.T., in "Aquatic Microbial Communities," (J. Cairns, ed.) p 161, Garland Publishing Inc., New York (1977).

910. Groeneweg, J., Khein, B., Mohn, F.L., Runkel, K.H. and Stengel, E., in "Algae Biomass," (G. Shelef and C.J. Soeder, eds.) p 255, Elsevier/North Holland Biomedical Press, Amsterdam (1980).

911. Balloni, W., Filipi, C. and Florenzano, G., in "Algae Biomass," (G. Shelef and C.J. Soeder, eds.) p 217, Elsevier/North Holland Biomedical Press, Amsterdam (1980).

912. Fujii, T., Honda, Y., Ando, A. and Yabuki, M., *Agric. Biol. Chem.*, *46*, 2209 (1982).

913. Mostecky, J., Barta, J., Nemec, M., Sara, R., Sliva,
     O., Stros, F. and Vernerova, J., Czech. Patent
     174,431, October 1978; CA *90*, 70610f.
914. Rychtera, M., *Dechema Monograph, 83,* 53 (1979).
915. Lines, K.K., U.S. Pat. 4,144,132, March 1979; CA *90*,
     166 582z.
916. Wheatley, A.D., Mitra, R.I. and Hawkes, H.A., *J. Chem.
     Tech. Biotechnol., 32,* 203 (1982).
917. Fedurina, M.D., *Gidroliz. Lesokhim. Prom. st.,* (7)
     14 (1978); CA *90*, 85 249u.
918. Ali, A.H., Hameed, K.M. and Abu-Tabikh, A.T., *Ber. Dtsch.
     Bot. Ges., 91,* 305 (1979).
919. Opoku, A.R. and Adoga, P.A., *Enzyme Microb. Technol., 2,*
     241 (1980).
920. Barker, T.W., Quinn, J.P. and Merchant, R., *Eur. J.
     Appl. Microbiol., 14,* 247 (1982).
921. Illanes, A. and Schaffeld, G., *Biotechnol. Lett., 5,*
     305 (1983).
922. Moebus, O and Teuber, M., *Kieler Milchwirtsch. Forsch.,
     31,* 297 (1979).
923. Weber, J., Glanser, M. and Ban S., in "Conf. Papers
     Third Int. Conf. Effluent Treatment in Biochem. Ind.,"
     London, November 1979 (pages are not numbered).
924. Ban, S., Glanser, M. and Smailagic, M., *Biotechnol.
     Bioeng., 21,* 1917 (1979).
925. Glanser, M., Ban, S. and Cooney, C., *Eur. J. Appl.
     Microbiol. Biotechnol., 13,* 54 (1981).
926. Therkildgen, M., *Biotechnol. Lett., 2,* 117 (1980).
927. Kargi, F. and Shuler, M.L., *Biotechnol. Lett., 3,* 409
     (1981).
927a. Poznanski, S., Benardski, W., Leman, J., Kowalewska, J.
     and Mieczkowski, M., *Agr. Wastes, 6,* 77 (1983).
928. Voegeli, H.F. and Cousminer, J.J., *Int. Biodeterior.
     Bull., 14,* 119 (1978).
929. Morinaga, Y., Chiba, M., Yamamoto, S., Takinami, K.
     and Hirose, Y., *Agr. Biol. Chem., 44,* 1267 (1980).
930. Hsu, C.C. and Thayer, D.W., *Appl. Environ. Microbiol.,
     44,* 339 (1982).
931. Kargi, F., Shuler, M.L., Vashon, R., Seeley, H.W.,
     Henry, A. and Austic, R.E., *Biotechnol. Bioeng., 22,*
     1567 (1980).
932. Kargi, F. and Shuler, M.L., *Agr. Wastes, 2,* 1 (1980).
933. Heddle, J.F., *Water Res., 13,* 581 (1979).
934. Evans, M.R., Hisset, R., Smith, M.P.W., Thacker, F.E.,
     Williams, A.G., *Agr. Wastes, 2,* 93 (1980).

935.   Poznanski, S., Andrzejewaska, J., Bednarski, W.,
       Tywonczuk, J. and Leman, J., *Przem. Ferment. Polny.*,
       21, 6 (1977).
936.   Brown, D.E., Boardman, T.D. and Reddington, S.W.,
       Agr. Wastes, 2, 185 (1980).
937.   Garraway, J.L., *Agr. Wastes, 4,* 131 (1982).
938.   Golcz, A., Oleszkiewics, A. and Jankowski, K., *Agr.
       Wastes, 4,* 285 (1982).
939.   Hissett, R., Deans, E.A. and Evans, M.R., *Agr. Wastes,
       4,* 477 (1982).
940.   Somerville, H.J., in "Mixed Culture Fermentations,"
       (M.E. Bushel and J.H. Slater, eds.) p 81, Academic
       Press, London (1981).
941.   Dallas-Newton, J. and Forster, C.F., *Environ. Technol.
       Letters, 1,* 31 (1980).
942.   Houtmeyers, J., van den Eynde, E., Poffe, R. and
       Verachtert, H., *Eur. J. Appl. Microbiol. Biotechnol.,
       9,* 63 (1980).
943.   Verachtert, H., van den Eynde, E., Poffe, R. and
       Houtmeyers, J., *Eur. J. Appl. Microbiol. Biotechnol.,
       9,* 137 (1980).
944.   Jacobson, S.N., O'Mara, N.L. and Alexander, M., *Appl.
       Environ. Microbiol., 40,* 917 (1980).
945.   Shimizu, N. and Odawara, Y., *J. Ferment. Technol., 59,*
       391 (1981).
946.   Vavilin, V.A. and Vasiliev, V.B., *Ecol. Modelling, 9,*
       57 (1980).
947.   Cocci, A.A., McKim, M.P., Landine, R.C. and Viraraghavan,
       T., *Agr. Wastes, 2,* 273 (1980).
948.   Seal, K.J. and Eggins, H.O.W., in "Food From Wastes,"
       (G.G. Berch, K.J. Parker and J.T. Worgan, eds.) p 58,
       Applied Science Publishers Ltd., London (1976).
949.   Thayer, D.W. and David, C.A., *App. Environ. Microbiol.,
       36,* 291 (1978).
950.   Thayer, D.W., Yang, S.P. and Ou, S.L., *Dev. Ind.
       Microbiol., 19,* 385 (1978).
951.   Wu, Y.Q., Jing, G.A., Fei, J.X. and Ma, D.L., *Wei
       Sheng Wu Hsueh Pao, 19,* 400 (1979); CA *92*, 92685c.
952.   De Leon, C.A. and Jason, L.M., *Acta Manilana Ser. A.,
       29*(19) 75 (1980).
953.   Chang, W.T.H., Hsu, W.H., Lai, M.N. and Chang, P.P.,
       Dev. Ind. Microbiol., 21, 313 (1980).
954.   Molina, O.E., Callieri, D.A., Perotti de Galvez, N.I.,
       Rev. lat-amer. Microbiol., 22, 123 (1980).
955.   Pierotti de Galvez, N.I. and Molina, O.E., *Biotechnol.
       Lett., 3,* 717 (1981).

956.   Choudhury, N., Dunn, N.W. and Gray, P.P., *Biotechnol. Lett., 3*, 493 (1981).

957.   Marsden, W.L., Morjanoff, P., Fox, D., Gray, P.O. and Dunn, N.W., *Proc. 5th Australian Biotechnol. Conf., 33*, (1982).

958.   Viesturi, U.E., Apsite, A.F., Laukevics, J.J., Ose, V.P. Bekers, M.J. and Tengerdy, R.P., *Biotechnol. Bioeng. Symp., 11*, 359 (1981).

959.   Ringpfeil, M., Beck, D., Hadeball, W., Kreuter, Th. and Heinritz, H.J., in "GIAM VI Global Impacts of Applied Microbiology," (S.O. Emejuaiwe, O. Ogunbo and S.O. Sanni, eds.) p 233, Academic Press, London (1981).

960.   de la Torre, M.M., *Ciencia y Desarrollo, 7*, (Mexico) (37), 111 (1981).

961.   de la Torre, M.M., *Conservation & Recycling, 5*, 41 (1982).

962.   Sarkar, C. and Trabhu, K.A., *J. Ferment. Technol., 60*, 297 (1982).

963.   Han, Y.W., *J. Ferment. Technol., 60*, 99 (1982).

964.   Wilson, D.U. and Thayer, D.W., *Dev. Ind. Microbiol., 23*, 337 (1982).

965.   Thouvenot, D., Abily, B. and Moreau, M., *Rev. Mycol., 42*, 347 (1978).

966.   Haug, R.T., in "Compost Engineering Principles and Practices," Ann Arbor Science, Ann Arbor, MI (1980).

967.   Hughes, E.G., in "Handbook of Organic Wastes Conversion," (M.W.M Bewick, ed.) p 108, Van Nostrand Reinhold Co., New York (1980).

968.   Goluecke, C.G., *Biocycle, 22*(4), 36 (1981).

969.   Colacicco, D., *Biocycle, 23*(5), 26 (1982).

970.   Crawford, J.H., *Process Biochem., 18*(1), 14 (1983).

971.   Kuster, E. and Safwat, M.S.A., *Zent. Bakteriol. II, 133*, 385 (1978).

972.   Griftis, C.L. and Mote, C.R., *Arkansas Farm Res., 27*(4), 3 (1978).

973.   Niese, G. and Neumeyer-Seekatz, S., *Z. Pflanzen. Bodenkd., 142*, 310 (1979).

974.   Goluecke, C.G., Lafrenz, D., Chaser, B. and Diaz, L.F., *Comp. Science/Land Utilization, 21*(4), 42 (1980).

975.   de Vleeschauwer, D., Verdonck, O. and de Boodt, M., *Resource Recovery Conservation, 5*, 285 (1980).

976.   Chang, C.S., Lai, F.S. and Miller, B.S., *Trans. Am. Soc. Agr. Eng., 23*, 709 (1980).

977.   Safwat, M.S.A., *Compost Science/Land Utilization, 21*, 27 (1980).

978.   Strom, P.F., Morris, M.L., Finstein, M.S., *Compost Science/Land Utilization, 21,* 44 (1980).

979.   Gouin, F.R. and Shanks, J.B., *Biocycle, 22,* 41 (1981).

980.   Park, Y.K. and Castro Gomez, R.J.H., *Sugar J., 45*(5), 14 (1982).

981.   Hyde, M.A. and Consolazio, G.A., *Biocycle, 23,* 58 (1982).

982.   Rathbun, B.L. and Shuler, M.L., *Biotechnol. Bioeng., 25,* 929 (1983).

983.   Mote, C.R. and Griffis, C.L., *Agr. Wastes, 4,* 65 (1982).

984.   MacGregor, S.T., Miller, F.C., Psarianos, K.M. and Finstein, M.S., *Appl. Env. Microbiol., 41,* 1321 (1980).

985.   Higgins, A.J., *Biocycle, 23,* 36 (1982).

986.   Miller, F.C., MacGregor, S.T., Psarianos, K.M., Cirello, J. and Finstein, M.S., *J. Water Poll. Control, 54,* 111 (1982).

987.   de Bertoldi, M., Vallini, G., Pera, A. and Zucconi, F., *Biocycle, 23,* 45 (1982).

988.   Mote, C.R. and Griffis, C.L., *Agr. Wastes, 1,* 191 (1979).

989.   Deschamps, A.M., Henno, P., Pernelle, C., Caignault, L. and Lebeault, J.M., *Biotechnol. Lett., 1,* 239 (1979).

990.   Whang, D.S. and Meenaghan, G.F., *Comp. Science/Land Utilization, 21,* 44 (1980).

991.   Psarianos, K.M., MacGregor, S.T., Miller, F.C. and Finstein, M.S., *Biocycle, 24*(2), 27 (1983).

992.   van Faassen, H.G. and van Dijk, H., in "Straw decay and its effect on disposal and utilization," (E. Grossbard, ed.) p 113, John Wiley & Sons, Chichester (1979).

993.   Smith, J.F. and Spencer, D.M., *Scientia Horticulturae, 5,* 23 (1976).

994.   Fermor, T.R., Smith, J.F. and Spencer, D.M., *J. Hort. Science, 54,* 137 (1979).

995.   Fermor, T.R. and Wood, D.A., in "Straw decay and its effect on disposal and utilization," (E. Grossbard, ed) p 105, John Wiley & Sons, Chichester (1979).

996.   Kleyn, J.G. and Wetzler, T.F., *Can. J. Microbiol., 27,* 748 (1981).

997.   Kuster, E. and Schmitten, I., in der *Eur. J. Appl. Microbiol. Biotechnol., 11,* 248 (1981).

998.   Millner, P.D., *Dev. Ind Microbiol., 23,* 61 (1982).

999.   Stutzenberger, F.J. and Bowden, M.W., *Biotechnol. Bioeng., 22,* 2443 (1980).

1000. Nolan, E.J., Hottzapple, M.T., Phillips, J.A., Staff, G.F. and Humphrey, A.E., in "Advances in Biotechnology," (M. Moo-Young and C.W. Robinson, eds.) Vol. II, p 33, Pergamon Press, Toronto (1981).

1001. Lamot, E. and Vandamme, E.J., *J. Chem. Technol. Biotechnol., 32,* 735 (1982).

1002. Berg, B. and Soderstrom, B., *Soil Biol. Biochem., 11,* 339 (1979).

1003. Sorensen, L.H., *Soil. Biol. Biochem., 11,* 23 (1979).

1004. Kjoller, A. and Struwe, S., *Soil Biol. Biochem., 12,* 425 (1980).

1005. Shortle, W.C., Monge, J.A. and Cowling, E.B., *Eur. J. For. Path., 8,* 293 (1978).

1006. Deschamps, A.M. and Lebeault, J.M., *Eur. J. For. Path., 10,* 316 (1980).

1007. Deschamps, A.M., Comtat, J., Nouvior, N. and Lebeault, J.M., *J. Gen. Appl. Microbiol., 28,* 275 (1982).

1008. Antheunisse, J., *J. Gen. Appl. Microbiol., 25,* 273 (1979); *26,* 167 (1980); *27,* 435 (1981).

1009. Amer, G.I. and Drew, S.W., *Ann. Rep. Ferment. Process, 4,* 67 (1980).

1010. Kirk, T.K., in "Trends in the Biology of Fermentation for Fuels and Chemicals," (A. Hollaender, ed.) p 131, Plenum Press, New York (1981).

1011. Zeikus, J.G., *Adv. Microbiol. Ecol., 5,* 211 (1981); in "Lignin Biodegradation," (T.K. Kirk; T. Higuchi and H. Chang, eds.) Vol. I, p 101, CRC Press, Boca Raton (1980).

1011a. Martin, J.P. and Haider, K., in "Lignin Biodegradation," (T.K. Kirk, T. Higuchi and H. Chang, eds.) Vol. I, p 77, CRC Press, Boca Raton (1980).

1012. Kirk, T.K. and Chang, H.M., *Enzyme Microb. Technol., 3,* 189 (1981).

1013. Wood, D.A. and Goodenough, P.W., *Arch. Microbiol., 114,* 161 (1973).

1014. Wood, D.A., *J. Gen. Microbiol., 117,* 327 (1980); *117,* 339 (1980).

1015. Wood, D.A., in "Straw decay and its effects on disposal and utilization, (E. Grossbard, ed.) p 95, John Wiley, Chichester (1979).

1016. Noguchi, A., Shimada, M. and Higuchi, T., *Holzforsch, 34,* 86 (1980).

1017. Shimada, M., in "Lignin Biodegradation," (T.K. Kirk, T. Higuchi and H. Chang, eds.) Vol. I, p 195, CRC Press, Boca Raton (1980).

1018. Leonowicz, A. and Grzycunowicz, K., *Enzyme Microb. Technol., 3,* 55 (1981).

1019. Settlif, E.C. and Eudy, W.W., in "Lignin Biodegradation," (T.K. Kirk, T. Higuchi and H. Chang, eds.) Vol. I, p 135, CRC Press, Boca Raton (1980).

1020. Petroski, R.J., Peczynska-Czoch, W. and Rosazza, J.P., *Appl. Environ. Microbiol., 40,* 1003 (1980).

1021.   Amer. G.I. and Drew, S.W., *Devel. Ind. Microbiol.*, *22*, 479 (1981).

1022.   Antai, S.P. and Crawford, D.L., *Eur. J. Appl. Microbiol. Biotechnol.*, *14*, 165 (1982).

1023.   Antai, S.P. and Crawford, D.L., *Appl. Environ. Micorbiol.* *42*, 378 (1981).

1024.   Kaufmann, U. and Wellendorf, H., *Eur. J. Forest Pathol.*, *10*, 90 (1980).

1025.   Wicklow, D.T., Detroy, R.W. and Adami, S., *Mycologia*, *72*, 1065 (1980).

1026.   Janshekar, H., Haltmeier, T. and Brown, C., *Eur. J. Appl. Microbiol. Biotechnol.*, *14*, 174 (1982).

1027.   Yang, H.H., Effland, M.J. and Kirk, T.K., *Biotechnol. Bioeng.*, *22*, 65 (1980).

1028.   Jeffries, T.W., Choi, S. and Kirk, T.K., *Appl. Environ. Microbiol.*, *42*, 290 (1981).

1029.   Leisola, M., Brown, C., Laurila, M., Ulmer, D. and Fiechter, A., *Eur. J. Appl. Microbiol.*, *15*, 180 (1982).

1030.   Forney, L.J., Reddy, C.A. and Pankratz, H.S., *Appl. Environ. Microbiol.*, *44*, 732 (1982).

1031.   Gold, M.H., Cheng, T.M. and Mayfield, M.B., *Appl. Environ. Microbiol.*, *44*, 996 (1982).

1032.   Reid, I., *Appl. Environ. Microbiol.*, *45*, 830 (1983); *45*, 838 (1983).

1033.   Flegel, T.W., Meevootisom, V. and Kiatpajan, S., *J. Ferment. Technol.*, *60*, 473 (1982).

1034.   Crawford, D.L., *Biotechnol. Bioeng. Symp.*, *11*, 275 (1981).

1035.   Deschamps, A.M., Gillie, J.P. and Lebeault, J.M., *Eur. J. Appl. Microbiol. Biotechnol.*, *13*, 222 (1981).

1036.   Glanser, M., Ban, S.N. and Cooney, C.L., *Process Biochem.*, *17*(5), 19 (1982).

1036a. Leisola, M., Ulmer, D. and Fiechter, A., *Eur. J. Appl. Microbiol. Biotechnol.*, *17*, 113 (1983); *17*, 117 (1983).

1037.   Chang, H., Chen, C.L. and Kirk, T.K., in "Lignin Biodegradation," (T.K. Kirk, T. Higuchi and H. Chang, eds.) Vol. I, p 215, CRC Press, Boca Raton (1980).

1038.   Haars, A. and Huttermann, A., *Naturwiss*, *67*, 39 (1980).

1039.   Vered, Y., Milstein, O., Flowers, H.M. and Gressel, J., *Eur. J. Appl. Microbiol. Biotechnol.*, *12*, 183 (1981).

1040.   Crawford, R.L., Robinson, L.E. and Cheh, A.M., in "Lignin Biodegradation," (T.K. Kirk, T. Higushi and H. Chang, eds.) Vol. I, p 61, CRC Press, Boca Raton (1980).

1041. Crawford, R.L., Robinson, L.E. and Foster, R.D., *Appl. Environ. Microbiol.*, *41*, 1112 (1981).

1042. Haider, K. and Trojanowski, J., in "Lignin Biodegradation," (T.K. Kirk, T. Higuchi and H. Chang, eds.) Vol. I, p 111, CRC Press, Boca Raton (1980).

1043. Kaplan, D.L. and Hartenstein, R., *Soil. Biol. Biochem.*, *12*, 65 (1980).

1044. Iwahara, Sh., in "Lignin Biodegradation," (T.K. Kirk, T. Higuchi and H. Chang, eds.), Vol. I, p 151, CRC Press, Boca Raton (1980).

1045. Higuchi, T., in "lignin Biodegradation," (T.K. Kirk, T. Higuchi and H. Chang, eds.) Vol. I, p 172, CRC Press, Boca Raton (1980).

1046. Katayama, T., Nakatsubu, F. and Higuchi, T., *Arch. Microbiol.*, *126*, 127 (1980).

1047. Iwahara, S., Nishihira, T., Jonori, T., Kuwahara, M. and Higuchi, T., *J. Ferment. Technol.*, *58*, 183 (1980).

1048. Norris, D., *Appl. Environ. Microbiol.*, *40*, 376 (1980).

1049. Deschamps, A.M., Mahoudeau, G. and Lebeault, J.M., *Eur. J. Appl. Microbiol. Biotechnol.*, *9*, 45 (1980).

1050. Deschamps, A.M., Mahoudeau, G., Conti, M. and Lebeault, J.M., *J. Ferment. Technol.*, *58*, 93 (1980).

1051. Deschamps, A.M. and Lebeault, J.M., in "Advances in Biotechnology," (M. Moo-Young and C.W. Robinson, eds.) Vol. II, p 639, Pergamon Press, Toronto (1981).

1052. Deschamps, A.M., Otuk, G. and Lebeault, J.M., *J. Ferment. Technol.*, *61*, 55 (1983).

1053. Rast, H.G., Engelhardt, G. and Wallnofer, P.R., *FEMS Microbiol. Lett.*, *7*, 1 (1980).

1054. Rast, H.G., Engelhardt, G., Ziegler, W. and Wallnofer, P.F., *FEMS Microbiol. Lett.*, *8*, 259 (1980).

1055. Odier, E., Janin, G. and Monties, B., *Appl. Environ. Microbiol.*, *41*, 337 (1981).

1056. Eggeling, L. and Sahm, H., *Arch. Microbiol.*, *126*, 141 (1980).

1057. Janshehar, H. and Fiechter, A., *Eur. J. Appl. Microbiol. Biotechnol.*, *14*, 47 (1982).

1057a. Kellens, R.D., Aseltine, M.S. and Church, D.C., *J. Anim. Science*, *53*, 1601 (1981).

1057b. Beszedits, S. and Lugowski, A., *Process Biochem.*, *18*(2), 35 (1983).

1058. Caswell, L.F., Fontenot, J.P. and Webb, K.E., *J. Anim. Sci.*, *46*, 547 (1978).

1059. D'Orso, G., Lanza, E., Scevra, V. and Galofaro, V., *Zootec. Nutr. Anim.*, *5*, 263 (1979).

1060. Jayal, M.M., Jain, V.K. and Pathak, N.N., *Agr. Wastes*, *3*, 157 (1981).

1061. Jakhmola, R.C., Kamra, D.N., Baruah, K.K. and Pathak, N.N., *Agr. Wastes, 5,* 95 (1983).

1062. Kutty, K.P.A. and Prasad, D.A., *Indian J. Anim. Sci., 50,* 189 (1980).

1063. Weiner, B.A., *Eur. J. Appl. Microbiol. Biotechnol., 16,* 39 (1982).

1064. Malossini, F., Martillotti, F. and Pace, V., *Ann. Ist. Sper. Zootec., 11,* 113 (1978).

1065. Malossini, F., Martillotti, F., Francia, U., Carretta, A. and Demichelis, F., *Zootec. Nutr. Anim., 5,* 81 (1979).

1066. Malossini, F., Romita, A., Carreta, A., diGiacomo, A. and Pace, V., *Zootec. Nutr. Anim., 6,* 143 (1980).

1067. Pace, V., de Michelis, F., Fontanelli, G. and Malossini, F., *Zootec. Nutr. Anim., 6,* 193 (1980).

1067a. Cornman, A.W., Lamm, W.D., Webb, K.E. and Fontenot, J.P., *J. Anim. Sciences, 52,* 1233 (1981).

1068. Prior, R.L. and Hashimoto, A.G., in "Fuel Gas Production from Biomass," (D.L. Wise, ed.) Vol. II, p 216, CRC Press, Boca Raton (1981).

1069. Marchaim, U. and Criden, J., in "Fuel Gas Production from Biomass," (D.L. Wise, ed.) Vol. I, p 95, CRC Press, Boca Raton (1981).

1070. Lizdas, D.J., Coe, W.B. and Turk, M.H., in "Fuel Gas Production from Biomass," (D.L. Wise, ed.) Vol. I, p 201, CRC Press, Boca Raton (1981).

1071. Hackett, W.F., Connors, W.J., Kirk, T.K. and Ziekus, J.G., *Appl. Environ. Microbiol., 33,* 43 (1977).

1072. MacCubbin, A.E. and Hodson, R.E., *Appl. Environ. Microbiol., 40,* 735 (1980).

1073. Colberg, P.J. and Young, L.Y., *Can. J. Microbiol., 28,* 886 (1982).

1073a. Akin, D.E., *Appl. Environ. Microbiol. 40,* 809 (1980).

1074. Akin, D.E., *J. Anim. Science, 48,* 701 (1979).

1075. Dinsdale, D., Morris, E.J. and Bacon, J.S.D., *Appl. Environ. Microbiol., 36,* 160 (1978).

1076. Akin, D.E. and Amos, H.E., *Appl. Environ. Microbiol., 37,* 332 (1979).

1076a. Orpin, C.G. and Letcher, A.J., *Current Microbiol., 3,* 121 (1979).

1077. Harbers, L.H. and Thouvenelle, M.L., *J. Anim. Science, 50,* 514 (1980).

1077a. Akin, D.E. and Burdick, D., *Crop Science, 21,* 577 (1981).

1078. Morris, E.J. and van Gylswyk, N.D., *J. Agric. Sci., 95,* 313 (1980).

1079. Akin, D.E., *Appl. Environ. Microbiol., 39,* 242 (1980).

1080. Cheng, K.J., Fay, J.P., Howarth, R.E. and Costerton,
       J.W., *Appl. Environ. Microbiol.*, *40*, 613 (1980).
1081. Cheng, K.J. and Costerton, J.W., *Can. J. Microbiol.*,
       26, 1104 (1980).
1082. Davis, C.L. and Clark, J.H., *Dev. Ind. Microbiol.*, *22*,
       247 (1981).
1083. Hobson, P.N., in "Microbial Polysaccharides and
       Polysaccharases," (R.C.W. Berkeley, G.W. Gooday and
       D.C. Ellwood, eds.) p 377, Academic Press, London
       (1979).
1084. Russell, J.B. and Hespell, R.B., *J. Dairy Science, 64*,
       1153 (1981).
1084a. Morrison, I.M., in "Straw decay and its effects on
       disposal and utilization," (E. Grossbard, ed.) p 237,
       John Wiley & Sons, Chichester (1979).
1085. Hespell, R.B., *Dev. Ind. Microbiol.*, *22*, 261 (1981).
1086. Stevenson, I.L., *Can. J. Microbiol.*, *24*, 1236 (1978).
1087. Chesson, A., Stewart, C.S. and Wallace, R.J., *Appl.
       Environ. Microbiol.*, *44*, 597 (1982).
1088. Collings, C.F. and Yokoyama, M.T., *Appl. Env. Microbiol.*,
       39, 566 (1980).
1089. Turner, K.W. and Robertson, A.M., *Appl. Env. Microbiol.*,
       38, 7 (1979).
1090. Stewart, C.S., Dinsdale, D., Chang E.J. and Paniagua,
       C., in "Straw decay and its effect on disposal and
       utilization," (E. Grossbard, ed.) p 123, John Wiley &
       Sons, Chichester (1979).
1091. Baldwin, R.L. and Denbam, S.L., *J. Anim. Science.*, *49*,
       1631 (1979).
1092. Tamminga, S., *J. Anim. Science, 49*, 1615 (1979).
1093. Mahadevan, S., Erfle, J.D. and Sauer, F.D., *J. Anim.
       Science, 50*, 723 (1980).
1094. Brock, F.M. Forsberg, C.W. and Buchnan-Smith, J.G.,
       Appl. Environ. Microbiol., *44*, 561 (1982).
1095. Broderick, G.A. and Balthrop, J.E., *J. Anim. Science,
       49*, 1101 (1979).
1096. Smith, R.H., *J. Anim. Science, 49*, 1604 (1979).
1097. Pisulewski, P.M., Okorie, A.U., Buttery, P.J., Havesign,
       W. and Lewis, D., *J. Sci. Food Agric.*, *32*, 759 (1981).
1098. Dawson, K.A., Preziosi, M.C. and Caldwell, D.R., *J.
       Bacteriol.*, *139*, 384 (1979).
1099. Chalupa, W., *Dev. Ind. Microbiol.*, *22*, 277 (1981).
1100. Jouany, J.P. and Senaud, J., *Ann. Zootech. 27*, 61 (1978).
1101. Wallace, R.J., Czerhawski, J.W. and Brechenridge, G.,
       British J. Nutr., *46*, 131 (1981).

1102. Stanier, G. and Davies, A., *British J. Nutr., 45,* 567 (1981).

1103. Horn, G.W., Mader, T.L., Ambruster, S.L. and Frahm, R.R., *J. Anim. Sci., 52,* 447 (1981).

1104. Thornton, J.H. and Owens, F.N., *J. Anim. Science, 52,* 628 (1981).

1105. Fuller, J.R. and Johnson, D.E., *J. Anim. Science, 53,* 1574 (1981).

1106. Dennis, S.M., Nagaraja, T.G. and Bartley, E.E., *J. Dairy Science, 64,* 2350 (1981).

1107. Henderson, C., *J. Gen. Microbiol., 119,* 485 (1980).

1108. Thomson, D.J., Beever, D.E., Latham, M.J., Sharpe, M.E. and Terry, R.A., *J. Agric. Science, 91,* 1 (1978).

1109. Van Nevel, C.J. and Demeyer, D.I., *Eur. J. Appl. Microbiol. Biotechnol., 7,* 111 (1979).

1109a. Hungate, R.E. and Stack, R.J., *Appl. Environ. Microbiol. 44,* 79 (1982).

1110. Miura, H., Horiguchi, M. and Matsumoto, T., *Appl. Environ. Microbiol., 40,* 294 (1980).

1111. Stanton, T.B. and Canale-Parola, E., *Arch. Microbiol., 127,* 145 (1980).

1112. Russell, J.B., Cotta, M.A. and Dombrowski, D.B., *Appl. Environ. Microbiol., 41,* 1394 (1981).

1113. Therion, J.J., Kistner, A. and Kornelios, J.H., *Appl. Environ. Microbiol., 44,* 428 (1982).

1114. Bauchop, T. and Mountfort, D.O., *Appl. Environ. Microbiol., 42,* 1103 (1981).

1115. Mountfort, D.O., Asher, R.A. and Bauchop, T., *Appl. Environ. Microbiol., 42,* 1103 (1981).

1116. Russell, J.B., *Appl. Environ. Microbiol., 45,* 1566 (1983).

1117. Hobson, P.N., in "Mixed Culture Fermentations," (M.E. Bushell and J.H. Slater, eds.) p 53, Academic Press, London (1981).

1118. Lane, A.G., *Proc. 5th Australian Biotechnol. Conf.,* 103 (1982).

1119. Khan, A.W., Saddler, J.N., Patel, G.B., Calvin, J.R. and Martin, S.M., *FEMS Microbiol. Lett., 7,* 47 (1980).

1120. Khan, A.W., *FEMS Microbiol. Lett., 9,* 233 (1980).

1121. Khan, A.W., Wall, D. and van den Berg, L. *Appl. Environ. Microbiol., 41,* 1214 (1981).

1122. Khan, A.W. and Mes-Hartree, M., *J. Appl. Bacteriol., 50,* 283 (1981).

1123. Khan, A.W. and Murray, W.D., *FEMS Microbiol. Lett., 13,* 377 (1982).

1124. Khan, A.W. and Murray, W.D., *Biotechnol. Lett.*, *4*, 177 (1982).

1125. Guiliano, C., Asther, M. and Khan, A.W., *Biotechnol. Lett.*, *5*, 395 (1983).

1126. Khan, A.W., Miller, S.S. and Murray, W.D., *Biotechnol. Bioeng.* *25*, 1571 (1983).

1127. Levy, P.F., Sanderson, J.E. and Wise, D.L., *Biotechnol. Bioeng. Symp.*, *11*, 239 (1981).

1128. Levy, P.F., Sanderson, J.E., Kispert, and Wise, D.L., *Enzyme Microb. Technol.*, *3*, 207 (1981).

1129. Joergensen, M.H., *Eur. J. Appl. Microbiol. Biotechnol.*, *6*, 181 (1978).

1130. Segers, L., Verstrynge, L. and Verstraete, W., *Biotechnol. Lett.*, *3*, 635 (1981).

1131. de la Torre, I. and Goma, G., *Biotechnol. Bioeng.*, *23*, 185 (1981).

1132. Gomez, G.J. and Viniegra, G., in "Advances in Biotechnology," (M. Moo-Young and C.W. Robinson, eds.) Vol. II, p 627, Pergamon Press, Toronto (1981).

1133. Datta, R., *Biotechnol. Bioeng.*, *23*, 61 (1981).

1134. Datta, R., *Biotechnol. Bioeng.*, *23*, 2167 (1981).

1135. Datta, R., *Biotechnol. Bioeng. Symp.*, *11*, 521 (1981).

1136. Playne, M.J., in "Advances in Biotechnology," (M. Moo-Young and C.W. Robinson, eds.) Vol. II, p 85, Pergamon Press, Toronto (1981).

1137. Clausen, E.C. and Gaddy, J.L., in "Advances in Biotechnology," (M. Moo-Young and C.W. Robinson, eds.) Vol. II, p 63, Pergamon Press, Toronto (1981).

1138. Clausen, E.C., Shah, R.B., Najafpour, G. and Gaddy, J.L., *Biotechnol. Bioeng. Symp.*, *12*, 67 (1982).

1139. Nishio, N., Kitaura, Sh. and Nagai, Sh., *J. Ferment. Technol.*, *60*, 423 (1982).

1140. Bekers, M.J., Upitis, A.A., Upite, D.J., Klintsare, A., Beker, V.F., Ose, V.P. and Ramnietse, V.E., in "Advances in Biotechnology," (M. Moo-Young and C.W. Robinson, eds.) Vol. II, p 357, Pergamon Press, Toronto (1981).

1141. Chesson, A., *J. Appl. Bacteriol.*, *48*, 1 (1980).

1142. Yoshihara, K. and Kobayashi, Y., *Agric. Biol. Chem.*, *46*, 109 (1982).

1143. Yoshihara, K. and Kobayashi, Y., *Biotechnol. Lett.*, *3*, 118 (1981).

1144. Thomas, P.C., Kelly, N.C., Chamberlain, D.G., *Proc. Nutrition Soc.*, *39*, 257 (1980).

1145. McDonald, P., "The Biochemistry of Silage," John Wiley & Sons, Chichester (1981).

1146.   Castle, M.E. and Watson, J.N., *Grass Forage Science,*
        34, 101 (1979).
1147.   Lessard, J.R., Erfle, J.D., Sauer, F.D. and Mohadevan,
        S., *J. Sci. Fd. Agric., 29,* 506 (1978).
1148.   Ohshima, M. and McDonald, P., *J. Sci. Fd. Agric., 29,*
        497 (1978).
1149.   Szigeti, J., *Acta Aliment., 8,* 25 (1979).
1150.   Ohshima, M., McDonald, P. and Acamovic, T., *J. Sci.
        Fd. Agric., 30,* 97 (1979).
1151.   Wiesemuller, W., Breite, L., Poppe, S. and Meier, H.,
        Tierernaehr. Fuetter, 11, 286 (1979); *11,* 301 (1979).
1152.   Siddons, R.C., Beever, D.E. and Kaiser, A.G., *J. Sci.
        Fd. Agric., 33,* 609 (1982).
1153.   Preston, T.R., Hinojosa, C. and Martinez, L., *Trop.
        Anim. Prod., 1,* 120 (1976).
1154.   de Gonzalez, E. and MacLeod, N.A., *Trop. Anim. Prod.,
        1,* 80 (1976).
1155.   Gill, M. and Nunoz, R., *Trop. Anim. Prod., 6,* 154 (1981).
1156.   Alli, I., Baker, B.E. and Garcia, G., *Anim. Feed
        Science Technol., 7,* 411 (1982).
1157.   Huber, J.T., Smith, N.E. and Stiles, J., *J. Anim.
        Science, 51,* 1387 (1980).
1158.   Halverson, A.W. and Emerick, R.J., *J. Agric. Food
        Chem., 30,* 474 (1982).
1159.   Andreis, H.J. and Destefano, R.P., *Sugar, J., 41*(5),
        13 (1978).

Index